THE DICTIONARY OF SODIUM, FATS, AND CHOLESTEROL

THE DICTIONARY OF
SODIUM, FATS, AND CHOLESTEROL

BY BARBARA KRAUS

GROSSET & DUNLAP
Publishers   New York

For Robert Alden and William E. Pecot, Jr.

Published simultaneously in Canada
Library of Congress catalog card number: 72-90848
ISBN 0-448-01371-1
First printing
Printed in the United States of America

CONTENTS

INTRODUCTION

This dictionary lists the total fat, saturated and unsaturated fatty acids, cholesterol, and sodium content of several thousands of food items. These nutrients have been receiving increasing attention by nutritionists and the medical profession because of their possible relationship to atherosclerosis (coronary heart disease). Of course, diet is not the only factor to be considered in heart disease. Other considerations are heredity, obesity, high blood pressure, blood cholesterol, blood lipids, cigarette smoking, lack of exercise, stress, and certain ailments such as diabetes.

Patients with coronary disease and high blood pressure are often placed on diets in which one or more of these nutrients is controlled. Since physicians now believe that the conditions leading to early heart attacks are the result of a lifetime of habits that predispose individuals to such attacks, they recommend that certain changes in dietary patterns be made also in early childhood.

Total Fat

The caloric intake from fat is often as much as 45% to 50% of the American diet. In the past 25 years, the consumption of fats and oils alone has increased, on the average, about 20%. In the same period, Americans have consumed over 18% more meat, which is also high in fat content. It is now frequently recommended that only about one-third of the calories be from fat, and of this amount, about two-thirds should be from polyunsaturated and monounsaturated fatty acids, amounting to approximately 10% to 15% of the diet.

Some fat in the diet is essential. It is used in the body to supply energy, yielding about nine calories per gram, more than twice as much as either protein or carbohydrate (the other energy-yielding nutrients in foods), which each supply about four calories per gram. Fat is also important in supplying essential fatty acids (those that cannot be synthesized by the body) and in the supply and utilization of fat soluble vitamins A, D, E, and K. Fat adds flavor and satiety value to otherwise bulky and bland diets.

Fatty Acids

Fats are composed of "saturated" and "unsaturated" fatty acids. In scientific terms, they are saturated if the carbon atoms contain all the hydrogen they can hold. If there is one double bond where hydrogen can be added, the fat is monounsaturated; if there are two or more, the fat is polyunsaturated. In layman's terms: if the fat is solid at room temperature it is saturated, if liquid, it is unsaturated. This is a very rough rule of thumb and there are exceptions to it. Most foods contain both kinds of fat in varying proportions. Oils from plant foods and fish contain the most abundant amounts of polyunsaturated fats. Coconut oil is an exception. It is high in saturated fatty acids and is solid at room temperature.

Meats, cheeses, eggs, and most animal products are high in saturated fatty acids. The listings in this book showing the values for unsaturated fatty acids include both mono and polyunsaturated fats. At this time, the available information for many foods is not adequate to give a breakdown between these two forms. New federal requirements for labeling commercial products will result in more complete information in the future, and later editions of this book will incorporate such values.

the values for saturated and unsaturated fats are rounded to the nearest

Since the values for saturated and unsaturated fats are rounded to the nearest whole number they may add up to more or less than the figure given for total fat. In many cases, the values for unsaturated fat were derived by the subtraction of the saturated from the total fats. In a very few cases, manufactures reported analyzed saturated and unsaturated fats which did not add up to the total fat shown, because the undetermined fatty acids were, in their opinion, not present in significant amounts.

Cholesterol

Cholesterol is one of the complex compounds known as sterols. It is an essential nutrient in normal metabolic processes and is synthesized in the body. It is also present in many foods of animal origin that we consume. Plant foods do not contain cholesterol in any significant amounts. Foods such as chocolate, cocoa, olive oil, coconut butter, and peanut butter are devoid of it since they are plant products.

Animal products are generally high in cholesterol. Organ and glandular meats such as brains, kidney, liver, sweetbreads, and heart are especially high in cholesterol. Egg yolk contains high concentrations of cholesterol, but it is absent in the white.

Cholesterol occurs in both the lean portions and the fatty portions of meat; the removal of fatty tissue if replaced by an equal amount of lean does not reduce the cholesterol intake. This is unlike the effect on the amount of fatty acids and must be taken into account in planning low cholesterol diets.

Sodium

Sodium is an essential element in the growth of animals. It occurs naturally in many foods but the principle source in diets is sodium chloride or ordinary table salt. Salt is used in the processing of many foods in freezing, canning, and other manufacturing methods. It is also added in the home in cooking and at the table.

For some people, the use of salt is excessive and low-sodium diets are prescribed by physicians. Low-sodium diets are used in the treatment of diseases such as cardiac failure, edema formation, and high blood pressure.

Summary

Always keep in mind that in making any drastic changes in dietary habits the advice and guidance of a competent physician should be sought. It is an essential of good health that the diet contain all the nutrients required by the individual and in adequate amounts. Reductions or increases in one or more of the essential nutrients may bring about adverse effects.

The tables given here are intended to provide basic information to help plan a varied and attractive diet within broad guidelines that are nutritionally adequate when one or more of the nutrients must be controlled.

ARRANGEMENT OF THIS BOOK

Foods are listed alphabetically by brand name or by the name of the food. The singular form is used for the entries, that is, blackberry instead of blackberries. Most items are listed individually although a few are grouped (see p. xi). For example, all candies are listed together so that if you are looking for *Mars* bar, you look first under Candy, then under *M* in alphabetical order. But, if you are looking for a breakfast food such as Oatmeal, you will find it under *O* in the main alphabet. Many cross references are included to assist you in finding items known by different names.

Under the main headings, it was often not possible nor even desirable to follow an alphabetical arrangement. For basic foods, such as apricots, the first entries are for the fresh product weighed with seeds as it is purchased in the store, then the fruit in small portions as they may be eaten or measured. These entries are followed by the processed products, canned (although it may actually be a bottle or a jar), dehydrated, dried, and frozen items. This basic plan, with adaptations where necessary, was followed for fruits, vegetables, and meats.

In almost all entries, where data were available, the U.S. Department of Agriculture figures are shown first. The Department values represent averages from several manufacturers and are shown for comparison with the values from individual companies or for use where particular brands are not available.

All brand-name products have been italicized and company names appear in parentheses.

Portions Used

The portion column is a most important one to read and note. Common household measures are used wherever possible. For some items, the amounts given are those commonly purchased in the store, such as one pound of meat, or a 15-ounce package of cake mix. These quantities can be divided into the number of servings used in the home and the nutritive values in each portion can then be readily determined. Any ingredients added in preparing such products must also be taken into account.

The smaller portions given are for foods as served or measured in moderate amounts, such as one-half cup of reconstituted juice, or four ounces of meat. Be sure to adjust the amount of the nutrients to the actual portions you use. For example, if you serve one cup of juice instead of one-half cup, multiply the amount of the nutrients shown for the smaller amount by two.

The size of portions you use is extremely important in controlling the intake of any nutrient. The amount of a nutrient is directly related to the weight of the food served. The weight of a volumetric measure, such as a cup or a pint may vary considerably depending on many factors; four ounces by weight may be very different from one-half cup or four fluid ounces. Ounces in the tables are always ounces by weight unless specified as fluid ounces, fractions of a cup, or other volumetric measure. Foods that are fluffy in texture such as flaked coconut and bean sprouts vary greatly in weight per cup, depending on how tightly they are packed. Such foods as canned green beans also vary when measured with and without liquid; for instance, canned beans with liquid weigh 4.2 ounces for one-half cup, but drained beans weigh 2.5 ounces for the same half cup. Check the weights of your serving portions regularly. Bear in mind that you can reduce or increase the intake of any nutrient by changing the serving size.

It was impossible to convert all the portions to a uniform basis. Some sources were able to report data only in terms of weights with no information on cup or other volumetric measures. We have shown small portions in quantities that might reasonably be expected to be served or measured in the home or institution.

You will find in the portion column the phrases "weighed with bone," and "weighed with skin and seeds." These descriptions apply to the products as you purchase them in the markets, but the nutritive values as shown are for the amount of edible food after you discard the bone, skin, seed, or other inedible part. The weight given in the "measure" or "quantity" column is to the nearest gram or fraction of an ounce.

Data on the composition of foods are constantly changing for many reasons. Better sampling and analytical methods, improvements in marketing procedures, and changes in formulas of mixed products may alter values for all of the nutrients. Weights of packaged foods are frequently changed. It is essential to read label information in order to be knowledgeable about these matters and to make intelligent use of food tables.

Other Nutrients

These tables are not intended as a dietary guide. Any drastic change from a normal mixed diet should be undertaken only under the guidance of a qualified physician. Do not forget that other nutrients — protein, carbohydrate, minerals, and vitamins — are extremely important in diet planning. From a nutritional viewpoint, perhaps the best advice that can be given is to eat a varied diet with all classes of food represented. Meat, fish, chicken, fats and oils, milk, vegetables, fruits, and grain products are all important sources of essential nutrients and some foods from each of these classes should be included in the diet every day.

If your doctor has recommended the control of one or more of the nutrients shown in these tables, you can choose foods from this book under the doctor's guidance that will fit his specifications and will provide a varied selection of products that are acceptable. Control of certain nutrients such as cholesterol or sodium does not condemn you to a monotonous diet. There is a rich and varied assortment of foods in our markets that will meet any medical requirements. Choose wisely and eat well.

Sources of Data

Values in this dictionary are based on publications issued by the U.S. Department of Agriculture and on data submitted by manufacturers and processors. The U.S. Department of Agriculture issues basic tables on food composition for use in the United States. The commercial products from U.S.D.A. publications represent average values obtained on products of more than one company. The figures designated as ''home recipe'' are based on recipes on file with the Department of Agriculture. Data on commercial products listed by brand name in this publication are based on values supplied by manufacturers and processors for their own individual products. Supermarket brand names, such as the A & P's *Ann Page*, or private labels could not be included in this book inasmuch as they are not usually analyzed under these trade names. Every care has been taken to interpret the data and the descriptions supplied by the companies as fully and as accurately as possible. Many values have been recalculated to different portions from those submitted, in order to bring about greater uniformity among similar items.

Analysis of foods to provide information on nutritive values are extremely expensive to conduct. Many small companies cannot afford to have their products analyzed and were unable to provide data or were able to provide only a portion of the data requested. Other companies have simply never gotten around to having the analyses done. New requirements for labeling nutritive values for products may provide information on additional items in the future. Wherever data were unavailable blank spaces were left which may be filled in by the reader at a later time.

Foods Listed by Groups

Foods in the following classes are reported together rather than as individual items in the main alphabet: baby food, bread, cake icing, cake icing mix, candy, cheese, cookies, cookie mix, crackers, gravy, salad dressing, and sauce.

BARBARA KRAUS

ABBREVIATIONS AND SYMBOLS

(USDA) = United States Department
      of Agriculture
    * = prepared as packaging directs[1]
    < = less than
    > = more than
    & = and
    " = inch
    + = values do not include amount
      of sodium found in the local
      water used in packaging
canned = bottles or jars as well as cans
  dia. = diameter
    fl. = fluid
  liq. = liquid

lb. = pound
med. = medium
oz. = ounce
pkg. = package
pt. = pint
qt. = quart
sq. = square
T. = tablespoon
Tr. = trace
tsp. = teaspoon
wt. = weight
mg. = milligram
gr. = gram

Italics or name in parentheses = registered trademark,®.
Where zero in parenthesis appears in the tabular column it means that (0) zero is imputed by author wherever there is a reasonable assumption that that nutrient is not present.

EQUIVALENTS

By Weight	By Volume
1 pound = 16 ounces	1 quart = 4 cups
1 ounce = 28.35 grams	1 cup = 8 fluid ounces
3.52 ounces = 100 grams	1 cup = ½ pint
1 milligram = .001 gram	1 cup = 16 tablespoons
	2 tablespoons = 1 fluid ounce
	1 tablespoon = 3 teaspoons
	1 pound butter = 4 sticks or 2 cups

[1]If the package directions call for whole or skim milk, the data given here are for whole milk, unless otherwise stated.

THE DICTIONARY OF SODIUM, FATS, AND CHOLESTEROL

Food and Description	Measure or Quantity	Sodium (mg.)	Total	—Fats in grams— Satu- rated	Unsatu- rated	Choles- terol (mg.)

A

ABALONE (USDA):
Raw, meat only	4 oz.		.6			
Canned	4 oz.		.3			

AC'CENT	¼ tsp. (1 gram)	129	0.			0

ACEROLA, fresh (USDA):
Fruit	½ lb. (weighed with seeds)	15	.6			0
Juice	½ cup (4.3 oz.)	4	.4			0

ALBACORE, raw, meat only
(USDA)	4 oz.	45	8.6	3.	5.	

ALCOHOLIC BEVERAGES (See individual listings)

ALEWIFE (USDA):
Raw, meat only	4 oz.		5.6			
Canned, solids & liq.	4 oz.		9.1			

ALLSPICE (Spice Islands):
Ground	1 tsp.	2				(0)
Whole	1 tsp.	1				(0)

ALMOND:
In shell:
(USDA)	4 oz. (weighed in shell)	2	31.4	2.	29.	0
(USDA)	1 cup (2.8 oz.)	2	21.7	2.	20.	0

Shelled:
Plain, unsalted:
Whole (USDA)	1 oz.	1	15.4	1.	14.	0
Whole (USDA)	1 cup (5 oz.)	6	77.0	6.	71.	0
Whole (USDA)	13-15 almonds (.6 oz.)	<1	9.5	<1.	9.	0
Chopped (USDA)	1 cup (4.5 oz.)	5	68.8	5.	64.	0
(Blue Diamond)	1 cup (5.6 oz.)	6	83.8	10.	74.	(0)

Blanched:
Salted (USDA)	1 cup (5.5 oz.)	311	90.6	8.	83.	0
Slivered (Blue Diamond)	1 cup (5.6 oz.)	6	86.2	10.	76.	(0)

(USDA): United States Department of Agriculture
*Prepared as Package Directs

(1)

Food and Description	Measure or Quantity	Sodium (mg.)	— Fats in grams —			Choles- terol (mg.)
			Total	Satu- rated	Unsatu- rated	
Chocolate-covered (See **CANDY**)						
Flavored (Blue Diamond) barbecue, cheese, French-fried, onion-garlic or smokehouse-style	1 oz.	56				(0)
Roasted, salted:						
(USDA)	1 oz.	56	16.4	1.	15.	0
(USDA)	1 cup (5.5 oz.)	311	90.6	8.	83.	0
Diced (Blue Diamond)	1 oz.	56				(0)
Dry (Flavor House)	1 oz.	56	16.4			(0)
Dry (Planters)	1 oz.	340	16.4	1.	15.	0
ALMOND MEAL, partially defatted (USDA)	1 oz.	2	5.2	Tr.	5.	0
ALPHA-BITS, oat cereal (Post)	1 cup (1 oz.)	150	1.1			0
AMARANTH, raw (USDA):						
Untrimmed	1 lb. (weighed untrimmed)		1.4			0
Trimmed	4 oz.		.6			0
AMBROSIA, chilled, bottled (Kraft)	4 oz.	115	2.4			(0)
A.M., fruit juice drink (Mott's)	½ cup		.1			(0)
ANCHOVY PASTE, canned (Crosse & Blackwell)	1 T. (.5 oz.)	1540	.5			
ANCHOVY, PICKLED, canned with or without added oil, not heavily salted (USDA)	1 oz.		2.9			
ANGEL FOOD CAKE, home recipe (USDA)[1]	1/12 of 8″ cake (1.4 oz.)	113	<.1			0
ANGEL FOOD CAKE MIX:						
Dry (USDA)	4 oz.	215	.2			0
*(USDA)[2]	1/12 of 10″ cake (1.9 oz.)	77	.1			0

(USDA): United States Department of Agriculture
*Prepared as Package Directs
[1]Made with sodium aluminum sulfate-type baking powder.
[2]Prepared with water, flavorings.

Food and Description	Measure or Quantity	Sodium (mg.)	Fats in grams			Cholesterol (mg.)
			Total	Saturated	Unsaturated	
*(Betty Crocker):						
1 step	$^1/_{16}$ of cake	172	.1			(0)
2 step	$^1/_{16}$ of cake	110	.1			(0)
Confetti	$^1/_{16}$ of cake		.1			(0)
Lemon custard	$^1/_{16}$ of cake	133	.1			(0)
Strawberry	$^1/_{16}$ of cake	148	.1			(0)
*(Duncan Hines)	$^1/_{12}$ of cake (2 oz.)	89	.1			0
*(Swans Down)	$^1/_{12}$ of cake (1.8 oz.)	95	.1			0
ANISE SEED (Spice Islands)	1 tsp.	<1				(0)
APPLE, any variety:						
Fresh (USDA):						
Eaten with skin	1 lb. (weighed with skin & core)	4	2.5			0
Eaten with skin	1 med., 2½" dia. (about 3 per lb.)	1	.8			0
Eaten without skin	1 lb. (weighed with skin & core)	4	1.2			0
Eaten without skin	1 med., 2½" dia. (about 3 per lb.)	1	.4			0
Pared, diced	1 cup (3.8 oz.)	1	.3			0
Pared, quartered	1 cup (4.3 oz.)	1	.4			0
Canned (See **APPLESAUCE**)						
Dehydrated:						
Uncooked (USDA)	1 oz.	2	.6			0
Cooked, sweetened (USDA)	½ cup (4.2 oz.)	1	.4			0
Dried:						
Uncooked (USDA)	1 cup (3 oz.)	4	1.4			0
Uncooked (Del Monte)	1 cup (3 oz.)	75	1.6			0
Cooked, unsweetened (USDA)	½ cup (4.3 oz.)	1	.6			0
Cooked, sweetened (USDA)	½ cup (4.9 oz.)	1	.6			0
Frozen, sweetened, slices, not thawed (USDA)	4 oz.	16	.1			0
APPLE BROWN BETTY, home recipe (USDA)[1]	1 cup (8.1 oz.)	352	8.0	2.	6.	
APPLE BUTTER:						
(USDA)	½ cup (5 oz.)	3	1.1			0

(USDA): United States Department of Agriculture
*Prepared as Package Directs
[1]Principal sources of fat: butter, bread crumbs.

Food and Description	Measure or Quantity	Sodium (mg.)	— Fats in grams —			Cholesterol (mg.)
			Total	Saturated	Unsaturated	
(USDA)	1 T. (.6 oz.)	<1	.1			0
(Bama)	1 T. (.6 oz.)	1	<.1			(0)
(Smucker's) spiced	1 T. (.6 oz.)	4.	.1			(0)
APPLE CAKE MIX, cinnamon:						
*(Betty Crocker) pudding cake	1/6 of cake	325	4.7			
*(Betty Crocker) upside down	1/9 of cake	226	10.1			
*(Duncan Hines)	1/12 of cake (2.7 oz.)	325	6.1			50
APPLE CIDER (USDA)	½ cup (4.4 oz.)	1	Tr.			0
APPLE DRINK, canned:						
(Del Monte)	6 fl. oz. (6.5 oz.)	32	Tr.			(0)
(Hi-C)	6 fl. oz. (6.3 oz.)	<1	Tr.			0
APPLE DUMPLING, frozen						
(Pepperidge Farm)	1 dumpling (3.3 oz.)	206	16.4			
APPLE FRITTER, frozen						
(Mrs. Paul's)	12 oz. pkg.		32.1			
APPLE JACKS, cereal						
(Kellogg's)	1 cup (1 oz.)	68	.2			(0)
APPLE JELLY, dietetic or low calorie:						
(Diet Delight)	1 T. (.6 oz.)	5	Tr.			(0)
(Kraft)	1 oz.	35	<.1			(0)
(S and W) *Nutradiet*	1 T. (.5 oz.)		<.1			(0)
(Slenderella)	1 T. (.6 oz.)	22	Tr.			(0)
(Tillie Lewis)	1 T. (.5 oz.)	3	Tr.			(0)
APPLE JUICE, canned:						
(USDA)	½ cup (4.4 oz.)	1	Tr.			0
(Heinz)	5½-fl.-oz. can	5	.2			(0)
(Mott's)	½ cup		Tr.			(0)
(Mott's) McIntosh	½ cup		Tr.			(0)
APPLE NECTAR (Mott's)	½ cup		<.1			(0)
APPLE PIE:						
Home recipe, 2 crusts (USDA):						
Made with lard & butter	1/6 of 9″ pie (5.6 oz.)	476	17.5	6.	11.	

(USDA): United States Department of Agriculture
*Prepared as Package Directs

Food and Description	Measure or Quantity	Sodium (mg.)	Total	Fats in grams — Satu- rated	Unsatu- rated	Choles- terol (mg.)
Made with vegetable shortening & butter	¹/₆ of 9″ pie (5.6 oz.)	476	17.5	5.	13.	Tr.
(Hostess)	4½ oz. pie	589	11.0			
(McDonald's)	1 serving (3 oz.)	395	15.4			
(Tastykake)	4-oz. pie		14.7			
French apple (Tastykake)	4½-oz. pie		18.4			
Frozen:						
Unbaked (USDA)[1]	5 oz.	251	11.8	3.	9.	
Baked (USDA)[1]	5 oz.	302	14.3	3.	11.	
(Banquet)	5-oz. serving		15.4			
(Morton)	¹/₆ of 20-oz. pie	229	10.8			
(Morton)	¹/₆ of 24-oz. pie	242	11.5			
(Morton)	⅛ of 46-oz. pie	287	16.5			
(Mrs. Smith's)	¹/₆ of 8″ pie (4.2 oz.)	302	14.6			
(Mrs. Smith's) old fashion	¹/₆ of 9″ pie (5.8 oz.)	547	22.7			
(Mrs. Smith's) golden deluxe	⅛ of 10″ pie (5.6 oz.)	395	19.1			
Dutch apple (Mrs. Smith's)	¹/₆ of 8″ pie (4.2 oz.)	284	14.2			
Tart (Mrs. Smith's)	¹/₆ of 8″ pie (4.2 oz.)	207	9.9			

APPLE PIE FILLING, canned:
Sweetened:

(Comstock)	½ cup (5.4 oz.)	9	<.1			
(Comstock) French	½ cup (5.4 oz.)	49	.3			
(Comstock-Greenwood) Pie, sliced	1-lb. 1-oz. can	9	1.4			
(Lucky Leaf)	8 oz.	120	.4			
Unsweetened:						
(Lucky Leaf)	8 oz.	5	.5			
(Wilderness)	21-oz. can	397				

APPLESAUCE, canned:
Sweetened:

(USDA)	½ cup (4.5 oz.)	3	.1			0
(Del Monte)	½ cup (4.6 oz.)	2	.2			0
(Hunt's)	5-oz. can	3	.2			(0)
(Mott's)	½ cup (4.5 oz.)		.1			(0)
(Mott's) Golden Delicious	½ cup (4.5 oz.)		.2			(0)
(Mott's) cinnamon, country style	½ cup (4.5 oz.)		.1			(0)
(Stokely-Van Camp)	½ cup (4.2 oz.)		.1			(0)
Unsweetened, dietetic or low calorie:						
(USDA)	½ cup (4.3 oz.)	2	.2			0

(USDA): United States Department of Agriculture
*Prepared as Package Directs
[1]Principal source of fat: vegetable shortening.

Food and Description	Measure or Quantity	Sodium (mg.)	— Fats in grams —			Cholesterol (mg.)
			Total	Saturated	Unsaturated	
(Blue Boy)	4 oz.	1	.1			(0)
(Diet Delight)	½ cup (4.4 oz.)	4	Tr.			(0)
(Mott's)	½ cup (4.4 oz.)		.2			(0)
(S and W) *Nutradiet,* low calorie	4 oz.	1	.1			(0)
(S and W) *Nutradiet,* unsweetened	4 oz.	1	<.1			
(Tillie Lewis)	½ cup (4.2 oz.)	<10	.2			(0)
*APPLESAUCE CAKE MIX, raisin (Duncan Hines)	¹/₉ of cake (2.7 oz.)	331	5.3			
APPLE SOFT DRINK, red, low calorie (Shasta)	6 fl. oz.	37	0.			0
APPLE TURNOVER, frozen (Pepperidge Farm)	1 turnover (3.3 oz.)	244	20.6			
APRICOT:						
Fresh (USDA):						
Whole	1 lb. (weighed with pits)	4	.9			0
Whole	3 apricots (about 12 per lb.)	1	.2			0
Halves	1 cup (5.5 oz.)	2	.3			0
Canned, regular pack, solids & liq.:						
Juice pack (USDA)	4 oz.	1	.2			0
Light syrup (USDA)	4 oz.	1	.1			0
Heavy syrup:						
Halves & syrup (USDA)	½ cup (4.4 oz.)	1	.1			0
Halves & syrup (USDA)	4 med. halves with 2 T. syrup (4.3 oz.)	1	.1			0
(Del Monte)	½ cup (4.4 oz.)	15	.4			0
(Hunt's)	½ cup (4.5 oz.)	1	.1			(0)
(Stokely-Van Camp)	½ cup (4.2 oz.)		.1			
Extra heavy syrup (USDA)	4 oz.	1	.1			0
Canned, unsweetened or low calorie:						
Water pack, halves & liq. (USDA)	4 oz.	1	.1			0
Water pack, halves & liq. (USDA)	½ cup (4.3 oz.)	1	.1			0
(Diet Delight)	½ cup (4.4 oz.)	5	<.1			(0)

(USDA): United States Department of Agriculture
*Prepared as Package Directs

Food and Description	Measure or Quantity	Sodium (mg.)	— Fats in grams —			Choles- terol (mg.)
			Total	Satu- rated	Unsatu- rated	
(S and W) *Nutradiet,* low calorie, whole	2 whole (3.5 oz.)	2	<.1			(0)
(S and W) *Nutradiet,* low calorie, halves	4 halves (3.5 oz.)	2	.1			(0)
(S and W) *Nutradiet,* unsweetened, halves	4 halves (3.5 oz.)	3	.1			(0)
(Tillie Lewis)	½ cup (4.3 oz.)	<10	.1			(0)
Dehydrated (USDA):						
Uncooked	4 oz.	37	1.1			0
Cooked, sugar added, solids & liq.	4 oz.	9	.2			0
Dried:						
Uncooked:						
(USDA)	1 lb.	118	2.3			0
(USDA)	14 large halves (½ cup or 2.8 oz.)	21	.4			0
(USDA)	10 small halves (¼ cup or 1.3 oz.)	10	.2			0
(USDA)	½ cup (2.3 oz.)	17	.3			0
(Del Monte)	½ cup (2.3 oz.)	2	.6			0
Cooked (USDA):						
Sweetened	½ cup with liq. (12-13 halves, 5.7 oz.)	11	.2			0
Unsweetened	½ cup with liq. (4.3 oz.)	10	.2			0
Frozen, sweetened, not thawed (USDA)	4 oz.	5	.1			0
APRICOT-APPLE JUICE & PRUNE (Sunsweet)	½ cup		<.1			(0)
APRICOT, CANDIED (USDA)	1 oz.		<.1			0
APRICOT LIQUEUR (Leroux) 60 proof	1 fl. oz.	<1	(0.)			(0)
APRICOT NECTAR, canned:						
Sweetened:						
(USDA)	½ cup (4.2 oz.)	Tr.	.1			0
(Del Monte)	½ cup (4.3 oz.)	8	.2			0
(Heinz)	5½-fl.-oz. can	5	.2			(0)
(Sunsweet)	½ cup		.1			(0)
Low calorie (S and W) *Nutradiet*	4 oz. (by wt.)	2	.1			(0)

(USDA): United States Department of Agriculture
*Prepared as Package Directs

⑦

Food and Description	Measure or Quantity	Sodium (mg.)	Fats in grams Total	Satu- rated	Unsatu- rated	Choles- terol (mg.)
APRICOT-ORANGE PIE						
(Tastykake)	4-oz. pie		15.6			
APRICOT PIE FILLING:						
(Comstock)	1 cup (10¾ oz.)	262	.2			
(Lucky Leaf)	8 oz.	182	.2			
APRICOT & PINEAPPLE NECTAR,						
(S and W) *Nutradiet*, unsweetened	4 oz. (by wt.)	2	.1			(0)
APRICOT & PINEAPPLE PRESERVE:						
Sweetened (Bama)	1 T. (.7 oz.)	2	<.1			(0)
Dietetic or low calorie:						
(Diet Delight)	1 T. (.6 oz.)	3	<.1			(0)
(S and W) *Nutradiet*	1 T. (.5 oz.)		<.1			(0)
(Tillie Lewis)	1 T. (.5 oz.)	3	Tr.			0
APRICOT PRESERVE, sweetened						
(Bama)	1 T. (.7 oz.)	2	<.1			(0)
APRICOT SOUR COCKTAIL						
(National Distillers) *Duet*, 12½% alcohol	2 fl. oz.	Tr.	0.			0
ARROWROOT (Spice Islands)	1 tsp.	Tr.				(0)
ARTICHOKE, Globe or French (See also **JERUSALEM ARTICHOKE**):						
Raw, whole (USDA)	1 lb. (weighed untrimmed)	78	.4			0
Boiled without salt, drained (USDA)	4 oz.	34	.2			0
Frozen, hearts (Birds Eye)	5-6 hearts (3 oz.)		.2			0
ASPARAGUS:						
Raw, whole spears (USDA)	1 lb. (weighed untrimmed)	5	.5			0
Boiled without salt, whole spears (USDA)	4 spears (½″ dia. at base, 2.1 oz.)	<1	.1			0
Boiled without salt, 1½″-2″ pieces, drained	1 cup (5.1 oz.)	1	.3			0

(USDA): United States Department of Agriculture
*Prepared as Package Directs

Food and Description	Measure or Quantity	Sodium (mg.)	Fats in grams Total	Satu- rated	Unsatu- rated	Choles- terol (mg.)
Canned, regular pack:						
Green:						
Spears & liq. (USDA)	4 oz.	268	.3			0
Spears & liq. (USDA)	1 cup (8.6 oz.)	576	.7			0
Spears only (USDA)	1 cup (7.6 oz.)	507	.9			0
Spears only (USDA)	6 med. spears (3.4 oz.)	227	.4			0
Liq. only (USDA)	2 T. (1.1 oz.)	71	Tr.			0
Cut spears & liq. (Green Giant)	½ of 10½-oz. can	492	.1			(0)
Spears & liq. (Green Giant)	⅓ of 15-oz. can	468	.1			(0)
Spears & liq. *LeSueur*	¼ of 1-lb. 3-oz. can	446	.1			(0)
Solids & liq. (Stokely-Van Camp)	½ cup (3.9 oz.)		.4			(0)
Drained solids (Del Monte)	1 cup (7.8 oz.)	417	.8			0
White:						
Spears & liq. (USDA)	4 oz.	268	.3			0
Spears & liq. (USDA)	1 cup (8.4 oz.)	564	.7			0
Spears & liq. (Del Monte)	1 cup (8 oz.)	865	.2			0
Spears only (USDA)	1 cup (7.6 oz.)	507	1.1			0
Spears only (USDA)	6 med. spears (3.4 oz.)	227	.5			0
Spears only (Del Monte)	1 cup (7.6 oz.)	821	.4			0
Liq. only (USDA)	2 T. (1.1 oz.)	71	Tr.			0
Canned, dietetic pack:						
Green:						
Spears & liq. (USDA)	4 oz.	3	.2			0
Spears & liq. (Blue Boy)	4 oz.	1	.2			(0)
Solids & liq. (Diet Delight)	4 oz.	7	.1			(0)
Solids & liq. (Tillie Lewis)	½ cup (4.2 oz.)	<10	.2			(0)
Drained solids (USDA)	4 oz.	3	.3			0
Liq. only (USDA)	4 oz.	3	Tr.			0
(S and W) *Nutradiet*, unseasoned	5 spears (3.5 oz.)	3	.1			(0)
White:						
Spears & liq. (USDA)	4 oz.	5	.2			0
Drained solids (USDA)	4 oz.	5	.2			0
Drained liq. (USDA)	4 oz.	5	Tr.			0
Frozen:						
Cuts & tips, not thawed (USDA)	4 oz.	2	.2			0
Cuts & tips, boiled without salt, drained (USDA)	4 oz.	1	.2			0
Cuts & tips, boiled without salt, drained (USDA)	½ cup (3.2 oz.)	<1	.2			0

(USDA): United States Department of Agriculture
*Prepared as Package Directs

Food and Description	Measure or Quantity	Sodium (mg.)	Total	Satu-rated	Unsatu-rated	Choles-terol (mg.)
				—Fats in grams—		
Cuts (Birds Eye)	½ cup (3.3 oz.)	2	.2			0
Cut spears in butter sauce (Green Giant)	⅓ of 9-oz. pkg.	353	2.6			
Spears, not thawed (USDA)	4 oz.	2	.2			0
Spears, boiled without salt, drained (USDA)	4 oz.	1	.2			0
Spears (Birds Eye)	⅓ of 10-oz. pkg.	2	.2			0
Spears with Hollandaise sauce (Birds Eye)	⅓ of 10-oz. pkg.	88	8.5			57
ASPARAGUS SOUP, Cream of, canned:						
Condensed (USDA)	8 oz. (by wt.)	1861	3.2			
*Prepared with equal volume water (USDA)	1 cup (8.5 oz.)	984	1.7			
*Prepared with equal volume milk (USDA)[1]	1 cup (8.5 oz.)	1046	5.8	2.	4.	
*(Campbell)	1 cup (8 oz.)	940	3.2	2.	1.	
AUNT JEMIMA SYRUP	¼ cup	2	0.			(0)
AVOCADO, peeled, pitted (USDA):						
All commercial varieties:						
Whole	1 lb. (weighed with seed & skin)	14	55.8	11.	45.	0
Diced	½ cup (2.6 oz.)	3	12.1	2	10.	0
Mashed	½ cup (4.1 oz.)	5	19.0	3.	16.	0
California varieties, mainly Fuerte:						
Whole	½ avocado (3⅛" dia.)	4	18.4	3.	15.	0
½" cubes	½ cup (2.7 oz.)	3	12.9	2.	11.	0
Florida varieties:						
Whole	½ avocado (3⅝" dia.)	6	16.7	3.	14.	0
½" cubes	½ cup (2.7 oz.)	3	8.4	2.	7.	0
***AWAKE** (Birds Eye)	½ cup (4.4 oz.)	5	.1			0
AYDS, vanilla or chocolate	1 piece (7 grams)		.6			2

B

BABY FOOD:

Apple:						
& apricot, junior (Beech-Nut)	7¾ oz.	28	.7			
& apricot, strained (Beech-Nut)	4¾ oz.	21	.4			

(USDA): United States Department of Agriculture
*Prepared as Package Directs
[1]Principal source of fat: milk.

Food and Description	Measure or Quantity	Sodium (mg.)	—Fats in grams—			Choles-terol (mg.)
			Total	Satu-rated	Unsatu-rated	
& cranberry, junior (Heinz)	7¾ oz.	13	.2			
& cranberry, strained (Heinz)	4¾ oz.	9	.1			
& honey, junior (Heinz)	7½ oz.	15	.2			
& honey with tapioca, strained (Heinz)	4½ oz.	7	Tr.			
& pear, junior (Heinz)	7¾ oz.	18	.2			
& pear, strained (Heinz)	4½ oz.	9	.1			
Dutch, dessert (Gerber):						
Junior	7⁸/₁₀ oz.	61	1.6			
Strained	4⁷/₁₀ oz.	36	1.3			
Apple-apricot juice, strained (Heinz)	4½ fl. oz.	3	.4			
Apple Betty (Beech-Nut):						
Junior	7¾ oz.	31	1.1			
Strained	4¾ oz.	21	.7			
Apple-cherry juice:						
Strained (Beech-Nut)	4¹/₅ fl. oz. (4.4 oz.)	1				
Strained (Gerber)	4¹/₅ fl. oz. (4.6 oz.)	2	.1			
Strained (Heinz)	4½ fl. oz.	3	.1			
Apple-grape juice:						
Strained (Beech-Nut)	4¹/₅ fl. oz. (4.4 oz.)	1				
Strained (Heinz)	4½ fl. oz.	4	Tr.			
Apple juice:						
Strained (Beech-Nut)	4¹/₅ fl. oz. (4.4 oz.)	1				
Strained (Gerber)	4¹/₅ fl. oz. (4.6 oz.)	3	.1			
Strained (Heinz)	4½ fl. oz.	2	.1			
Apple pie (Heinz):						
Junior	7¾ oz.	8	2.2			
Strained	4¾ oz.	6	1.4			
Apple-pineapple juice, strained (Heinz)	4½ fl. oz.	3	.1			
Apple-prune & honey (Heinz):						
Junior	7½ oz.		.2			
With tapioca, strained	4½ oz.		.1			
Apple-prune juice, strained (Heinz)	4½ fl. oz.	6	Tr.			
Applesauce:						
Junior (Beech-Nut)	7¾ oz.	33	.7			
Junior (Gerber)	7⁸/₁₀ oz.	4	.5			
Junior (Heinz)	7¾ oz.	18	.2			
Strained (Beech-Nut)	4¾ oz.	17	.3			
Strained (Gerber)	4⁷/₁₀ oz.	2	.2			
Strained (Heinz)	4½ oz.	11	.2			

(USDA): United States Department of Agriculture
*Prepared as Package Directs

(11)

Food and Description	Measure or Quantity	Sodium (mg.)	—Fats in grams—			Cholesterol (mg.)
			Total	Saturated	Unsaturated	
& apricots, junior (Gerber)	7⁸/₁₀ oz.	3	.7			
& apricots, junior (Heinz)	7¾ oz.	15	.3			
& apricots, strained (Gerber)	4⁷/₁₀ oz.	2	.2			
& apricots, strained (Heinz)	4¾ oz.	8	.1			
& cherries, junior (Beech-Nut)	7¾ oz.	18	1.1			
& cherries, strained (Beech-Nut)	4¾ oz.	13	.9			
& pineapple, junior (Gerber)	7⁸/₁₀ oz.	4	.3			
& pineapple, strained (Gerber)	4⁷/₁₀ oz.	2	.3			
& raspberries, junior (Beech-Nut)	7¾ oz.	28	1.1			
& raspberries, strained (Beech-Nut)	4¾ oz.	25	.5			
Apricot with tapioca:						
Junior (Beech-Nut)	7¾ oz.	112	.4			
Junior (Gerber)	7⁸/₁₀ oz.	70	.1			
Junior (Heinz)	7¾ oz.	11	.3			
Strained (Beech-Nut)	4¾ oz.	52	.1			
Strained (Gerber)	4⁷/₁₀ oz.	42	.1			
Strained (Heinz)	4¾ oz.	15	.5			
Banana:						
Strained (Heinz)	4½ oz.	7	.3			
Pie, junior (Heinz)	7¾ oz.	10	2.3			
Pie, strained (Heinz)	4¾ oz.	6	1.4			
& pineapple, junior (Heinz)	7¾ oz.	14	.2			
& pineapple, strained (Heinz)	4¾ oz.	12	.1			
& pineapple with tapioca:						
Junior (Beech-Nut)	7¾ oz.	120	.2			
Junior (Gerber)	7⁸/₁₀ oz.	214	.3			
Strained (Beech-Nut)	4¾ oz.	95	.5			
Strained (Gerber)	4⁷/₁₀ oz.	40	.3			
Dessert, junior (Beech-Nut)	7¾ oz.	237	.2			
Pudding, junior (Gerber)	7⁸/₁₀ oz.	273	1.6			
With tapioca:						
Strained (Beech-Nut)	4¾ oz.	95	.3			
Strained (Gerber)	4⁷/₁₀ oz.	40	.3			
Bean, green:						
Junior (Beech-Nut)	7¼ oz.	252	.2			
Strained (Beech-Nut)	4½ oz.	146	.3			
Strained (Gerber)	4½ oz.	146	.1			
Strained (Heinz)	4½ oz.	151	.3			
Creamed with bacon, junior (Gerber)	7½ oz.	652	5.4			
In butter sauce, junior (Beech-Nut)	7¼ oz.	195	1.8			

(USDA): United States Department of Agriculture
*Prepared as Package Directs

Food and Description	Measure or Quantity	Sodium (mg.)	Total	Satu-rated	Unsatu-rated	Choles-terol (mg.)
In butter sauce, strained (Beech-Nut)	4½ oz.	118	1.2			
With potatoes & ham, casserole, toddler (Gerber)	6¹/₅ oz.	742	6.0			
Beef:						
Junior (Beech-Nut)	3½ oz.	162	3.9			
Junior (Gerber)	3½ oz.	154	4.0			
Strained (Beech-Nut)	3½ oz.	162	5.4			
Strained (Gerber)	3½ oz.	180	4.1			
Beef & beef broth (Heinz):						
Junior	3½ oz.	157	4.4			
Strained	3½ oz.	146	4.1			
Beef & beef heart, strained (Gerber)	3½ oz.	149	3.7			
Beef dinner:						
Junior (Beech-Nut)	4½ oz.	125	7.2			
Strained (Beech-Nut)	4½ oz.	165	8.4			
& noodles, junior (Beech-Nut)	7½ oz.	273	6.4			
& noodles, junior (Gerber)	7½ oz.	222	2.1			
& noodles, strained (Beech-Nut)	4½ oz.	205	3.6			
& noodles, strained (Gerber)	4½ oz.	163	1.1			
& noodles, strained (Heinz)	4½ oz.	170	1.2			
With vegetables:						
Junior (Gerber)	4½ oz.	279	4.6			
Strained (Gerber)	4½ oz.	194	4.7			
Strained (Heinz)	4¾ oz.	126	5.4			
With vegetables & cereal, junior (Heinz)	4¾ oz.	124	5.8			
Beef lasagna, toddler (Gerber)	6¹/₅ oz.	768	4.1			
Beef liver, strained (Gerber)	3½ oz.	160	3.2			
Beef liver soup, strained (Heinz)	4½ oz.	115	.6			
Beef stew, toddler (Gerber)	6¹/₅ oz.	871	2.3			
Beet:						
Strained (Gerber)	4½ oz.	205	.1			
Strained (Heinz)	4½ oz.	98	.2			
Blueberry buckle (Gerber):						
Junior	7⁸/₁₀ oz.	101	.2			
Strained	4½ oz.	57	.1			
Butterscotch pudding (Gerber):						
Junior	7½ oz.	254	2.8			
Strained	4½ oz.	157	3.1			
Caramel pudding (Beech-Nut):						
Junior	7¾ oz.	112	.9			

(USDA): United States Department of Agriculture
*Prepared as Package Directs

⑬

Food and Description	Measure or Quantity	Sodium (mg.)	— Fats in grams —			Choles- terol (mg.)
			Total	Satu- rated	Unsatu- rated	
Strained	4¾ oz.	84	.7			
Carrot:						
Junior (Beech-Nut)	7½ oz.	269	.2			
Junior (Gerber)	7½ oz.	259	.4			
Junior (Heinz)	7¾ oz.	248	.3			
Strained (Beech-Nut)	4½ oz.	163	.1			
Strained (Gerber)	4½ oz.	156	.2			
Strained (Heinz)	4½ oz.	217	.2			
& pea, junior (Gerber)	7½ oz.	213	.7			
In butter sauce (Beech-Nut):						
Junior	7½ oz.	301	1.5			
Strained	4½ oz.	182	.8			
Cereal, dry:						
Barley (Gerber)	3 T. (7 grams)	1	.3			
Barley, instant (Heinz)	2 T.	19	.1			
High protein (Gerber)	3 T. (7 grams)	1	.4			
High protein, instant (Heinz)	2 T.	15	.2			
Hi-protein (Beech-Nut)	1 oz.	24	1.4			
Mixed (Beech-Nut)	1 oz.	26	1.3			
Mixed (Gerber)	3 T. (7 grams)	1	.3			
Mixed (Heinz)	2 T.	7	.2			
Mixed, honey (Beech-Nut)	1 oz.	29	1.4			
Mixed, with banana (Gerber)	3 T. (7 grams)	12	.3			
Oatmeal (Beech-Nut)	1 oz.	21	2.0			
Oatmeal (Gerber)	3 T. (7 grams)	1	.6			
Oatmeal, honey (Beech-Nut)	1 oz.	34	1.8			
Oatmeal, instant (Heinz)	2 T.	9	.3			
Oatmeal, with banana (Gerber)	3 T. (7 grams)	13	.4			
Rice (Beech-Nut)	1 oz.	26	1.6			
Rice (Gerber)	3 T. (7 grams)	1	.3			
Rice, honey (Beech-Nut)	1 oz.	20	1.1			
Rice, instant (Heinz)	2 T.	30	.1			
Rice, with strawberry (Gerber)	3 T. (7 grams)	5	.4			
Cereal, or mixed cereal:						
With applesauce & banana:						
Junior (Gerber)	7⁸/₁₀ oz.	196	.6			
Strained (Gerber)	4⁷/₁₀ oz.	134	.9			
Strained (Heinz)	4¾ oz.	8	.5			
With egg yolks & bacon:						
Junior (Beech-Nut)	7½ oz.	229	12.3			
Junior (Gerber)	7½ oz.	284	8.8			
Junior (Heinz)	7½ oz.	872	8.7			
Strained (Beech-Nut)	4½ oz.	198	7.3			

(USDA): United States Department of Agriculture
*Prepared as Package Directs

Food and Description	Measure or Quantity	Sodium (mg.)	—Fats in grams—			Choles-terol (mg.)
			Total	Satu-rated	Unsatu-rated	
Strained (Gerber)	4½ oz.	172	5.0			
Strained (Heinz)	4½ oz.	413	6.7			
With fruit, strained (Beech-Nut)	4¾ oz.	142	.3			
High protein with apple & banana, strained (Heinz)	4¾ oz.	20	.6			
Oatmeal, with applesauce & banana, junior (Gerber)	7⁸/₁₀ oz.	306	1.5			
Oatmeal with applesauce & banana, strained (Gerber)	4⁷/₁₀ oz.	196	1.2			
Oatmeal, with fruit, strained (Beech-Nut)	4¾ oz.	204	.9			
Rice, with applesauce & banana, strained (Gerber)	4⁷/₁₀ oz.	163	.8			
Cheese:						
Cottage, creamed with pineapple:						
Junior (Beech-Nut)	7¾ oz.	322	2.2			
Strained (Gerber)	4⁷/₁₀ oz.	195	6.0			
Cottage, creamed with pineapple juice, strained (Beech-Nut)	4¾ oz.	154	1.3			
Cottage, dessert, with pineapple (Gerber):						
Junior	7⁸/₁₀ oz.	301	2.6			
Strained	4½ oz.	175	1.5			
Cottage, with banana (Heinz):						
Junior	7¾ oz.	23	.5			
Strained	4½ oz.	14	.3			
Cherry vanilla pudding (Gerber):						
Junior	7⁸/₁₀ oz.	110	1.6			
Strained	4⁷/₁₀ oz.	69	1.3			
Chicken:						
Junior (Beech-Nut)	3½ oz.	162	4.9			
Junior (Gerber)	3½ oz.	206	8.2			
Strained (Beech-Nut)	3½ oz.	142	5.2			
Strained (Gerber)	3½ oz.	168	8.5			
Chicken & chicken broth (Heinz):						
Junior	3½ oz.	134	5.7			
Strained	3½ oz.	122	6.4			
Chicken dinner:						
Junior (Beech-Nut)	4½ oz.	221	4.6			
Strained (Beech-Nut)	4½ oz.	191	5.0			
Noodle:						
Junior (Beech-Nut)	7½ oz.	199	2.1			

(USDA): United States Department of Agriculture
*Prepared as Package Directs

Food and Description	Measure or Quantity	Sodium (mg.)	Fats in grams Total	Satu- rated	Unsatu- rated	Choles- terol (mg.)
Junior (Gerber)	7½ oz.	320	1.0			
Junior (Heinz)	7½ oz.	213	4.6			
Strained (Beech-Nut)	4½ oz.	131	1.2			
Strained (Gerber)	4½ oz.	147	.9			
Strained (Heinz)	4½ oz.	139	2.5			
With vegetables:						
Junior (Beech-Nut)	7½ oz.	191	1.3			
Junior (Gerber)	4½ oz.	262	5.7			
Junior (Heinz)	4¾ oz.	145	7.7			
Strained (Beech-Nut)	4½ oz.	128	1.0			
Strained (Gerber)	4½ oz.	172	5.3			
Strained (Heinz)	4¾ oz.	138	7.3			
Chicken soup:						
Junior (Heinz)	7½ oz.	210	3.4			
Strained (Heinz)	4½ oz.	116	2.1			
Cream of, junior (Gerber)	7½ oz.	293	2.1			
Cream of, strained (Gerber)	4½ oz.	177	1.3			
Chicken stew, toddler (Gerber)	6 oz.	716	4.1			
Chicken sticks:						
Junior (Beech-Nut)	2½ oz.	203	11.1			
Junior (Gerber)	2½ oz.	300	9.6			
Cookie, animal-shaped (Gerber)	1 cookie (6 grams)	28	1.0			
Cookie, assorted (Beech-Nut)	½ oz.	49	1.8			
Corn, creamed:						
Junior (Gerber)	7½ oz.	214	.5			
Junior (Heinz)	7½ oz.	232	.8			
Strained (Beech-Nut)	4½ oz.	118	.9			
Strained (Gerber)	4½ oz.	129	.3			
Strained (Heinz)	4½ oz.	103	.5			
Custard:						
Junior (Beech-Nut)	7¾ oz.	272	3.9			
Junior (Heinz)	7¾ oz.	270	5.9			
Strained (Beech-Nut)	4½ oz.	150	2.2			
Strained (Heinz)	4½ oz.	84	3.0			
Chocolate, junior (Gerber)	7⁸/₁₀ oz.	221	3.3			
Chocolate, strained (Beech-Nut)	4½ oz.	133	2.0			
Chocolate, strained (Gerber)	4½ oz.	129	2.1			
Vanilla, junior (Gerber)	7½ oz.	229	3.5			
Vanilla, strained (Gerber)	4½ oz.	129	1.6			
Egg yolk:						
Strained (Beech-Nut)	3⅓ oz.	94	15.6			
Strained (Gerber)	3³/₁₀ oz.	168	16.7			
Strained (Heinz)	3¼ oz.	110	15.9			

(USDA): United States Department of Agriculture
*Prepared as Package Directs

16

Food and Description	Measure or Quantity	Sodium (mg.)	—Fats in grams—			Choles- terol (mg.)
			Total	Satu- rated	Unsatu- rated	
& bacon, strained (Beech-Nut)	3⅓ oz.	206	13.3			
& ham, strained (Gerber)	3³/₁₀ oz.	312	16.1			
Fruit (Heinz):						
Mixed, & honey, junior	7½ oz.	10	.2			
Mixed, & honey, with tapioca, strained	4½ oz.	5	.2			
Fruit dessert:						
Junior (Heinz)	7¾ oz.	14	0.			
Strained (Heinz)	4½ oz.	13	.1			
Tropical, junior (Beech-Nut)	7¾ oz.	191	.4			
With tapioca:						
Junior (Beech-Nut)	7¾ oz.	116	.7			
Junior (Gerber)	7⁸/₁₀ oz.	91	.3			
Strained (Beech-Nut)	4¾ oz.	77	.1			
Strained (Gerber)	4⁷/₁₀ oz.	51	.4			
Fruit juice:						
Mixed, strained (Beech-Nut)	4¹/₅ fl. oz. (4.4 oz.)	1	.2			
Mixed, strained (Gerber)	4¹/₅ fl. oz. (4.6 oz.)	3	.3			
Ham:						
Junior (Gerber)	3½ oz.	204	6.0			
Strained (Beech-Nut)	3½ oz.	183	6.6			
Strained (Gerber)	3½ oz.	200	6.2			
Ham dinner:						
Junior (Beech-Nut)	4½ oz.	224	6.9			
Strained (Beech-Nut)	4½ oz.	228	8.2			
With vegetables:						
Junior (Gerber)	4½ oz.	263	3.6			
Junior (Heinz)	4¾ oz.	297	7.1			
Strained (Gerber)	4½ oz.	224	3.8			
Strained (Heinz)	4¾ oz.	263	7.0			
Lamb:						
Junior (Beech-Nut)	3½ oz.	173	4.8			
Junior (Gerber)	3½ oz.	201	3.9			
Strained (Beech-Nut)	3½ oz.	142	4.5			
Strained (Gerber)	3½ oz.	164	4.0			
& noodles, junior (Beech-Nut)	7½ oz.	231	6.4			
Lamb & lamb broth (Heinz):						
Junior	3½ oz.	158	4.6			
Strained	3½ oz.	116	3.5			
Liver with liver broth, strained (Heinz)	3½ oz.	170	2.7			
Macaroni:						
Alphabets & beef casserole, toddler (Gerber)	6¹/₅ oz.	881	3.1			

(USDA): United States Department of Agriculture
*Prepared as Package Directs

(17)

Food and Description	Measure or Quantity	Sodium (mg.)	—Fats in grams— Total	Satu- rated	Unsatu- rated	Choles- terol (mg.)
& bacon, junior (Beech-Nut)	7½ oz.	299	10.6			
& beef with vegetables, junior (Beech-Nut)	7½ oz.	131	5.9			
With tomato, beef & bacon:						
Junior (Gerber)	7½ oz.	265	3.0			
Junior (Heinz)	7½ oz.	293	4.4			
Strained (Gerber)	4½ oz.	137	2.4			
Strained (Heinz)	4½ oz.	166	3.0			
With tomato sauce, beef & bacon dinner, strained (Beech-Nut)	4½ oz.	166	5.6			
Meat sticks, junior (Beech-Nut)	2½ oz.	279	9.9			
Meat sticks, junior (Gerber)	2½ oz.	326	7.5			
Noodles & beef, junior (Heinz)	7½ oz.	234	2.8			
Orange-apple juice, strained:						
(Beech-Nut)	4¹/₅ fl. oz. (4.4 oz.)	1	.2			
(Gerber)	4¹/₅ fl. oz. (4.6 oz.)	3	.5			
Orange-apple-banana juice, strained:						
(Gerber)	4¹/₅ fl. oz. (4.6 oz.)	4	.3			
(Heinz)	4½ fl. oz.	4	.2			
Orange-apricot juice, strained:						
(Beech-Nut)	4¹/₅ fl. oz. (4.4 oz.)	1	.4			
(Gerber)	4¹/₅ fl. oz. (4.6 oz.)	2	.2			
(Heinz)	4½ fl. oz.	2	.1			
Orange-banana juice, strained (Beech-Nut)	4¹/₅ fl. oz. (4.4 oz.)	1	.5			
Orange juice, strained:						
(Beech-Nut)	4¹/₅ fl. oz. (4.4 oz.)	1	.4			
(Gerber)	4¹/₅ fl. oz. (4.6 oz.)	3	.5			
(Heinz)	4½ fl. oz.	1	.2			
Orange-pineapple dessert, strained (Beech-Nut)	4¾ oz.	130	.7			
Orange-pineapple juice, strained:						
(Beech-Nut)	4¹/₅ fl. oz. (4.4 oz.)	1	.4			
(Gerber)	4¹/₅ fl. oz. (4.6 oz.)	2	.2			
(Heinz)	4½ fl. oz.	2	Tr.			
Orange pudding, strained:						
(Gerber)	4⁷/₁₀ oz.	129	1.2			
(Heinz)	4½ oz.	79	.4			
Pea:						
Strained (Beech-Nut)	4½ oz.	129	.5			
Strained (Gerber)	4½ oz.	129	.4			

(USDA): United States Department of Agriculture
*Prepared as Package Directs

Food and Description	Measure or Quantity	Sodium (mg.)	— Fats in grams —			Choles- terol (mg.)
			Total	Satu- rated	Unsatu- rated	
Pea, creamed (Heinz):						
Junior	7¾ oz.	205	3.8			
Strained	4½ oz.	88	2.3			
Pea, in butter sauce (Beech-Nut):						
Junior	7¼ oz.	269	2.5			
Strained	4½ oz.	164	1.7			
Peach:						
Junior (Beech-Nut)	7¾ oz.	31	.4			
Junior (Gerber)	$7^8/_{10}$ oz.	12	.7			
Junior (Heinz)	7½ oz.	10	.3			
Strained (Beech-Nut)	4¾ oz.	15	.3			
Strained (Gerber)	$4^7/_{10}$ oz.	3	.1			
Strained (Heinz)	4½ oz.	7	.3			
Peach cobbler (Gerber):						
Junior	$7^8/_{10}$ oz.	38	.3			
Strained	$4^7/_{10}$ oz.	23	.3			
Peach & honey (Heinz):						
Junior	7½ oz.	11	.1			
With tapioca, strained	4½ oz.	9	.2			
Peach Melba (Beech-Nut):						
Junior	7¾ oz.	177	.9			
Strained	4¾ oz.	51	.4			
Peach pie (Heinz):						
Junior	7¾ oz.	13	2.2			
Strained	4¾ oz.	8	1.3			
Pear:						
Junior (Beech-Nut)	7½ oz.	19	0.			
Junior (Gerber)	$7^8/_{10}$ oz.	4	.4			
Junior (Heinz)	7¾ oz.	2	.3			
Strained (Beech-Nut)	4½ oz.	19	.3			
Strained (Gerber)	$4^7/_{10}$ oz.	3	.1			
Strained (Heinz)	4½ oz.	3	.3			
Pear & pineapple:						
Junior (Beech-Nut)	7½ oz.	32	.2			
Junior (Gerber)	$7^8/_{10}$ oz.	4	.6			
Junior (Heinz)	7¾ oz.	3	.2			
Strained (Beech-Nut)	4½ oz.	31	.1			
Strained (Gerber)	$4^7/_{10}$ oz.	3	.2			
Strained (Heinz)	4¾ oz.	2	.2			
Pineapple dessert, strained (Beech-Nut)	4¾ oz.	163	.1			
Pineapple-grapefruit juice drink, strained (Gerber)	$4^1/_5$ fl. oz. (4.6 oz.)	3	.1			

(USDA): United States Department of Agriculture
*Prepared as Package Directs

Food and Description	Measure or Quantity	Sodium (mg.)	Fats in grams Total	Satu- rated	Unsatu- rated	Choles- terol (mg.)
Pineapple juice, strained (Heinz)	4½ fl. oz.	2	.2			
Pineapple-orange dessert (Heinz):						
Junior	7¾ oz.	46	.4			
Strained	4½ oz.	31	.2			
Pineapple pie (Heinz):						
Junior	7¾ oz.	15	3.3			
Strained	4¾ oz.	16	1.8			
Plum with tapioca:						
Junior (Beech-Nut)	7¾ oz.	90	.4			
Junior (Gerber)	7⁸/₁₀ oz.	12	.4			
Strained (Beech-Nut)	4¾ oz.	47	.3			
Strained (Gerber)	4⁷/₁₀ oz.	6	.2			
Strained (Heinz)	4½ oz.	8	.1			
Pork:						
Junior (Beech-Nut)	3½ oz.	178	5.9			
Junior (Gerber)	3½ oz.	241	6.1			
Strained (Beech-Nut)	3½ oz.	170	6.4			
Strained (Gerber)	3½ oz.	218	6.0			
Pork with pork broth, strained (Heinz)	3½ oz.	98	4.0			
Potatoes, creamed, with ham, toddler (Gerber)	6 oz.	750	9.3			
Pretzel (Gerber)	1 piece (5 grams)	30	.1			
Prune-orange juice:						
Strained (Beech-Nut)	4¹/₅ fl. oz. (4.4 oz.)	2	.1			
Strained (Gerber)	4¹/₅ fl. oz. (4.6 oz.)	5	.2			
Strained (Heinz)	4½ fl. oz.	3	.1			
Prune with tapioca:						
Junior (Beech-Nut)	7¾ oz.	81	.2			
Junior (Gerber)	7⁸/₁₀ oz.	44	.5			
Strained (Beech-Nut)	4¾ oz.	59	.1			
Strained (Gerber)	4⁷/₁₀ oz.	26	.2			
Strained (Heinz)	4¾ oz.	6	.2			
Raspberry cobbler (Gerber):						
Junior	7⁸/₁₀ oz.	101	.2			
Strained	4½ oz.	62	.1			
Similac:						
Advance	1 fl. oz. (1 oz.)	12	.5			<1
Isomil	1 fl. oz. (1 oz.)	9	1.0			0
*Powder	1 fl. oz. (1 oz.)	11	1.0			Tr.
Ready-to-feed	1 fl. oz. (1 oz.)	9	1.0			Tr.
Spaghetti & meat balls, toddler (Gerber)	6¹/₅ oz.	813	1.2			

(USDA): United States Department of Agriculture
*Prepared as Package Directs

Food and Description	Measure or Quantity	Sodium (mg.)	—Fats in grams— Total	Satu- rated	Unsatu- rated	Choles- terol (mg.)
Spaghetti, tomato sauce & beef:						
Junior (Beech-Nut)	7½ oz.	352	6.8			
Junior (Gerber)	7½ oz.	483	1.6			
Junior (Heinz)	7½ oz.	272	4.4			
Spaghetti, tomato sauce & meat, strained (Heinz)	4½ oz.	48	2.9			
Spinach, creamed:						
Junior (Gerber)	7½ oz.	320	2.2			
Strained (Gerber)	4½ oz.	150	1.0			
Strained (Heinz)	4½ oz.	112	1.5			
Split pea with bacon, junior (Gerber)	7½ oz.	267	5.4			
Split pea, vegetables & bacon:						
Junior (Heinz)	7½ oz.	334	8.9			
Strained (Heinz)	4½ oz.	213	4.8			
Split pea, vegetables & ham, junior (Beech-Nut)	7½ oz.	390	2.8			
Squash:						
Junior (Beech-Nut)	7½ oz.	254	.4			
Junior (Gerber)	7½ oz.	214	.5			
Strained (Beech-Nut)	4½ oz.	148	.4			
Strained (Gerber)	4½ oz.	129	.2			
Strained (Heinz)	4½ oz.	108	.3			
In butter sauce (Beech-Nut):						
Junior	7½ oz.	267	1.7			
Strained	4½ oz.	147	1.0			
Sweet potato:						
Junior (Beech-Nut)	7¾ oz.	241	.2			
Junior (Gerber)	7⁸/₁₀ oz.	222	.2			
Strained (Beech-Nut)	4½ oz.	223	1.3			
Strained (Gerber)	4⁷/₁₀ oz.	135	.1			
Strained (Heinz)	4½ oz.	40	.1			
In butter sauce (Beech-Nut):						
Junior	7¾ oz.	250	1.5			
Strained (Beech-Nut)	4½ oz.	202	.9			
Teething biscuit (Gerber)	1 piece (.4 oz.)	60	.6			
Teething ring, honey (Beech-Nut)	½ oz.	101	.9			
Tuna with noodles, strained (Heinz)	4½ oz.	176	.5			
Turkey:						
Junior (Beech-Nut)	3½ oz.	199	4.5			
Junior (Gerber)	3½ oz.	166	4.7			
Strained (Beech-Nut)	3½ oz.	169	5.1			

(USDA): United States Department of Agriculture
*Prepared as Package Directs

Food and Description	Measure or Quantity	Sodium (mg.)	Total	Satu-rated	Unsatu-rated	Choles-terol (mg.)
			—Fats in grams—			
Strained (Gerber)	3½ oz.	180	8.1			
Turkey dinner:						
Junior (Beech-Nut)	4½ oz.	225	3.2			
Strained (Beech-Nut)	4½ oz.	250	4.0			
With rice:						
Junior (Gerber)	7½ oz.	316	1.5			
Strained (Beech-Nut)	4½ oz.	175	.8			
Strained (Gerber)	4½ oz.	129	1.2			
With rice & vegetables, junior						
(Beech-Nut)	7½ oz.	199	1.1			
With vegetables:						
Junior (Gerber)	4½ oz.	322	4.5			
Strained (Gerber)	4½ oz.	195	4.1			
Strained (Heinz)	4¾ oz.	149	3.3			
Tutti frutti dessert (Heinz):						
Junior	7¾ oz.	79	.5			
Strained	4½ oz.	69	.5			
Veal:						
Junior (Beech-Nut)	3½ oz.	158	3.7			
Junior (Gerber)	3½ oz.	166	4.2			
Strained (Beech-Nut)	3½ oz.	169	6.1			
Strained (Gerber)	3½ oz.	177	4.0			
Veal dinner:						
Junior (Beech-Nut)	4½ oz.	145	8.2			
Strained (Beech-Nut)	4½ oz.	177	6.9			
With vegetables:						
Junior (Gerber)	4½ oz.	289	1.6			
Junior (Heinz)	4¾ oz.	150	4.7			
Strained (Gerber)	4½ oz.	161	1.8			
Strained (Heinz)	4¾ oz.	167	2.7			
Veal & veal broth (Heinz):						
Junior	3½ oz.	200	3.8			
Strained	3½ oz.	143	3.6			
Vegetables:						
Garden, strained (Beech-Nut)	4½ oz.	134	.3			
Garden, strained (Gerber)	4½ oz.	142	.3			
Mixed, junior (Gerber)	7½ oz.	286	.3			
Mixed, junior (Heinz)	7½ oz.	208	.1			
Mixed, strained (Gerber)	4½ oz.	218	.1			
Vegetables & bacon:						
Junior (Beech-Nut)	7½ oz.	337	8.3			
Junior (Gerber)	7½ oz.	362	5.2			
Junior (Heinz)	7½ oz	289	8.4			

(USDA): United States Department of Agriculture
*Prepared as Package Directs

Food and Description	Measure or Quantity	Sodium (mg.)	—Fats in grams—			Choles- terol (mg.)
			Total	Satu- rated	Unsatu- rated	
Strained (Beech-Nut)	4½ oz.	137	5.1			
Strained (Gerber)	4½ oz.	183	3.9			
Strained (Heinz)	4½ oz.	170	2.4			
Vegetables & beef:						
Junior (Beech-Nut)	7½ oz.	337	6.4			
Junior (Gerber)	7½ oz.	310	3.3			
Junior (Heinz)	7½ oz.	270	2.0			
Strained (Beech-Nut)	4½ oz.	157	4.1			
Strained (Gerber)	4½ oz.	173	2.6			
Strained (Heinz)	4½ oz.	226	2.0			
Vegetables & chicken (Gerber):						
Junior	7½ oz.	227	.6			
Strained	4½ oz.	137	.8			
Vegetables, dumplings, beef & bacon (Heinz):						
Junior	7½ oz.	253	6.5			
Strained	4½ oz.	182	4.3			
Vegetables, egg noodles & chicken (Heinz):						
Junior	7½ oz.	210	4.9			
Strained	4½ oz.	134	2.5			
Vegetables, egg noodles & turkey, junior (Heinz)	7½ oz.	206	3.2			
Vegetables & ham:						
Junior (Heinz)	7½ oz.	238	5.8			
Strained (Beech-Nut)	4½ oz.	259	3.1			
With bacon, junior (Gerber)	7½ oz.	445	4.0			
With bacon, strained (Gerber)	4½ oz.	298	2.6			
With bacon, strained (Heinz)	4½ oz.	198	4.1			
Vegetables & lamb:						
Junior (Beech-Nut)	7½ oz.	290	5.1			
Junior (Gerber)	7½ oz.	273	2.9			
Junior (Heinz)	7½ oz.	251	2.5			
Strained (Gerber)	4½ oz.	184	1.7			
Strained (Heinz)	4½ oz.	149	1.3			
Vegetables & liver:						
Junior (Beech-Nut)	7½ oz.	184	1.1			
Strained (Beech-Nut)	4½ oz.	186	.5			
With bacon, junior (Gerber)	7½ oz.	498	2.0			
With bacon, strained (Gerber)	4½ oz.	262	3.6			
Vegetables & turkey:						
Junior (Gerber)	7½ oz.	224	.6			
Strained (Gerber)	4½ oz.	157	.5			

(USDA): United States Department of Agriculture
*Prepared as Package Directs

| Food and Description | Measure or Quantity | Sodium (mg.) | — Fats in grams — | | | Choles- terol (mg.) |
			Total	Satu- rated	Unsatu- rated	
Toddler, casserole (Gerber)	6¹/₅ oz.	991	5.6			
Vegetable soup:						
Junior (Beech-Nut)	7½ oz.	189	.2			
Junior (Heinz)	7½ oz.	225	1.3			
Strained (Beech-Nut)	4½ oz.	163	.1			
Strained (Heinz)	4½ oz.	292	.8			

BAC ONION:
(Lawry's)	1 pkg. (3½ oz.)		8.2			
(Lawry's)	1 tsp. (4 grams)		.3			

BACO NOIR BURGUNDY WINE
(Great Western) 12.5% alcohol	3 fl. oz.	36	0.			0

BAC*OS (General Mills)
*BAC*OS* (General Mills)	1 T.	341	1.3			

BACON, cured:
Raw:						
(USDA) sliced	1 lb.	3084	314.3	101.	213.	
(USDA) sliced	1 oz.	193	19.6	6.	13.	
(USDA) slab	1 lb. (weighed with rind)	2900	295.5	95.	200.	
(Hormel) Black Label	1 piece (.8 oz.)	270	12.9			
(Hormel) *Range Brand*	1 piece (1.6 oz.)	832	28.1	9.	15.	34
(Wilson)	1 oz.	193	17.7	6.	11.	20
Broiled or fried, crisp, drained:						
(USDA) thin slice	1 slice (5 grams)	51	2.6	<1.	2.	
(USDA) medium slice	1 slice (8 grams)	77	3.9	1.	3.	
(USDA) thick slice	1 slice (.4 oz.)	123	6.2	2.	4.	
(Oscar Mayer) 11–14 slices per lb. raw	1 slice (.4 oz.)	209	6.3	2.	4.	6
(Oscar Mayer) 18–26 slices per lb. raw	1 slice (6 grams)	114	3.4	1.	2.	3
(Oscar Mayer) 25–30 slices per ¾ lb. raw	1 slice (4 grams)	76	2.3	<1.	1.	2
Canned (USDA)	3 oz.		60.8	20.	41.	

BACON BITS:
(Wilson)	1 oz.	778	10.2	4.	6.	27
Imitation:						
(Durkee)	1 tsp. (2 grams)	229	.4			
(French's)	1 tsp. (2 grams)	40	.3			
(McCormick)	1 tsp. (2 grams)	14	.4			

(USDA): United States Department of Agriculture
*Prepared as Package Directs

Food and Description	Measure or Quantity	Sodium (mg.)	—Fats in grams—			Choles-terol (mg.)
			Total	Satu-rated	Unsatu-rated	
BACON, CANADIAN:						
Unheated:						
(USDA)	1 oz.	538	4.1	1.	3.	
(Oscar Mayer)	1-oz. slice	343	2.3			
(Wilson)	1 oz.	293	2.3			18
Broiled or fried, drained (USDA)	1 oz.	724	5.0	2.	3.	
BAGEL (USDA):						
Egg	3″ dia. (1.9 oz.)		2.0			
Water	3″ dia. (1.9 oz.)		2.0			
BAKING POWDER:						
Regular:						
Phosphate (USDA)	1 tsp. (5 grams)	386	Tr.			0
SAS (USDA)	1 tsp. (4 grams)	405	Tr.			0
Tartrate (USDA)	1 tsp. (4 grams)	270	Tr.			0
(Calumet) SAS	1 tsp. (4 grams)	276				0
(Royal) tartrate	1 tsp. (4 grams)	250	Tr.			0
Low sodium, commercial (USDA)	1 tsp. (4 grams)	<1	Tr.			0
BAKON DELITES (Wise):						
Regular	½-oz. bag	295	4.3			
Barbecue flavor	½-oz. bag	352	4.1			
BAMBOO SHOOT, raw (USDA):						
Untrimmed	½ lb. (weighed untrimmed)		.2			0
Trimmed	4 oz.		.3			0
BANANA:						
Common:						
Fresh:						
Whole (USDA)	1 lb. (weighed with skin)	3	.6			0
Small size (USDA)	4.9-oz. banana (7¾″ x 1¹¹/₃₂″)	1	.2			0
Small size (Del Monte)	1 peeled banana (3.5 oz.)	2	.5			0
Medium size (USDA)	6.2-oz. banana (8¾″ x 1¹³/₃₂″)	1	.2			0
Large size (USDA)	7-oz. banana (9¾″ x 1⁷/₁₆″)	1	.3			0

(USDA): United States Department of Agriculture
*Prepared as Package Directs

25

Food and Description	Measure or Quantity	Sodium (mg.)	Fats in grams Total	Satu-rated	Unsatu-rated	Choles-terol (mg.)
Chunks (USDA)	1 cup (5 oz.)	1	.3			0
Mashed (USDA)	1 cup (2 med., 7.8 oz.)	2	.4			0
Sliced (USDA)	1 cup (1¼ med., 5.1 oz.)	1	.3			0
Dehydrated (USDA):						
Flakes	½ cup (1.8 oz.)	2	.4			0
Powder	1 oz.	1	.2			0
Red, fresh, whole (USDA)	1 lb. (weighed with skin)	3	.6			0
Red, fresh, peeled (USDA)	4 oz.	1	.2			0

BANANA, BAKING (See PLANTAIN)

BANANA CAKE MIX:

*(Betty Crocker) layer	¹/₁₂ of cake	244	5.7			
*(Duncan Hines)	¹/₁₂ of cake (2.7 oz.)	339	6.1			50

BANANA PIE, cream or custard:

Home recipe (USDA)	¹/₆ of 9″ pie (5.4 oz.)	295	14.1			
(Tastykake)	4-oz. pie		15.2			
Frozen (Banquet)	2½-oz. serving		8.7			
Frozen (Morton)	¹/₆ of 14.4-oz. pie	125	9.3			
Frozen (Mrs. Smith's)	¹/₆ of 8″ pie (2.3 oz.)	95	11.8			

BANANA PUDDING, canned

(Del Monte)	5-oz. can	251	5.3			

BANANA PUDDING & PIE MIX:

*Instant (Jell-O)	½ cup (5.3 oz.)	406	4.7			13
*Instant (Royal)	½ cup (5.1 oz.)	350	4.2			14
*Regular (Jell-O)	½ cup (5.2 oz.)	224	4.6			13
*Regular (Royal)	½ cup (5.1 oz.)	170	4.4			14

BARBADOS CHERRY
(See ACEROLA)

BARBECUE DINNER MIX

(Hunt's) Skillet[1]	2-lb. 1-oz. pkg.	5813	30.7	8.	22.	

BARBECUE SAUCE (See SAUCE, Barbecue)

(USDA): United States Department of Agriculture
*Prepared as Package Directs
[1]Principal source of fat: vegetable shortening.

Food and Description	Measure or Quantity	Sodium (mg.)	—Fats in grams—			Choles-terol (mg.)
			Total	Satu-rated	Unsatu-rated	
BARBECUE SEASONING (French's)	1 tsp. (2 grams)	70	.3			
BARLEY, pearled, dry:						
Light:						
(USDA)	¼ cup (1.8 oz.)	2	.5			0
(Albers)	¼ cup		.5			(0)
(Quaker Scotch)	¼ cup (1.7 oz.)	4	.6			(0)
Pot or Scotch (USDA)	2 oz.		.6			0
BARRACUDA, raw, meat only						
(USDA)	4 oz.		2.9			
BASS (USDA):						
Black sea:						
Raw, whole	1 lb. (weighed whole)	120	2.1			
Baked, home recipe[1]	4 oz.		17.9			
Smallmouth & largemouth, raw:						
Whole	1 lb. (weighed whole)		3.7			
Meat only	4 oz.		2.9			
Striped:						
Raw, whole	1 lb. (weighed whole)		5.3			
Raw, meat only	4 oz.		3.1			
Oven-fried[2]	4 oz.		9.6			
White, raw, whole	1 lb. (weighed whole)		4.1			
White, raw, meat only	4 oz.		2.6			
BASIL (Spice Islands)	1 tsp.	Tr.				(0)
BAVARIAN PIE FILLING, canned						
(Lucky Leaf)	8 oz.	268	10.8			
BAVARIAN-STYLE BEANS &						
SPAETZLE, frozen (Birds Eye)	⅓ of 10-oz. pkg.	411	8.4			0
BAY LEAF (Spice Islands)	1 med. leaf	Tr.				(0)

(USDA): United States Department of Agriculture
*Prepared as Package Directs
[1]Prepared with bacon, butter, onion, celery & bread cubes.
[2]Prepared with milk, bread crumbs, butter & salt.

Food and Description	Measure or Quantity	Sodium (mg.)	Total	Satu- rated	Unsatu- rated	Choles- terol (mg.)
				— Fats in grams —		

BEAN, BAKED:

Food and Description	Measure or Quantity	Sodium (mg.)	Total	Satu-rated	Unsatu-rated	Choles-terol (mg.)
Canned with brown sugar sauce:						
(B & M) red kidney bean	1 cup (8 oz.)	780	8.9			
(B & M) yellow eye bean	1 cup (8 oz.)	965	10.6			
(Homemaker's) red kidney bean	1 cup (8 oz.)	999	8.4			
Canned in molasses sauce:						
(Heinz)	1 cup (9¼ oz.)	1072	1.1			
& brown sugar sauce (Campbell)	1 cup	651	7.7			
Canned with pork:						
(Campbell) Home Style	1 cup	848	4.0			
(Hunt's) *Snack Pack*[1]	5-oz. can	561	.7			
(Van Camp)	1 cup (7.7 oz.)		6.0			
Canned with pork & molasses sauce:						
(USDA)[2]	1 cup (9 oz.)	969	12.0	5.	7.	
(B & M) Michigan Pea, New England-style	1 cup (7.9 oz.)	811	7.8			
(Heinz) Boston-style	1 cup (8¾ oz.)	754	4.4			
(Homemaker's) Michigan Pea, New England-style	1 cup (8 oz.)	840	7.3			
Canned with pork & tomato sauce:						
(USDA)[2]	1 cup (9 oz.)	1181	6.6	3.	4.	
(Campbell)	1 cup	1134	3.0			
(Heinz)	1 cup (9¼ oz.)	1148	4.5			
Canned with tomato sauce:						
(USDA)	1 cup (9 oz.)	862	1.3			0
(Heinz) *Campside*	1 cup (9½ oz.)	1307	9.6			
(Heinz) Vegetarian	1 cup (9¼ oz.)	1184	1.4			(0)
(Van Camp)	1 cup (8.1 oz.)		1.2			
BEAN, BARBECUE (Campbell)	1 cup	1089	3.4			
BEAN, BAYO, dry (USDA)	4 oz.	28	1.7			0
BEAN 'N BEEF (Campbell)	1 cup	1211	6.1			
BEAN, BLACK, dry (USDA)	4 oz.	28	1.7			0
BEAN, BROWN, dry (USDA)	4 oz.	28	1.7			0
BEAN, CALICO, dry (USDA)	4 oz.	11	1.4			0

(USDA): United States Department of Agriculture
*Prepared as Package Directs
[1]Principal source of fat: bacon.
[2]Principal source of fat: pork.

Food and Description	Measure or Quantity	Sodium (mg.)	—Fats in grams—			Choles- terol (mg.)
			Total	Satu- rated	Unsatu- rated	
BEAN & FRANKFURTER, canned:						
(USDA)	1 cup (9 oz.)	1374	18.1			
(Campbell) in tomato & molasses						
sauce	1 cup	1143	16.1			
(Heinz)	1 can (8¾ oz.)	1121	19.8			
(Van Camp) *Beanie-Weenee*	1 cup (7.8 oz.)		15.6			
BEAN & FRANKFURTER						
DINNER, frozen:						
(Banquet):						
Meat compartment	6¼ oz.		32.3			
Apple compartment	2½ oz.		.3			
Cornbread compartment	2 oz.		3.0			
Complete dinner	10 ¾-oz. dinner		35.7			
(Morton)	12-oz. dinner	881	17.9			
(Swanson)	11½-oz. dinner	1085	28.7			
BEAN, GREAT NORTHERN						
(See **BEAN, WHITE**)						
BEAN, GREEN or SNAP:						
Fresh (USDA):						
Whole	1 lb. (weighed					
	untrimmed)	28	.8			0
1½" to 2" pieces	½ cup (1.8 oz.)	4	.1			0
French-style	½ cup (1.4 oz.)	3	<.1			0
Boiled without salt, drained,						
whole (USDA)	½ cup (2.2 oz.)	2	.1			0
Boiled without salt, drained,						
1½" to 2" pieces (USDA)	½ cup (2.4 oz.)	3	.1			0
Canned, regular pack:						
Solids & liq. (USDA)	½ cup (4.2 oz.)	283	.1			0
Drained solids, whole (USDA)	4 oz.	268	.2			0
Drained solids, cut (USDA)	½ cup (2.5 oz.)	165	.1			0
Drained liq. (USDA)	4 oz.	268	.1			0
Cut, solids & liq. (Comstock-						
Greenwood)	½ cup (3 oz.)	200	.2			(0)
Solids & liq. (Del Monte)	½ cup (4 oz.)	440	.3			0
Whole, solids & liq. (Green Giant)	¼ of 16-oz. can	415	.1			(0)
Cut, solids & liq. (Green Giant)	¼ of 16-oz. can	415	.1			(0)
Solids & liq. (Green Giant)						
French style	½ of 8-oz. can	415	.1			(0)

(USDA): United States Department of Agriculture
*Prepared as Package Directs

Food and Description	Measure or Quantity	Sodium (mg.)	Fats in grams — Total	Satu- rated	Unsatu- rated	Choles- terol (mg.)
Solids & liq. (Stokely-Van Camp)	½ cup (3.9 oz.)		.1			(0)
Drained, whole or cut (Del Monte)	½ cup (2.5 oz.)	274	.2			0
Seasoned, solids & liq. (Del Monte)	½ cup (4 oz.)	526	.2			0
Seasoned, drained solids (Del Monte)	½ cup (2.5 oz.)	325	.2			0
Seasoned, drained liq. (Del Monte)	4 oz.	523	.1			0
With bacon (Comstock-Greenwood)	4 oz.	778	.9			
With mushroom, solids & liq. (Comstock-Greenwood)	4 oz.	473	.2			
Canned, dietetic pack:						
Solids & liq. (USDA)	4 oz.	2	.1			0
Drained solids (USDA)	4 oz.	2	.1			0
Drained liq. (USDA)	4 oz.	2	.1			0
Cut, solids & liq. (Blue Boy)	4 oz.	9	.1			(0)
Cut (S and W) *Nutradiet*, unseasoned	4 oz.	1	.1			(0)
Solids & liq. (Diet Delight)	½ cup (4.2 oz.)	4	.1			(0)
Solids & liq. (Tillie Lewis)	½ cup (4.2 oz.)	<10	.1			0
Frozen:						
Whole (Birds Eye)	⅓ of 9-oz. pkg.	1	.1			0
Cut, not thawed (USDA)	10-oz. pkg.	3	.3			0
Cut, boiled without salt, drained (USDA)	4 oz.	1	.1			0
Cut, boiled without salt, drained (USDA)	½ cup (2.8 oz.)	<1	<.1			0
Cut (Birds Eye)	⅓ of 9-oz. pkg.	1	.1			0
French-style, not thawed (USDA)	10-oz. pkg.	6	.3			0
French-style, boiled, drained (USDA)	½ cup (2.8 oz.)	2	<.1			0
French-style (Birds Eye)	⅓ of 9-oz. pkg.	2	.1			0
French-style with sliced mushrooms (Birds Eye)	⅓ of 9-oz. pkg.	153	.1			0
French-style, with toasted almonds (Birds Eye)	½ cup (3 oz.)	360	2.8			0
In butter sauce, cut or French-style (Green Giant)	⅓ of 9-oz. pkg.	361	1.7			
In mushroom sauce, casserole (Green Giant)	⅓ of 12-oz. pkg.	414	1.9			

(USDA): United States Department of Agriculture
*Prepared as Package Directs

Food and Description	Measure or Quantity	Sodium (mg.)	Fats in grams Total	Satu- rated	Unsatu- rated	Choles- terol (mg.)
With onions & bacon						
(Green Giant)	⅓ of 9-oz. pkg.	327	1.5			
In mushroom sauce, cut						
(Green Giant)	⅓ of 10-oz. pkg.	321	.6			
BEAN, ITALIAN (See **BROADBEAN)**						
BEAN, KIDNEY or RED:						
Dry:						
(USDA)	4 oz.	11	1.7			0
(USDA)	½ cup (3.3 oz.)	9	1.4			0
Cooked without salt (USDA)	½ cup (3.3 oz.)	3	.5			0
Canned:						
Solids & liq. (USDA)	½ cup (4.5 oz.)	4	.5			0
Red kidney & chili gravy						
(Nalley's)	4 oz.		2.3			
BEAN, LIMA, young:						
Raw, whole (USDA)	1 lb. (weighed in pod)	4	.9			0
Raw, without shell (USDA)	1 lb. (weighed shelled)	9	2.3			0
Boiled without salt, drained (USDA)	½ cup (3 oz.)	<1	.4			0
Canned, regular pack:						
Solids & liq. (USDA)	½ cup (4.4 oz.)	293	.4			0
Drained solids (USDA)	½ cup (3.1 oz.)	205	.3			0
Drained liq. (USDA)	4 oz.	268	Tr.			0
Drained solids (Del Monte)	½ cup (3.1 oz.)	292	.2			0
Seasoned, drained solids (Del Monte)	½ cup (3.1 oz.)	266	.2			0
Solids & liq. (Stokely-Van Camp)	½ cup (4.1 oz.)		.4			(0)
With ham (Nalley's)	4 oz.		4.8			
Canned, dietetic pack:						
Solids & liq., low sodium (USDA)	4 oz.	5	.3			0
Drained solids, low sodium (USDA)	4 oz.	5	.3			0
Solids & liq., unseasoned (Blue Boy)	4 oz.	5	.3			(0)

(USDA): United States Department of Agriculture
*Prepared as Package Directs

31

Food and Description	Measure or Quantity	Sodium (mg.)	—Fats in grams— Total	Satu- rated	Unsatu- rated	Choles- terol (mg.)
Frozen:						
Baby butter beans (Birds Eye)	⅓ of 10-oz. pkg.	367	.2			0
Baby limas:						
Not thawed (USDA)	4 oz.	167	.2			0
Boiled, drained (USDA)	½ cup (3 oz.)	111	.2			0
(Birds Eye)	½ cup (3.3 oz.)	138	.2			0
In butter sauce (Green Giant)	⅓ of 10-oz. pkg.	439	2.8			
Tiny (Birds Eye)	½ cup (2.5 oz.)	104	.1			0
Fordhooks:						
Not thawed (USDA)	4 oz.	146	.1			0
Boiled, drained (USDA)	½ cup (3 oz.)	85	<.1			0
(Birds Eye)	⅓ of 10-oz. pkg.	121	.1			0
BEAN, LIMA, mature:						
Dry:						
Baby (USDA)	½ cup (3.4 oz.)	4	1.5			0
Large (USDA)	½ cup (3.1 oz.)	4	1.4			0
Boiled without salt, drained (USDA)	½ cup (3.4 oz.)	2	.6			0
BEAN, MUNG, dry (USDA)	½ cup (3.7 oz.)	6	1.4			0
BEAN, NAVY or PEA (See **BEAN, WHITE**)						
BEAN, PINTO:						
Dry (USDA)	4 oz.	11	1.4			0
Dry (USDA)	½ cup (3.4 oz.)	10	1.2			0
*(Uncle Ben's) including broth	¾ cup (6 oz.)	16	.6			0
BEAN, RED (See **BEAN, KIDNEY** or **BEAN, RED MEXICAN**)						
BEAN, RED MEXICAN, dry (USDA)	4 oz.	11	1.4			0
BEAN SALAD, canned, solids & liq.:						
(Comstock-Greenwood)	4 oz.	536	2.1			
(Hunt's) *Snack Pack*[1]	5-oz. can	547	1.2	Tr.	<1.	
(Le Sueur)	¼ of 1-lb. 1-oz. can	458	.8			

(USDA): United States Department of Agriculture
*Prepared as Package Directs
[1]Principal source of fat: cottonseed oil.

Food and Description	Measure or Quantity	Sodium (mg.)	Total	Fats in grams Satu- rated	Unsatu- rated	Choles- terol (mg.)
BEAN, SEMI-MATURE, shelled,						
in brine, drained (B & M)	½ of 8¾-oz. can	320	.7			(0)
BEAN SOUP, canned:						
*(Manischewitz)	1 cup		1.9			
*With bacon (Campbell)	1 cup	852	4.8	1.	4.	
With pork, condensed						
(USDA)	8 oz. (by wt.)	1830	10.4			
*With pork, prepared with						
equal volume water (USDA)	1 cup (8.8 oz.)	1008	5.8			
With smoked ham (Heinz)						
Great American	1 cup (8¾ oz.)	1215	6.8			
*With smoked pork (Heinz)	1 cup (8½ oz.)	1080	5.7			
BEAN SOUP, BLACK, canned:						
*(Campbell)	1 cup	816	1.6	Tr.	1.	
(Crosse & Blackwell)	6½ oz. (½ can)		.6			
***BEAN SOUP, LIMA,** canned						
(Manischewitz)	1 cup		1.6			
BEAN SOUP, NAVY, dehydrated						
(USDA)	1 oz.		.3			
BEAN SPROUT:						
Mung:						
Raw (USDA)	½ lb.	12	.4			0
Raw (USDA)	½ cup (1.6 oz.)	2	<.1			0
Boiled without salt,						
drained (USDA)	½ cup (2.2 oz.)	2	.1			0
Soy:						
Raw (USDA)	½ lb.		3.2			0
Raw (USDA)	½ cup (1.9 oz.)		.8			0
Boiled without salt,						
drained (USDA)	4 oz.		1.6			0
Canned (Hung's)	8 oz.		1.0			(0)
BEAN, WHITE, dry:						
Raw:						
Great Northern (USDA)	½ cup (3.1 oz.)	17	1.4			0
Navy or pea (USDA)	½ cup (3.7 oz.)	20	1.7			0
All other white (USDA)	1 oz.	5	.5			0

(USDA): United States Department of Agriculture
*Prepared as Package Directs

Food and Description	Measure or Quantity	Sodium (mg.)	—Fats in grams—			Choles- terol (mg.)
			Total	Satu- rated	Unsatu- rated	
Cooked without salt:						
Great Northern (USDA)	½ cup (3 oz.)	6	.5			0
Navy or pea (USDA)	½ cup (3.4 oz.)	7	.6			0
All other white (USDA)	4 oz.	8	.7			0
BEAN, YELLOW or WAX:						
Raw, whole (USDA)	1 lb. (weighed untrimmed)	28	.8			0
Boiled without salt, drained (USDA)	4 oz.	3	.2			0
Boiled without salt, drained 1″ pieces (USDA)	½ cup (2.9 oz.)	2	.2			0
Canned, regular pack:						
Solids & liq. (USDA)	½ cup (4.2 oz.)	283	.2			0
Drained solids (USDA)	½ cup (2.2 oz.)	146	.2			0
Drained liq. (USDA)	4 oz.	268	.1			0
Cut, solids & liq. (Comstock-Greenwood)	½ cup (3 oz.)	200	.2			(0)
Solids & liq. (Del Monte)	½ cup (4 oz.)	438	.2			0
Drained solids (Del Monte)	½ cup (2.5 oz.)	275	.2			0
Cut, solids & liq. (Green Giant)	½ of 8.5-oz. can	428	.1			(0)
Solids & liq. (Stokely-Van Camp)	½ cup (4.1 oz.)		.2			(0)
Canned, dietetic pack:						
Solids & liq. (USDA)	4 oz.	2	.1			0
Drained solids (USDA)	4 oz.	2	.1			0
Drained liq. (USDA)	4 oz.	2	.1			0
Solids & liq. (Blue Boy)	4 oz.	2	.1			(0)
Frozen:						
Cut, not thawed (USDA)	4 oz.	1	.1			0
Boiled, drained (USDA)	4 oz.	1	.1			0
Cut (Birds Eye)	⅓ of 9-oz. pkg.	14	.1			0
BEAVER, roasted (USDA)	4 oz.		15.5			
BEECHNUT:						
Whole (USDA)	4 oz. (weighed in shell)		34.6	3.	32.	0
Shelled (USDA)	1 oz. (weighed shelled)		14.2	1.	13.	0

(USDA): United States Department of Agriculture
*Prepared as Package Directs

Food and Description	Measure or Quantity	Sodium (mg.)	Total	Satu- rated	Unsatu- rated	Choles- terol (mg.)
				—Fats in grams—		

BEEF. Values for beef cuts are given below for "lean and fat" and for "lean only." Beef purchased by the consumer at the retail store usually is trimmed to about one-half inch layer of fat. This is the meat described as "lean and fat." If all the fat that can be cut off with a knife is removed, the remainder is the "lean only." These cuts still contain flecks of fat known as "marbling" distributed through the meat. Cooked meats are medium done. Choice grade cuts (USDA):

Brisket:

Raw, lean & fat	1 lb. (weighed with bone)	248	112.4	54.	58.	259
Raw, lean & fat	1 lb. (weighed without bone)	295	133.8	64.	69.	308
Raw, lean only	1 lb.	295	37.2	18.	20.	295
Braised:						
Lean & fat	4 oz.	68	39.5	19.	21.	107
Lean only	4 oz.	68	11.9	6.	6.	103
Chuck:						
Raw, lean & fat	1 lb. (weighed with bone)	248	75.0	36.	39.	259
Raw, lean & fat	1 lb. (weighed without bone)	295	88.9	43.	46.	308
Raw, lean only	1 lb.	295	33.6	18.	15.	295
Braised or pot-roasted:						
Lean & fat	4 oz.	68	27.1	13.	14.	107
Lean only	4 oz.	68	10.8	5.	6.	103
Dried (See BEEF, CHIPPED)						
Fat, separable, raw	1 oz.		21.6	10.	11.	21
Fat, separable, cooked	1 oz.		22.1	10.	12.	

Filet Mignon. There are no data available on its composition. For dietary estimates, the data for sirloin steak, lean only, afford the closest approximation.

Flank:

Raw, 100% lean	1 lb.	295	25.9	12.	13.	295
Braised, 100% lean	4 oz.	68	8.3	4.	4.	103

(USDA): United States Department of Agriculture
*Prepared as Package Directs

Food and Description	Measure or Quantity	Sodium (mg.)	Fats in grams — Total	Satu- rated	Unsatu- rated	Choles- terol (mg.)
Foreshank:						
Raw, lean & fat	1 lb. (weighed with bone)	156	34.8	17.	18.	163
Simmered:						
Lean & fat	4 oz.	68	19.4	9.	10.	107
Lean only	4 oz.	68	6.6	3.	3.	103
Ground:						
Regular:						
Raw	1 lb.	118	96.2	48.	48.	308
Raw	1 cup (8 oz.)	104	47.9	23.	25.	154
Broiled	4 oz.	53	23.0	11.	12.	107
Lean:						
Raw	1 lb.	236	45.4	22.	23.	295
Raw	1 cup (8 oz.)	118	22.6	11.	11.	147
Broiled	4 oz.	54	12.8	6.	7.	103
Heel of round:						
Raw, lean & fat	1 lb.	295	64.4	31.	34.	308
Raw, lean only	1 lb.	295	21.3	9.	12.	295
Roasted:						
Lean & fat	4 oz.	68	18.3	9.	10.	107
Lean only	4 oz.	68	6.5	3.	3.	103
Hindshank:						
Raw, lean & fat	1 lb. (weighed with bone)	136	48.9	23.	25.	142
Raw, lean & fat	1 lb. (weighed without bone)	295	106.1	51.	55.	308
Raw, lean only	1 lb.	295	20.9	10.	11.	295
Simmered:						
Lean & fat	4 oz.	68	31.9	15.	17.	107
Lean only	4 oz.	68	6.7	3.	4.	103
Neck:						
Raw, lean & fat	1 lb. (weighed with bone)	237	57.7	28.	30.	247
Pot-roasted:						
Lean & fat	4 oz.	68	22.3	11.	12.	107
Lean only	4 oz.	68	8.3	4.	4.	103
Oxtail, raw	1 lb. (weighed with bone)	73	7.9	4.	4.	73
Oxtail, raw	1 lb. (weighed without bone)	295	31.8	15.	17.	295
Plate:						
Raw, lean & fat	1 lb. (weighed with bone)	263	150.6	72.	78.	275

(USDA): United States Department of Agriculture
*Prepared as Package Directs

Food and Description	Measure or Quantity	Sodium (mg.)	— Fats in grams —			Choles- terol (mg.)
			Total	Satu- rated	Unsatu- rated	
Raw, lean & fat	1 lb. (weighed without bone)	295	169.2	81.	88.	308
Raw, lean only	1 lb.	295	37.2	18.	20.	295
Simmered:						
Lean & fat	4 oz.	68	48.5	23.	25.	107
Lean only	4 oz.	68	11.9	6.	6.	103
Rib roast:						
Raw, lean & fat	1 lb. (weighed with bone)	271	156.1	75.	81.	284
Raw, lean & fat	1 lb. (weighed without bone)	295	169.6	81.	88.	308
Raw, lean only	1 lb.	295	52.6	27.	25.	295
Roasted:						
Lean & fat	4 oz.	68	44.7	21.	23.	107
Lean only	4 oz.	68	15.2	7.	8.	103
Lean only, chopped	1 cup (4.5 oz.)	77	17.2	8.	9.	116
Lean only, diced	1 cup (5 oz.)	86	19.2	9.	10.	130
Round:						
Raw, lean & fat	1 lb. (weighed with bone)	286	53.9	26.	28.	299
Raw, lean & fat	1 lb. (weighed without bone)	295	55.8	27.	29.	308
Raw, lean only	1 lb.	295	21.3	9.	12.	295
Broiled:						
Lean & fat	4 oz.	68	17.5	8.	9.	107
Lean only	4 oz.	68	6.9	3.	4.	103
Rump:						
Raw, lean & fat	1 lb. (weighed with bone)	251	97.4	47.	50.	262
Raw, lean & fat	1 lb. (weighed without bone)	295	114.8	55.	60.	308
Raw, lean only	1 lb.	295	34.0	15.	19.	295
Roasted:						
Lean & fat	4 oz.	68	31.0	15.	16.	107
Lean only	4 oz.	68	10.5	5.	5.	103
Steak, club:						
Raw, lean & fat	1 lb. (weighed with bone)	229	132.1	63.	69.	259
Raw, lean & fat	1 lb. (weighed without bone)	295	157.9	76.	82.	308
Raw, lean only	1 lb.	295	46.7	22.	25.	295
Broiled:						
Lean & fat	4 oz.	68	46.0	22.	24.	107

(USDA): United States Department of Agriculture
*Prepared as Package Directs

Food and Description	Measure or Quantity	Sodium (mg.)	— Fats in grams —			Choles- terol (mg.)
			Total	Satu- rated	Unsatu- rated	
Lean only	4 oz.	68	14.7	7.	8.	103
One 8-oz. steak (weighed without bone before cooking) will give you:						
Lean & fat	5.9 oz.	100	67.4	32.	35.	156
Lean only	3.4 oz.	58	12.5	6.	7.	87
Steak, porterhouse:						
Raw, lean & fat	1 lb. (weighed with bone)	268	148.8	71.	78.	281
Broiled:						
Lean & fat	4 oz.	68	47.9	23.	25.	107
Lean only	4 oz.	68	11.9	6.	6.	103
One 16-oz. steak (weighed with bone before cooking) will give you:						
Lean & fat	10.2 oz.	173	121.5	59.	63.	271
Lean only	5.9 oz.	100	17.4	8.	9.	151
Steak, ribeye, broiled:						
One 10-oz. steak (weighed without bone before cooking) will give you:						
Lean & fat	7.3 oz.	124	81.6	39.	42.	195
Lean only	3.8 oz.	64	14.3	7.	7.	97
Steak, sirloin, double-bone:						
Raw, lean & fat	1 lb. (weighed with bone)	242	108.4	52.	56.	253
Raw, lean & fat	1 lb. (weighed without bone)	295	132.0	63.	69.	308
Raw, lean only	1 lb.	295	33.6	18.	15.	295
Broiled:						
Lean & fat	4 oz.	68	39.3	19.	20.	107
Lean only	4 oz.	68	10.8	5.	6.	103
One 16-oz. steak (weighed with bone before cooking) will give you:						
Lean & fat	8.9 oz.	151	87.4	42.	45.	237
Lean only	5.9 oz.	100	15.8	8.	8.	151
One 12-oz. steak (weighed with bone before cooking) will give you:						
Lean & fat	6.6 oz.	113	65.2	31.	34.	177
Lean only	4.4 oz.	74	11.8	6.	6.	113

(USDA): United States Department of Agriculture
*Prepared as Package Directs

Food and Description	Measure or Quantity	Sodium (mg.)	— Fats in grams —			Choles- terol (mg.)
			Total	Satu- rated	Unsatu- rated	
Steak, sirloin, hipbone:						
Raw, lean & fat	1 lb. (weighed with bone)	251	149.3	72.	78.	262
Raw, lean & fat	1 lb. (weighed without bone)	295	176.0	84.	92.	308
Raw, lean only	1 lb.	295	44.9	21.	24.	295
Broiled:						
Lean & fat	4 oz.	68	50.9	24.	26.	107
Lean only	4 oz.	68	14.2	7.	7.	103
Steak, sirloin, wedge & round- bone:						
Raw, lean & fat	1 lb. (weighed with bone)	274	112.3	54.	58.	287
Raw, lean & fat	1 lb. (weighed without bone)	295	121.1	58.	63.	308
Raw, lean only	1 lb.	295	25.9	12.	14.	295
Broiled:						
Lean & fat	4 oz.	68	36.3	17.	19.	107
Lean only	4 oz.	68	8.7	4.	5.	103
Steak, T-bone:						
Raw, lean & fat	1 lb. (weighed with bone)	263	149.1	72.	78.	275
Broiled:						
Lean & fat	4 oz.	68	49.0	24.	26.	107
Lean only	4 oz.	68	11.7	6.	6.	103
One 16-oz. steak (weighed with bone before cooking) will give you:						
Lean & fat	9.8 oz.	167	120.1	58.	63.	261
Lean only	5.5 oz.	94	16.1	8.	8.	142
BEEFAMATO COCKTAIL						
(Mott's)	½ cup		.3			
BEEFARONI, canned (Chef Boy- Ar-Dee)	⅓ of 40-oz. can	1371	6.4			
BEEF BOUILLON/BROTH, cubes or powder (See also **BEEF SOUP**):						
(Croyden House) instant	1 tsp. (5 grams)	490	.1	Tr.	0.	
(Herb-Ox)	1 cube (4 grams)	900	.1			
(Herb-Ox) instant	1 packet (4 grams)	1000	.1			
(Maggi)	1 cube (4 grams)	735	.2			

(USDA): United States Department of Agriculture
*Prepared as Package Directs

39

Food and Description	Measure or Quantity	Sodium (mg.)	Fats in grams — Total	Satu- rated	Unsatu- rated	Choles- terol (mg.)
(Maggi) instant	1 tsp. (4 grams)	809	.2			
(Steero)	1 cube (4 grams)		.2			
(Wyler's)	1 cube (4 grams)		.2			
(Wyler's) instant	1 tsp.		.2			
(Wyler's) no salt added	1 cube (4 grams)	10	.3			
BEEF & CABBAGE, casserole, frozen						
(Mrs. Paul's)	12-oz. pkg.		29.1			
BEEF, CHIPPED:						
Uncooked:						
(USDA)	2 oz. (about ⅓ cup)	2451	3.6	2.	2.	
(USDA)	½ cup (2.9 oz.)	3526	5.2	2.	3.	
(Armour Star)	1 oz.		.7			
Cooked, creamed, home recipe						
(USDA)[1]	1 cup (8.6 oz.)	1754	25.2	15.	11.	66
Cooked (Oscar Mayer) thin sliced	1 slice (5 grams)	74	.2			
Canned, creamed (Swanson)	1 cup	1908	10.0			
Frozen, creamed (Banquet)	5-oz. bag		4.1			
BEEF, CHOPPED or						
DICED, canned:						
(Armour Star)	12-oz. can		92.2			
(Hormel)	12-oz. can		71.1			
Freeze dry (Wilson) *Campsite:*						
Dry	2 oz.	437	8.3	4.	4.	113
*Reconstituted	4 oz.	304	6.2	3.	3.	86
BEEF, CORNED (See **CORNED BEEF**)						
BEEF DINNER, frozen:						
(Banquet):						
Corn compartment	1.9 oz.		1.2			
Meat compartment	6.3 oz.		6.2			
Potato compartment	2.8 oz.		.9			
Complete dinner	11-oz. dinner		8.3			
(Morton)	11-oz. dinner	938	9.4			
(Swanson)	11½-oz. dinner	790	13.5			
(Swanson) 3-course	15-oz. dinner	1403	21.5			
Beef steak & carrots						
(Weight Watchers)	10-oz. luncheon		31.5			

(USDA): United States Department of Agriculture
*Prepared as Package Directs
[1]Principal sources of fat: milk, butter & beef.

Food and Description	Measure or Quantity	Sodium (mg.)	Total	—Fats in grams— Satu-rated	Unsatu-rated	Choles-terol (mg.)
Beef steak & cauliflower (Weight Watchers)	11-oz. luncheon		31.2			
Chopped (Banquet):						
Meat compartment	3.9 oz.		19.8			
Potato compartment	2.8 oz.		.7			
Corn compartment	2.3 oz.		1.2			
Complete dinner	9-oz. dinner		21.7			
Chopped sirloin (Swanson)	10-oz. dinner	978	19.9	7.	13.	
Chopped (Weight Watchers)	18-oz. dinner		54.4			
Pot roast, includes whole oven-browned potatoes, peas & corn (USDA)	10 oz.	736	9.1	6.	3.	
Sliced (Morton) 3-course	1-lb. 1-oz. dinner	1476	20.7			
BEEF & EGGPLANT, casserole, frozen (Mrs. Paul's)	12-oz. pkg.		26.0			
BEEF GOULASH:						
Canned (Heinz)	8½-oz. can	1070	12.3			
Seasoning mix (Lawry's)	1.7-oz. pkg.		1.3			
BEEF & GREEN PEPPER, casserole, frozen (Mrs. Paul's)	12-oz. pkg.		21.9			
BEEF, GROUND, seasoning mix:						
(Durkee)	1⅛-oz. pkg.	8852	.7			
With onions (French's)	1⅛-oz. pkg.		.3			
BEEF HASH, ROAST:						
Canned (Hormel) *Mary Kitchen*	7½ oz.	1327	27.1	10.	13.	62
Frozen (Stouffer's)	11½-oz. pkg.		26.1			
BEEF JERKY (General Mills)	1 piece (¼ oz.)	203	.7			
BEEF PATTIES:						
Canned, freeze dry (Wilson)						
Campsite:						
Dry	2-oz. can	1311	14.4	7.	7.	161
*Reconstituted	4 oz.	1043	11.6	6.	6.	130
Frozen (Morton House):						
& Burgundy sauce	4¹/₆-oz. serving	637	8.9	4.	4.	30
& Italian sauce	4¹/₆-oz. serving	832	8.9	4.	4.	30
& Mexican sauce	4¹/₆-oz. serving	654	9.0	4.	4.	30

(USDA): United States Department of Agriculture
*Prepared as Package Directs

Food and Description	Measure or Quantity	Sodium (mg.)	—Fats in grams—			Cholesterol (mg.)
			Total	Saturated	Unsaturated	
BEEF PIE:						
Baked, home recipe (USDA)[1]	4¼" pie (8 oz. before baking)	645	32.9	9.	24.	48
Baked, home recipe (USDA)[1]	⅓ of 9" pie (7.4 oz.)	596	30.4	8.	22.	44
Frozen:						
Commercial, unheated (USDA)[1]	1 pie (7.6 oz.)	791	21.4	6.	15.	39
(Banquet)	8-oz. pie		19.9			
(Banquet)	2-lb 4-oz. pie		56.3			
(Morton)	8-oz. pie	1051	18.2			
(Stouffer's)	10-oz. pkg.	1807	35.0			
(Swanson)	8-oz. pie	1060	24.1			
(Swanson) deep dish	16-oz. pie	1922	37.9			
BEEF, POTTED (USDA)	1 oz.		5.4			
BEEF PUFFS, frozen (Durkee)	1 piece (.5 oz.)		4.5			
BEEF, ROAST, canned:						
(USDA)	4 oz.		15.0	7.	8.	
(Wilson) *Tender Made*	4 oz.	492	2.9	1.	2.	77
BEEF, SLICED, with barbecue sauce						
(Banquet)	5-oz. bag		5.0			
BEEF SOUP, canned:						
*(Campbell)	1 cup	795	2.3	<1.	1.	
(Campbell) *Chunky*	1 cup	896	6.6			
*Barley (Manischewitz)	1 cup		2.7			
Bouillon, condensed (USDA)	8 oz. (by wt.)	1480	0.			
*Bouillon, prepared with equal volume water (USDA)	1 cup (8.5 oz.)	782	0.			
Broth:						
Condensed (USDA)	8 oz. (by wt.)	1480	0.			
*Prepared with equal volume water (USDA)	1 cup (8.5 oz.)	782	0.			
(Swanson)	1 cup	680	.8			
*Cabbage (Manischewitz)	1 cup		1.3			
Consommé:						
Condensed (USDA)	8 oz. (by wt.)	1480	0.			
*Prepared with equal volume water (USDA)	1 cup (8.5 oz.)	782	0.			
*(Campbell)	1 cup	642	0.			

(USDA): United States Department of Agriculture
*Prepared as Package Directs
[1]Principal sources of fat: vegetable shortening & beef.

Food and Description	Measure or Quantity	Sodium (mg.)	—Fats in grams—			Choles- terol (mg.)
			Total	Satu- rated	Unsatu- rated	
Noodle:						
Condensed (USDA)	8 oz. (by wt.)	1734	5.0			
*Prepared with equal volume water (USDA)	1 cup (8.5 oz.)	917	2.6			
*(Campbell)	1 cup	792	2.2	<1.	2.	7
*(Heinz)	1 cup (8.5 oz.)	972	2.9			
*(Manischewitz)	1 cup		1.6			
*Curly, with chicken (Campbell)	1 cup	1099	2.5			
With dumplings (Heinz) *Great American*	1 cup (8¾ oz.)	967	4.3			
Sirloin burger (Campbell) *Chunky*	1 cup	930	7.0			
*Vegetable (Manischewitz)	1 cup		1.3			
*With broth (Campbell)	1 cup	760	0.			
BEEF SOUP MIX:						
*Barley (Wyler's)	6 fl. oz.		.8			
Broth (Lipton) *Cup-a-Soup*	1 pkg. (8 grams)	899	.1	Tr.	0.	
Noodle:						
(USDA)[1]	1 oz.	672	2.1	<1	2	
*(USDA)	1 cup (8.1 oz.)	401	1.1			
(Lipton) *Cup-a-Soup*	1 pkg. (.4 oz.)	892	.3	Tr.	Tr.	5
*With vegetable (Lipton)	1 cup	1258	1.3	Tr.	<1.	17
*(Wyler's)	6 fl. oz.		.5			
BEEF STEAK, freeze dry, canned (Wilson) *Campsite:*						
Dry	2-oz. can	1267	4.7	2.	3.	169
*Reconstituted	4 oz.	1043	3.9	2.	2.	137
BEEF STEW:						
Home recipe, made with lean beef chuck (USDA)[2]	1 cup (8.6 oz.)	91	10.5	5.	6.	64
Canned:						
(USDA)	1 cup (8.6 oz.)	1007	7.6			34
(Armour Star)	24-oz. can		32.6			
(B & M)	1 cup (7.9 oz.)	1010	4.5			
(Dinty Moore)	8 oz.	887	9.8	4.	5.	36
(Heinz)	8½-oz. can	1272	10.7			
(Nalley's)	8 oz.		10.4			
(Swanson)	1 cup	882	5.7			
(Van Camp)	½ cup (4.6 oz.)		4.2			

(USDA): United States Department of Agriculture
*Prepared as Package Directs
[1]Principal sources of fat: vegetable shortening & egg.
[2]Principal source of fat: beef.

Food and Description	Measure or Quantity	Sodium (mg.)	Total	Fats in grams Satu- rated	Unsatu- rated	Choles- terol (mg.)
(Wilson)	15½-oz. can	1850	13.2	7.	7.	75
Dietetic (Claybourne)	8-oz. can	66	23.8			
Dietetic (Slim-ette)	8-oz. can	132	5.6			
Freeze dry (Wilson) *Campsite:*						
Dry	4½-oz. can	3424	25.3	12	14	
*Reconstituted	16 oz.	2676	25.4	12	14	
Meatball (Hormel)	1-lb. 8-oz. can		41.5			
Frozen:						
Buffet (Banquet)	2-lb. pkg.		25.0			
BEEF STEW SEASONING MIX:						
(Durkee)	1¾-oz. pkg.	6953	.6			
(French's)	1 pkg. (1⅞ oz.)	4400	.5			
*(Kraft)	1 oz.	52	4.0			
(Lawry's)	1 pkg. (1³/₅ oz.)		1.3			
BEEF STOCK BASE (French's)	1 tsp. (4 grams)		Tr.			
BEEF STROGANOFF:						
Canned (Hormel)	1-lb. can		39.0			
Mix (Chef Boy-Ar-Dee)	6⅔-oz. pkg.	1067	7.9			
Mix (Hunt's) *Skillet*[1]	1-lb. 2-oz. pkg.	3984	40.7	21.	19.	
Mix, dinner (Jeno's)						
Add 'n Heat	40-oz. pkg.	95.3				
*Mix, *Noodle-Roni*	4 oz.	446	2.7			
Seasoning mix (Lawry's)	1½-oz. pkg.		.5			
BEEF & ZUCCHINI, casserole, frozen (Mrs. Paul's)	12-oz. pkg.		21.0			
BEER, canned:						
Regular:						
(USDA) 4.5% alcohol	12 fl. oz.	25	0.			0
Budweiser, 4.9% alcohol	12 fl. oz.	20–40	0.			0
Budweiser, 3.9% alcohol	12 fl. oz.	20–40	0.			0
Busch Bavarian, 4.9% alcohol	12 fl. oz.	20–40	0.			0
Busch Bavarian, 3.9% alcohol	12 fl. oz.	20–40	0.			0
Michelob, 4.9% alcohol	12 fl. oz.	20–40	0.			0
Pabst Blue Ribbon	12 fl. oz.	7	(0.)			(0)
Schlitz	12 fl. oz.	24	(0.)			(0)
Low carbohydrate:						
Gablinger's, 4.5% alcohol	12 fl. oz.	21	0.			(0)
Meister Brau Lite, 4.6% alcohol	12 fl. oz.	Tr.	0.			(0)

(USDA): United States Department of Agriculture
*Prepared as Package Directs
[1]Principal sources of fat: sour cream, cottonseed oil & egg.

Food and Description	Measure or Quantity	Sodium (mg.)	Total	Satu-rated	Unsatu-rated	Choles-terol (mg.)
BEER, NEAR, *Kingsbury*						
(Heileman) 0.4% alcohol	12 fl. oz.	41				0
BEET:						
Raw (USDA)	1 lb. (weighed with skins, part tops)	133	.2			0
Raw (USDA)	1 lb. (weighed with skins, without tops)	190	.3			0
Raw, diced (USDA)	½ cup (2.4 oz.)	40	<.1			0
Boiled without salt:						
Whole, drained (USDA)	2 beets (2″ dia., 3.5 oz.)	43	.1			0
Diced, drained (USDA)	½ cup (3 oz.)	37	<.1			0
Sliced, drained (USDA)	½ cup (3.6 oz.)	44	.1			0
Canned, regular pack:						
Solids & liq. (USDA)	½ cup (4.3 oz.)	290	.1			0
Drained solids, whole (USDA)	½ cup (2.8 oz.)	188	<.1			0
Drained solids, diced (USDA)	½ cup (2.9 oz.)	194	<.1			0
Drained solids, sliced (USDA)	½ cup (3.1 oz.)	208	<.1			0
Drained liq. (USDA)	4 oz.	268	Tr.			0
Solids & liq., sliced (Comstock-Greenwood)	½ cup (2.6 oz.)	177	<.1			(0)
Solids & liq. (Del Monte)	½ cup (4 oz.)	316	.1			0
Solids & liq., tiny, whole (Le Sueur)	¼ of 1-lb. can	437	<.1			(0)
Solids & liq. (Stokely-Van Camp)	½ cup (4.1 oz.)		.1			(0)
Drained solids (Del Monte)	½ cup (3.1 oz.)	246	.1			0
Harvard, solids & liq. (Comstock-Greenwood)	4 oz.		<.1			(0)
Pickled, solids & liq. (Comstock-Greenwood)	4 oz.		<.1			(0)
Pickled, solids & liq. (Del Monte)	½ cup (4 oz.)	378	.2			0
Pickled, drained solids (Del Monte)	½ cup (3.1 oz.)	291	.4			0
Pickled, drained liq. (Del Monte)	4 oz.	378	Tr.			0
Canned, dietetic pack:						
Solids & liq. (USDA)	4 oz.	52	Tr.			0
Drained solids (USDA)	4 oz.	52	.1			0
Drained liq. (USDA)	4 oz.	52	Tr.			0
Whole (Blue Boy)	10 small (3.5 oz.)	44	<.1			(0)
Diced, solids & liq. (Blue Boy)	4 oz.	50	<.1			(0)

(USDA): United States Department of Agriculture
*Prepared as Package Directs

Food and Description	Measure or Quantity	Sodium (mg.)	—Fats in grams—			Choles- terol (mg.)
			Total	Satu- rated	Unsatu- rated	
Diced, solids & liq. (Tillie Lewis)	½ cup (4.3 oz.)	55	Tr.			(0)
Sliced (Blue Boy)	10 slices (3.5 oz.)	33	<.1			(0)
Sliced (S and W) *Nutradiet,* unseasoned	4 oz.	45	.8			(0)
Frozen, sliced, in orange flavor glaze (Birds Eye)	⅓ of 10-oz. pkg.	129	.1			0
BEET GREENS (USDA):						
Raw, whole	1 lb. (weighed untrimmed)	330	.8			0
Boiled without salt, leaves & stems, drained	½ cup (2.6 oz.)	55	.1			0
BELL PEPPER, dried:						
Green (Spice Islands)	1 tsp.	6				(0)
Red (Spice Islands)	1 tsp.	39				(0)
BEVERAGE (See individual listings)						
BIF (Wilson) canned luncheon meat	3 oz.	935	23.8	12.	12.	56
BIG MAC (McDonald's)	1 hamburger (6.5 oz.)	1064	31.9			
BIG WHEEL (Hostess)	1 cake (1.4 oz.)	101	8.7			
BIRCH BEER, soft drink (Yukon Club)	6 fl. oz.	14	0.			0
BISCUIT:						
Baking powder, home recipe (USDA)[1]:						
Made with regular flour & lard[2]	1-oz. biscuit (2″ dia.)	175	4.8	2.	3.	
Made with regular flour & vegetable shortening[3]	1-oz. biscuit (2″ dia.)	175	4.8	1.	4.	
Made with self-rising flour & lard[2]	1-oz. biscuit (2″ dia.)	185	4.9	2.	3.	
Made with self-rising flour & vegetable shortening[3]	1-oz. biscuit (2″ dia.)	185	4.9	1.	4.	
Egg (Stella D'oro):						
Dietetic	1 piece (.4 oz.)	3	1.0			

(USDA): United States Department of Agriculture
*Prepared as Package Directs
[1]Made with sodium aluminum sulfate-type baking powder.
[2]Principal source of fat: lard.
[3]Principal source of fat: vegetable shortening.

Food and Description	Measure or Quantity	Sodium (mg.)	— Fats in grams —			Choles- terol (mg.)
			Total	Satu- rated	Unsatu- rated	
Regular	1 piece (.4 oz.)		1.0			
Roman	1 piece (1.1 oz.)		5.2			
Sugared	1 piece (.5 oz.)		1.1			
BISCUIT DOUGH:						
Frozen, commercial (DA)[1]	1 oz.	258	3.4	<1.	2.	
Refrigerated:						
Commercial (USDA)[1]	1 oz.	246	1.8	<1.	2.	
*(Borden)	1 biscuit (.8 oz.)		1.3			
*Big 10's (Borden)	1 biscuit (1 oz.)		4.0			
*Buttered Up (Borden)	1 biscuit (1 oz.)		4.6			
*Gem (Borden)	1 biscuit (1 oz.)		3.2			
BISCUIT MIX:						
Dry, with enriched flour (USDA)[1]	1 oz.	369	3.6	<1.	3	
*Baked from mix, with added milk (USDA)[1]	1-oz. biscuit	276	2.6	<1.	2	
Bisquick (Betty Crocker)	1 cup	1475	17.2			
BITTER LEMON, soft drink:						
(Canada Dry) bottle or can	6 fl. oz.	13+	0.			0
(Hoffman)	6 fl. oz.	25+	0.			0
(Schweppes)	6 fl. oz.	30	0.	0		
BITTER ORANGE, soft drink						
(Schweppes)	6 fl. oz.	5+	0.			0
BITTERS (Angostura)						
45% alcohol	1 tsp. (5 grams)	Tr.				(0)
BLACKBERRY:						
Fresh (includes boysenberry, dewberry, youngberry):						
With hulls (USDA)	1 lb. (weighed untrimmed)	4	3.9			0
Hulled (USDA)	½ cup (2.6 oz.)	<1	.7			0
Canned, regular, solids & liq. (USDA):						
Juice pack	4 oz.	1	.9			0
Light syrup	4 oz.	1	.7			0
Heavy syrup	½ cup (4.6 oz.)	1	.8			0
Extra heavy syrup	4 oz.	1	.7			0

(USDA): United States Department of Agriculture
*Prepared as Package Directs
[1]Principal source of fat: vegetable shortening.

Food and Description	Measure or Quantity	Sodium (mg.)	— Fats in grams —			Choles- terol (mg.)
			Total	Satu- rated	Unsatu- rated	
Canned, water pack, solids & liq. (USDA)	½ cup (4.3 oz.)	1	.7			0
Canned, low calorie, solids & liq. (S and W) *Nutradiet*	4 oz.	2	.1			
Frozen (USDA):						
Sweetened, not thawed	4 oz.	1	.3			0
Unsweetened, not thawed	4 oz.	1	.3			0
BLACKBERRY JELLY, low calorie:						
(Slenderella)	1 T. (.7 oz.)	16	Tr.			(0)
& apple (Kraft)	1 oz.	37	<.1			(0)
BLACKBERRY JUICE, canned, unsweetened (USDA)	½ cup (4.3 oz.)	1	.7			0
BLACKBERRY PIE:						
Home recipe, 2-crust, made with lard (USDA)[1]	¹/₆ of 9″ pie (5.6 oz.)	423	17.4	6.	11.	
Home recipe, 2-crust, made with vegetable shortening (USDA)[2]	¹/₆ of 9″ pie (5.6 oz.)	423	17.4	5.	13.	
(Tastykake)	4-oz. pie		14.6			
Frozen (Banquet)	5-oz. serving		15.3			
BLACKBERRY PIE FILLING, canned:						
(Comstock)	1 cup (10¾ oz.)	320	<.1			
(Lucky Leaf)	8 oz.	276	.4			
BLACKBERRY PRESERVE or JAM:						
Sweetened (Bama)	1 T. (.7 oz.)	1	<.1			(0)
Low calorie:						
(Diet Delight)	1 T. (.6 oz.)	4	Tr.			(0)
(S and W) *Nutradiet*	1 T. (.5 oz.)		<.1			(0)
BLACK-EYED PEA, frozen (See also **COWPEA**):						
Not thawed (USDA)	10-oz. pkg.	142	1.1			0

(USDA): United States Department of Agriculture
*Prepared as Package Directs
[1]Principal sources of fat: lard, butter.
[2]Principal sources of fat: vegetable shortening, butter.

Food and Description	Measure or Quantity	Sodium (mg.)	—Fats in grams— Total	Satu-rated	Unsatu-rated	Choles-terol (mg.)
Boiled, drained (USDA)	½ cup (3 oz.)	33	.3			0
(Birds Eye)	½ cup (2.5 oz.)	35	.3			0

BLANCMANGE (See **VANILLA PUDDING**)

BLOOD PUDDING or SAUSAGE

(USDA)	1 oz.		10.5	4.	7.	

BLOODY MARY MIX (Bar-Tender's)

	1 serving (.3 oz.)	510	<.1			(0)

BLUEBERRY:

Fresh, whole (USDA)	1 lb. (weighed untrimmed)	4	2.1			0
Fresh, trimmed (USDA)	½ cup (2.6 oz.)	<1	.4			0
Canned, solids & liq. (USDA):						
Syrup pack, extra heavy	½ cup (4.4 oz.)	1	.2			0
Water pack	½ cup (4.3 oz.)	1	.2			0
Frozen:						
Sweetened, solids & liq. (USDA)	½ cup (4 oz.)	1	.3			0
Unsweetened, solids & liq. (USDA)	½ cup (2.9 oz.)	<1	.4			0
Quick thaw (Birds Eye)	½ cup (5 oz.)	1	.6			0

BLUEBERRY PIE:

Home recipe, 2 crust (USDA):						
Made with lard[1]	¹/₆ of 9″ pie (5.6 oz.)	423	17.1	6.	11.	
Made with vegetable shortening[2]	¹/₆ of 9″ pie (5.6 oz.)	423	17.1	5.	12.	
(Tastykake)	4-oz. pie		14.6			
Frozen:						
(Banquet)	5-oz. serving		14.4			
(Morton)	¹/₆ of 20-oz. pie	236	10.9			
(Morton)	⅛ of 46-oz. pie	317	16.7			
(Mrs. Smith's)	¹/₆ of 8″ pie (4.2 oz.)	302	14.2			
(Mrs. Smith's) old fashion	¹/₆ of 9″ pie (5.8 oz.)	470	23.2			

(USDA): United States Department of Agriculture
*Prepared as Package Directs
[1]Principal sources of fat: lard, butter.
[2]Principal sources of fat: vegetable shortening, butter.

Food and Description	Measure or Quantity	Sodium (mg.)	— Fats in grams —			Cholesterol (mg.)
			Total	Saturated	Unsaturated	
(Mrs. Smith's) golden deluxe	⅛ of 10″ pie (5.6 oz.)	410	18.4			
Tart (Pepperidge Farm)	1 pie tart (3 oz.)	188	14.9			
BLUEBERRY PIE FILLING, canned:						
(Comstock)	1 cup (10¾ oz.)	354	.7			
(Lucky Leaf)	8 oz.	298	.6			
(Wilderness)	21-oz. can	200				
BLUEBERRY PRESERVE or JAM, sweetened:						
(Bama)	1 T. (.7 oz.)	2	<.1			(0)
(Smucker's)	1 T. (.7 oz)	6	Tr.			(0)
BLUEBERRY TURNOVER, frozen (Pepperidge Farm)	1 turnover (3.3 oz.)	258	20.0			
BLUEFISH (USDA):						
Raw, whole	1 lb. (weighed whole)	171	7.6			
Raw, meat only	4 oz.	84	3.7			
Baked or broiled[1]	4.4-oz. piece (3½″ x 3″ x ½″)	130	6.5			
Fried[2]	5.3-oz. piece (3½″ x 3″ x ½″)	219	14.7			
BOCKWURST (USDA)	1 oz.		6.7	3.	4.	
BOLOGNA:						
All meat (USDA)	1 oz.		6.5			
All meat, very thin slice (USDA)	½-oz. slice (3″ x ⅛″)		3.0			
With cereal (USDA)	1 oz.		5.8			
All meat (Armour Star)	1 oz.		9.6			
All meat (Hormel)	1 oz.	266	7.7	2.	5.	14
All meat (Oscar Mayer):						
8–10 slices per ¾ lb.	1 slice (1.3 oz.)	342	11.0	5.	6.	14
8 slices per ½ lb.	1 slice (1 oz.)	252	8.1	3.	5.	11
10 slices per ½ lb.	1 slice (.8 oz.)	207	6.7	3.	4.	9
Garlic	.8-oz. slice	224	6.7			
Wisconsin made, coarse	1 oz.	221	7.4			

(USDA): United States Department of Agriculture
*Prepared as Package Directs
[1]Prepared with butter or margarine.
[2]Prepared with egg, milk or water, & bread crumbs.

Food and Description	Measure or Quantity	Sodium (mg.)	—Fats in grams—			Choles-terol (mg.)
			Total	Satu-rated	Unsatu-rated	
Wisconsin made, fine	1 oz.	221	7.9			
Coarse ground (Hormel)	1 oz.	306	6.2	2.	3.	14
Fine ground (Hormel)	1 oz.	306	7.3	3.	4.	16
German brand (Oscar Mayer)	.8-oz. slice	242	4.4			
Lebanon, pure beef (Oscar Mayer)	.8-oz. slice	269	2.5			
Pure beef (Oscar Mayer):						
8 slices per ¾ lb.	1 slice (1.3 oz.)	332	11.0	5	6	16
10 slices per ½ lb.	1 slice (.8 oz.)	201	6.7	3	4	10
(Vienna)	1 oz.					
(Wilson)	1 oz.	312	7.9	3.	5.	17

BONITO, raw (USDA):

Food and Description	Measure or Quantity	Sodium (mg.)	Total	Satu-rated	Unsatu-rated	Choles-terol (mg.)
Whole	1 lb. (weighed whole)		19.2			
Meat only	4 oz.		8.3			

BORSCHT (Manischewitz)

	1 cup		<.1			

BOSCO (Best Foods)[1]

	1 T. (.7 oz.)	33	.3			0

BOSTON BROWN BREAD (See
  BREAD)

BOSTON CREAM PIE:

Home recipe (USDA)[2]	$^1/_{12}$ of 8″ pie (2.4 oz.)	128	6.5			
*Mix (Betty Crocker)	⅛ of pie	451	7.6			

BOUILON CUBE (See also
  individual flavors) (USDA)
flavor not indicated

	1 cube (approx. ½″, 4 grams)	960	.1			

BOUQUET GARNI:

For beef (Spice Islands)	1 tsp.	<1				
For soup (Spice Islands)	1 tsp.	5				

BOURBON WHISKEY (See
  DISTILLED LIQUOR)

(USDA): United States Department of Agriculture
*Prepared as Package Directs
[1]Principal source of fat: cocoa.
[2]Made with sodium aluminum sulfate-type baking powder.

Food and Description	Measure or Quantity	Sodium (mg.)	— Fats in grams —			Cholesterol (mg.)
			Total	Saturated	Unsaturated	

BOYSENBERRY:
Fresh (See **BLACKBERRY,** fresh)
Canned, unsweetened or low calorie:
  Water pack, solids & liq.

Food and Description	Measure or Quantity	Sodium (mg.)	Total	Saturated	Unsaturated	Cholesterol (mg.)
(USDA)	4 oz.	1	.1			0
Solids & liq. (S and W) *Nutradiet*	4 oz.	2	.1			(0)
Frozen, not thawed (USDA):						
Sweetened	4 oz.	1	.3			0
Unsweetened	4 oz.	1	.3			0
BOYSENBERRY PIE, frozen:						
(Banquet)	5-oz. serving		15.0			
(Morton)	¹/₆ of 20-oz. pie	231	10.7			
BOYSENBERRY PIE FILLING						
(Comstock)	1 cup (10¾ oz.)	238	.2			
BOYSENBERRY PRESERVE or JAM, low calorie:						
(S and W) *Nutradiet*	1 T. (.5 oz.)		<.1			(0)
(Tillie Lewis)	1 T. (.5 oz.)	2	Tr.			0
BRAINS, all animals, raw (USDA)	4 oz.	142	9.8			2268
BRAN (USDA):						
With added sugar & defatted wheat germ	1 oz.	139	.5			0
With added sugar & malt extract	1 oz.	301	.9			0
BRAN BREAKFAST CEREAL:						
Plain:						
All-Bran (Kellogg's)	½ cup (1 oz.)	287	.7			(0)
Bran Buds (Kellogg's)	⅓ cup (1 oz.)	63	.7			(0)
40% bran flakes (USDA)	½ cup (.6 oz.)	162	.3			0
40% bran flakes (Kellogg's)	¾ cup (1 oz.)	141	.5			(0)
40% bran flakes (Post)	⅔ cup (1 oz.)	190	.3			0
100% bran (Nabisco)	½ cup (1 oz.)	187	1.0			(0)
Raisin bran flakes:						
(USDA)	½ cup (.9 oz.)	200	.4			0
(Kellogg's)	¾ cup (1 oz.)	143	.5			(0)
(Post)	½ cup (1 oz.)	136	.3			0
Cinnamon (Post)	½ cup (1 oz.)	136	.3			0

(USDA): United States Department of Agriculture
*Prepared as Package Directs

Food and Description	Measure or Quantity	Sodium (mg.)	—Fats in grams—			Choles-terol (mg.)
			Total	Satu-rated	Unsatu-rated	

BRANDY (See **DISTILLED LIQUOR**)

BRANDY, BLACKBERRY

(Leroux) 70 proof	1 fl. oz.	2	(0.)			(0)

BRATWURST (Oscar Mayer)	1 oz.	221	8.5			

BRAUNSCHWEIGER:

(USDA)	2 slices (2″ x ¼″, .7 oz.)		5.5	2.	4.	
(Oscar Mayer)	1 oz.	310	10.2			
(Wilson)	1 oz.	293	7.8	3.	5.	55
Liver cheese (Oscar Mayer)	1 slice (6 per ½ lb.)	459	8.4			

BRAZIL NUT (USDA):

Whole	1 lb. (weighed in shell)	2	145.6	29.	117.	0
Whole	1 cup (14 nuts, 4.3 oz. with shell)	1	39.2	8.	31.	0
Shelled	½ cup (2.5 oz.)	<1	46.8	9.	38.	0
Shelled	4 nuts (.6 oz.)	<1	11.7	2.	9.	0

BREAD (listed by type or brand name; toasting does not affect these nutritive values, only weight):

Boston brown (USDA)	1.7-oz. slice (3″ x ¾″)	120	.6			
Cheese, party (Pepperidge Farm)	1 slice (6 grams)	39	.4			
Cornbread (See **CORNBREAD**)						
Corn & Molasses (Pepperidge Farm)	.9-oz. slice	116	.6			
Cracked-wheat:						
(USDA)	.8-oz. slice	122	.5			
(USDA) 20 slices to 1 lb.	.9-oz. slice	132	.6			
(Pepperidge Farm)	.9-oz. slice	151	1.0			
Honey (Wonder)	.8-oz. slice	137	.7			
Daffodil Farm (Wonder)	.8-oz. slice	140	.6			
Date-nut loaf (Thomas')	1.1-oz. slice	150	1.5			
Finn Crisp	1 piece (6 grams)		<.1			
French:						
(USDA) 20 slices to 1 lb.[1]	.8-oz. slice	133	.7	Tr.	<1.	

(USDA): United States Department of Agriculture
*Prepared as Package Directs
[1]Principal source of fat: vegetable shortening.

Food and Description	Measure or Quantity	Sodium (mg.)	Total	Fats in grams — Satu-rated	Unsatu-rated	Choles-terol (mg.)
(Pepperidge Farm)	1″ slice (1.1 oz.)	173	1.2			
Glutogen Gluten (Thomas')	.5-oz. slice	80	.1			0
Honey Wheatberry (Pepperidge Farm)	1.1-oz. slice	172	1.1			
Italian:						
(USDA) 20 slices to 1 lb.	.8-oz. slice	135	.2			
(Pepperidge Farm)	1″ slice (1.2 oz.)	179	1.3			
King's Bread (Wasa)	3.5-oz. piece	400	2.0			
Low sodium, 1-lb. loaf (Van de Kamp's)	.8-oz. slice	7				
Natural Health (Arnold)	.9-oz. slice		1.9			
Oatmeal:						
(Arnold)	.8-oz. slice		1.2			
(Pepperidge Farm)	.9-oz. slice	166	1.0			
Profile, dark (Wonder)	.8-oz. slice	129	.7			
Profile, light (Wonder)	.8-oz. slice	143	.7			
Protogen Protein (Thomas')	.7-oz. slice	120	.1			0
Pumpernickel:						
(USDA) 20 slices to 1 lb.	.8-oz. slice	131	.3			
(Arnold) Jewish	1.4-oz. slice		1.2			
(Pepperidge Farm) family	1.2-oz. slice	239	.8			
(Pepperidge Farm) party	1 slice (8 grams)	58	.2			
(Wonder)	.8-oz. slice	148	.2			
Raisin:						
(USDA) 18 slices to 1 lb.[1]	.9-oz. slice	91	.7	Tr.	Tr.	
Cinnamon (Pepperidge Farm)	.9-oz. slice	66	1.6			
Cinnamon (Thomas')	.8-oz. slice	105	.5			0
Cinnamon (Wonder)	.8-oz slice	83	.5			
Orange (Arnold)	.9-oz. slice		1.8			
Tea (Arnold)	.9-oz. slice		1.9			
Rite Diet (Thomas')	.7-oz. slice	110	.3			
Roman Meal	.8-oz. slice	126	.7			
Rye:						
Light, 18 slices to 1 lb. (USDA)	.9-oz. slice	139	.3			
(Arnold) Melba thin, Jewish	.6-oz. slice		.6			
(Arnold) Jewish, seeded or unseeded	1.2-oz. slice		1.4			
(Pepperidge Farm) family	1.2-oz. slice	232	.8			
(Pepperidge Farm) party	1 slice (6 grams)	86	.2			
(Pepperidge Farm) seedless	1.2-oz. slice	235	.8			

(USDA): United States Department of Agriculture
*Prepared as Package Directs
[1]Prepared with vegetable shortening & nonfat dry milk.

Food and Description	Measure or Quantity	Sodium (mg.)	Fats in grams — Total	Satu- rated	Unsatu- rated	Choles- terol (mg.)
Ry-King (Wasa):						
Brown	1 piece (.4 oz.)	47	.2	Tr.	Tr.	
Golden	1 piece (10 grams)	35	.2	Tr.	Tr.	
Lite	1 piece (8 grams)	34	.1	Tr.	Tr.	
Seasoned	1 piece (9 grams)	49	.2			
(Wonder)	.8-oz. slice	134	.5			
(Wonder) *Beefsteak*	1.1-oz. slice	213	.7			
(Wonder) *Beefsteak,* family	.8-oz. slice	147	.6			
Salt-rising (USDA)	.9-oz. slice	66	.6	Tr.	Tr.	
Toaster cake (See **TOASTER CAKE**)						
Vienna, 20 slices to 1 lb.						
(USDA)	.8-oz. slice	133	.7	Tr.	Tr.	
Wheat (Wonder):						
Golden	.9-oz. slice	140	.7			
Home Pride	.9-oz. slice	131	.9			
Wheat germ (Pepperidge Farm)	.9-oz. slice	144	.6			
White, enriched or unenriched:						
Prepared with 1-2% nonfat dry milk (USDA)[1]	.8-oz. slice	117	.7	Tr.	<1.	
Prepared with 3-4% nonfat dry milk (USDA)[1]	.8-oz. slice	117	.7	Tr.	<1.	
Prepared with 5-6% nonfat dry milk (USDA)[1]	.8-oz. slice	114	.9	Tr.	<1.	
(Arnold) Melba thin, diet-slice	.5-oz. slice		1.0			
(Arnold) sandwich, soft, 1½-lb. loaf	.9-oz. slice		1.6			
(Arnold) small family	.8-oz. slice		1.5			
(Arnold) toasting	1.1-oz. slice		1.8			
(Pepperidge Farm) large loaf	1-oz. slice	151	1.4			
(Pepperidge Farm) large loaf, Calif. only	.8-oz. slice	131	1.2			
(Pepperidge Farm) sandwich	.8-oz. slice	126	1.2			
(Pepperidge Farm) toasting	1.2-oz. slice	230	.9			
(Pepperidge Farm) very thin slice, East	.5-oz. slice	74	.3			
(Pepperidge Farm) very thin slice, Midwest	.6-oz. slice	79	.4			
(Thomas')	.9-oz. slice	190	.6			
(Wonder)	.9-oz. slice	148	.7			
Brick Oven (Arnold) 1-lb. loaf	.8-oz. slice		1.5			
Brick Oven (Arnold) 30-oz. loaf	1.1-oz. slice		1.8			

(USDA): United States Department of Agriculture
*Prepared as Package Directs
[1]Principal source of fat: vegetable shortening.

Food and Description	Measure or Quantity	Sodium (mg.)	—Fats in grams—			Choles-terol (mg.)
			Total	Satu-rated	Unsatu-rated	
Brick Oven, golden (Arnold)						
2-lb. loaf	1-oz. slice		1.9			
English Tea Loaf						
(Pepperidge Farm)	.9-oz. slice	146	1.5			
Hearthstone (Arnold) 1-lb. loaf	.9-oz. slice		1.5			
Hearthstone (Arnold) 2-lb. loaf	1.1-oz. slice		1.8			
Home Pride (Wonder)	.9-oz. slice	145	1.0			
Whole-wheat:						
Made with 2% nonfat dry						
milk (USDA)[1]	.9-oz. slice	132	.8	Tr.	<1.	
Made with water (USDA)[1]	.8-oz. slice	122	.6	<1	Tr.	
Made with water (USDA)[1]	.9-oz. slice	132	.6	Tr.	Tr.	
(Arnold) Melba thin, diet slice	.6-oz. slice		1.1			
(Arnold) small family	.8-oz. slice		1.7			
(Pepperidge Farm)	.9-oz. slice	107	1.2			
(Thomas') 100%	.9-oz. slice	140	.6			0
(Wonder)	.8-oz. slice	136	.9			
Brick Oven (Arnold) 1-lb. loaf	.8-oz. slice		1.7			
Brick Oven (Arnold) 2-lb. loaf	1.1-oz. slice		2.1			

BREAD, CANNED:

Food and Description	Measure or Quantity	Sodium (mg.)	Total	Satu-rated	Unsatu-rated	Choles-terol (mg.)
Banana nut (Dromedary)	½″ slice (1 oz.)	212	2.0			
Brown, plain (B & M)	½″ slice (1.5 oz.)	209	Tr.			
Brown with raisins (B & M)	½″ slice (1.6 oz.)	201	.3			
Chocolate nut (Dromedary)	½″ slice (1 oz.)	151	2.4			
Date & nut (Crosse & Blackwell)	½″ slice (1 oz.)		.5			
Date & nut (Dromedary)	½″ slice (1 oz.)	155	2.1			
Orange nut (Dromedary)	½″ slice (1 oz.)	133	2.0			

BREAD CRUMBS:

Food and Description	Measure or Quantity	Sodium (mg.)	Total	Satu-rated	Unsatu-rated	Choles-terol (mg.)
Dry, grated (USDA)	1 cup (3.5 oz.)	736	4.6	1.	4.	
Dry, grated (USDA)[1]	1 T. (6 grams)	47	.3	Tr.	Tr.	
(Buitoni)	4 oz.		5.0			
(Old London)	1 cup (4.5 oz.)		3.6			
Seasoned (Contadina)	1 cup (4.1 oz.)	3695	2.2			

BREAD DOUGH, frozen (Morton)   1 oz.   141   1.3

BREAD PUDDING with raisins,
home recipe (USDA)[2]   1 cup (9.3 oz.)   557   16.2   8.   8.   170

(USDA): United States Department of Agriculture
*Prepared as Package Directs
[1]Principal source of fat: vegetable shortening.
[2]Principal sources of fat: milk, butter, eggs & bread crumbs.

Food and Description	Measure or Quantity	Sodium (mg.)	— Fats in grams —			Choles- terol (mg.)
			Total	Satu- rated	Unsatu- rated	

BREAD STICK:

Cheese (Keebler)	1 piece (3 grams)	25	.1			
Dietetic (Stella D'oro)	1 piece (9 grams)	3	1.0			
Garlic (Keebler)	1 piece (3 grams)	28	.1			
Onion (Stella D'oro)	1 piece (.4 oz.)		1.1			
Regular (Stella D'oro)	1 piece (10 grams)		1.0			
Salt:						
(USDA)[1]	1 piece (3 grams)	50	<.1	Tr.	Tr.	
Vienna type (USDA)[1]	1 piece (3 grams)	47	<.1	Tr.	Tr.	
(Keebler)	1 piece (3 grams)	30	.1			
Sesame (Keebler)	1 piece (3 grams)	41	.1			
Sesame (Stella D'oro)	1 piece (9 grams)		1.6			

BREAD STUFFING MIX:

Dry (USDA)[1]	1 cup (2.5 oz.)	945	2.7	<1.		2.
*Crumb type, prepared with water & fat (USDA)[2]	1 cup (5 oz.)	1263	30.7	16.	15.	
*Moist type, prepared with water, egg & fat (USDA)[3]	1 cup (7.2 oz.)	1023	26.0	14.	12.	
Corn bread (Pepperidge Farm)	8-oz. pkg.	4142	9.2			
Cube (Pepperidge Farm)	7-oz. pkg.	3389	4.0			
Herb seasoned (Pepperidge Farm)	8-oz. bag	3931	3.0			
Seasoned (Uncle Ben's)						
Stuff 'n Such	6-oz. pkg.	3381	2.9			0
*Seasoned (Uncle Ben's)						
Stuff 'n Such, no added butter	½ cup (2.9 oz.)	647	.6			0
*(Uncle Ben's) *Stuff 'n Such*, with added butter	½ cup (3.3 oz.)	747	8.6			

BREADFRUIT, fresh (USDA):

Whole	1 lb. (weighed untrimmed)	52	1.0			0
Peeled & trimmed	4 oz.	17	.3			0

BROADBEAN:

Immature seed (USDA)	1 lb. (weighed in pod)	6	.6			0
Immature seed (USDA)	1 oz. (without pod)	1	.1			0
Mature seed, dry (USDA)	1 oz.		.5			0

(USDA): United States Department of Agriculture
*Prepared as Package Directs
[1]Principal source of fat: vegetable shortening.
[2]Principal sources of fat: vegetable shortening & butter.
[3]Principal sources of fat: butter, vegetable shortening & egg.

Food and Description	Measure or Quantity	Sodium (mg.)	—Fats in grams—			Choles- terol (mg.)
			Total	Satu- rated	Unsatu- rated	
Canned, regular pack, drained solids (Del Monte)	½ cup (2.5 oz.)	368	<.1			(0)
Frozen, Italian bean (Birds Eye)	⅓ of 9-oz. pkg.	36	.1			0
Frozen, Italian bean, tomato sauce (Green Giant)	⅓ of 10-oz. pkg.	340	.9			

BROCCOLI:

Raw, whole (USDA)	1 lb. (weighed untrimmed)	42	.8			0
Raw, large leaves removed (USDA)	1 lb. (weighed partially trimmed)	53	1.1			0
Boiled without salt, drained (USDA):						
Whole stalk	1 stalk (6.3 oz.)	18	.5			0
½" pieces	½ cup (2.8 oz.)	18	.2			0
Frozen:						
Chopped or cut:						
Not thawed (USDA)	10-oz. pkg.	48	.8			0
Boiled without salt, drained (USDA)	1⅜ cups (10-oz. pkg.)	25	.8			0
(Birds Eye)	⅓ of 10-oz. pkg.	56	.3			0
& noodle with sour cream sauce (Green Giant)	⅓ of 10-oz. pkg.	425	4.7			
In cheese sauce (Green Giant)	⅓ of 10-oz. pkg.	331	2.1			
Spears:						
Not thawed (USDA)	10-oz. pkg.	37	.6			0
Boiled without salt, drained (USDA)	½ cup (3.3 oz.)	11	.2			0
(Birds Eye)	⅓ of 10-oz. pkg.	51	.2			0
Baby spears (Birds Eye)	⅓ of 10-oz. pkg.	12	.2			0
In butter sauce (Green Giant)	⅓ of 10-oz. pkg.	444	2.2			
In Hollandaise sauce (Birds Eye)	⅓ of 10-oz. pkg.	96	8.5			57

BROTH & SEASONING (See also individual kinds):

Maggi	1 T. (.6 oz.)	923	0.			
Golden (George Washington)	1 packet (4 grams)	1093	<.1			
Rich Brown (George Washington)	1 packet (4 grams)	1212	<.1			

BROTWURST (Oscar Mayer)	3-oz. link	663	23.8			

(USDA): United States Department of Agriculture
*Prepared as Package Directs

Food and Description	Measure or Quantity	Sodium (mg.)	Total	—Fats in grams— Satu- rated	Unsatu- rated	Choles- terol (mg.)
BROWNIE (See **COOKIE**)						
BROWN 'n SEASON (Adolph's)	1 tsp. (4 grams)	21	<.1	Tr.	Tr.	0
BRUSSELS SPROUT:						
Raw (USDA)	1 lb.	58	1.7			0
Boiled without salt, 1¼"–1½" dia., drained (USDA)	1 cup (7–8 sprouts, 5.5 oz.)	16	.6			0
Frozen:						
Not thawed (USDA)	10-oz. pkg.	45	.6			0
Boiled without salt, drained (USDA)	4 oz.	16	.2			0
Au gratin, casserole (Green Giant)	⅓ of 10-oz. pkg.	439	2.4			
Baby sprouts (Birds Eye)	½ cup (3.3 oz.)	15	.2			0
In butter sauce (Green Giant)	⅓ of 10-oz. pkg.	421	2.6			
BUCKWHEAT:						
Flour (See **FLOUR**)						
Groats:						
(Pocono)	1 oz.		.7			(0)
Wolff's Kasha (Birkett)	1 oz.		.5			0
Whole-grain (USDA)	1 oz.		.7			0
BUC WHEATS, cereal (General Mills)	1 cup	264	.5			(0)
BUFFALOFISH, raw (USDA):						
Whole	1 lb. (weighed whole)	76	6.1			
Meat only	4 oz.	59	4.8			
BULGUR (from hard red winter wheat) (USDA):						
Dry	1 lb.		6.8			0
Canned, seasoned	1 cup (4.8 oz.)	621	4.5			
Canned, unseasoned	4 oz.	679	.8			0
BULLHEAD, raw (USDA):						
Whole	1 lb. (weighed whole)		1.4			
Meat only	4 oz.		1.8			

(USDA): United States Department of Agriculture
*Prepared as Package Directs

59

Food and Description	Measure or Quantity	Sodium (mg.)	— Fats in grams —			Cholesterol (mg.)
			Total	Saturated	Unsaturated	

BULLOCK'S-HEART (See **CUSTARD APPLE**)

BUN (See **ROLL**)

BURBOT, raw (USDA):
Whole	1 lb. (weighed whole)		.6			
Meat only	4 oz.		1.0			

BURGUNDY WINE:
(Gold Seal) 12% alcohol	3 fl. oz.	3	0.			(0)
(Great Western) 12.5% alcohol	3 fl. oz.	36	0.			0

BURGUNDY WINE, SPARKLING
(Gold Seal) 12% alcohol	3 fl. oz.	3	0.			(0)

BUTTER:
Salted:						
(USDA)	¼ lb. (1 stick, ½ cup)	1119	92.0	52.	40.	284
(USDA)	1 T. (⅛ stick, .5 oz.)	138	11.3	6.	5.	35
(USDA)	1 pat (1″ x 1″ x ⅓″, 5 grams)	49	4.0	2.	2.	12
(Breakstone)	1 T. (.5 oz.)	95	11.0			29
(Sealtest)	1 T. (.5 oz.)	117	12.1			
Whipped (USDA)	2.7 oz. (1 stick, ½ cup)	750	61.6	35.	27.	190
Whipped (USDA)	1 T. (⅛ stick, 9 grams)	89	7.3	4.	3.	22
Whipped (USDA)	1 pat (1¼″ x 1¼″ x ⅓″, 4 grams)	39	3.2	2.	1.	10
Whipped (Breakstone)	1 T. (9 grams)	64	7.4			19
Whipped (Sealtest)	1 T. (9 grams)	74	7.6			
Unsalted:						
(USDA)	¼ lb. (1 stick, ½ cup)	11	92.0	52.	40.	284
(USDA)	1 T. (⅛ stick, .5 oz.)	1	11.3	6.	5.	35
(USDA)	1 pat (1″ x 1″ x ⅓″, 5 grams)	<1	4.0	2.	2.	12
(Breakstone)	1 T. (.5 oz.)	<1	11.0			29
(Sealtest)	1 T. (.5 oz.)	1	12.1			
Whipped (USDA)	½ cup (1 stick, 2.7 oz.)	8	61.6	35.	27.	190
Whipped (USDA)	1 T. (⅛ stick, 9 grams)	<1	7.3	4.	3.	22

(USDA): United States Department of Agriculture
*Prepared as Package Directs

Food and Description	Measure or Quantity	Sodium (mg.)	— Fats in grams —			Choles- terol (mg.)
			Total	Satu- rated	Unsatu- rated	
Whipped (USDA)	1 pat (1¼″ x 1¼″ x ⅓″, 4 grams)	<1	3.2	2.	1.	10
Whipped (Breakstone)	1 T. (9 grams)	<1	7.4			19

BUTTER BEAN (See **BEAN, LIMA**)

**BUTTER BRICKLE LAYER CAKE
   MIX* (Betty Crocker)   ¹/₁₂ of cake   5.7

BUTTERFISH, raw (USDA):
   Gulf:

Whole	1 lb. (weighed whole)		6.7			
Meat only	4 oz.		3.3			

   Northern:

Whole	1 lb. (weighed whole)		23.6			
Meat only	4 oz.		11.6			

BUTTERMILK (See **MILK**)

BUTTERNUT (USDA):

Whole	1 lb. (weighed in shell)		38.9			0
Shelled	4 oz.		69.4			0

BUTTER OIL or dehydrated butter
   (USDA)   1 cup (7.2 oz.)   203.0 112.   91.

BUTTERSCOTCH MORSELS
   (Nestlé's)   6-oz. pkg.   52.7

BUTTERSCOTCH PIE:

Home recipe, made with lard (USDA)[1]	¹/₆ of 9″ pie (5.4 oz.)	325	16.7	6.	11.	
Home recipe, made with vegetable shortening (USDA)[2]	¹/₆ of 9″ pie (5.4 oz.)	325	16.7	5.	12.	
Frozen, cream (Banquet)	2½-oz. serving		7.9			

**BUTTERSCOTCH PIE FILLING
   MIX** (See **BUTTERSCOTCH
   PUDDING MIX**)

(USDA): United States Department of Agriculture
*Prepared as Package Directs
[1]Principal sources of fat: lard, butter.
[2]Principal sources of fat: vegetable shortening, butter.

Food and Description	Measure or Quantity	Sodium (mg.)	—Fats in grams— Total	Satu- rated	Unsatu- rated	Choles- terol (mg.)
BUTTERSCOTCH PUDDING:						
Chilled (Breakstone)	5-oz. container	250	13.3			0
Chilled (Sealtest)	4 oz.	298	3.3			
Canned:						
(Betty Crocker)	½ cup	204	4.9			
(Del Monte)	5-oz. can	264	5.3			
(Hunt's)	5-oz. can	242	12.4	2.	10.	
(Thank You)	½ cup (4.5 oz.)		4.7			
BUTTERSCOTCH PUDDING or PIE MIX:						
Sweetened:						
*Instant (Jell-O)	½ cup (5.3 oz.)	406	4.7			13
*Instant (Royal)	½ cup (5.1 oz.)	380	5.1			14
*Regular (Jell-O)	½ cup (5.2 oz.)	224	4.6			13
*Regular (Royal)	½ cup (5.1 oz.)	260	5.2			14
*Low calorie (D-Zerta)	½ cup (4.6 oz.)	142	4.5			13
***B-V** (Wilson)	1 tsp. (7 grams)	978	0.			

C

Food and Description	Measure or Quantity	Sodium (mg.)	Total	Satu- rated	Unsatu- rated	Choles- terol (mg.)
CABBAGE:						
White (USDA):						
Raw:						
Whole	1 lb. (weighed untrimmed)	72	.7			0
Coarsely shredded or sliced	1 cup (2.5 oz.)	14	.1			0
Finely shredded or chopped	1 cup (3.2 oz.)	18	.2			0
Wedge	3½″ x 4½″ wedge (3.5 oz.)	20	.2			0
Boiled without salt, until tender:						
Shredded, small amount of water, drained	½ cup (2.6 oz.)	10	.1			0
Wedges, in large amount of water, drained	½ cup (3.2 oz.)	12	.2			0
Dehydrated	1 oz.	54	.5			0
Red, raw (USDA):						
Whole	1 lb. (weighed untrimmed)	93	.7			0
Coarsely shredded	1 cup (2.5 oz.)	18	.1			0
Red, canned (Comstock-Greenwood)	4 oz.		Tr.			(0)

(USDA): United States Department of Agriculture
*Prepared as Package Directs

Food and Description	Measure or Quantity	Sodium (mg.)	—Fats in grams—		Choles- terol (mg.)
			Total	Satu- rated	Unsatu- rated

Food and Description	Measure or Quantity	Sodium (mg.)	Total	Satu- rated	Unsatu- rated	Choles- terol (mg.)
Savoy, raw (USDA):						
Whole	1 lb. (weighed untrimmed)	79	.7			0
Coarsely shredded	1 cup (2.5 oz.)	15	.1			0
CABBAGE, CHINESE or CELERY, raw (USDA):						
Whole	1 lb. (weighed untrimmed)	101	.4			0
1″ pieces, leaves with stalk	½ cup (1.3 oz.)	9	Tr.			0
CABBAGE, SPOON or WHITE MUSTARD or PAKCHOY (USDA):						
Raw	1 lb. (weighed untrimmed)	112	.9			0
Boiled without salt, drained	½ cup (3 oz.)	15	.2			0
CACTUS COOLER, soft drink						
(Canada Dry) bottle or can	6 fl. oz.	13+	0.			(0)
CAKE. Most cakes are listed elsewhere by kind of cake such as **ANGEL FOOD** or **CHOCOLATE** or brand name such as **YANKEE DOODLES.** Those listed below are made with sodium aluminum sulfate-type baking powder. (USDA):						
Plain, home recipe:						
Without icing:						
Made with butter[1]	¹/₉ of 9″ sq. cake (3″ x 3″ x 1″, 3 oz.)	258	10.9	6.	5.	
Made with vegetable shortening[2]	¹/₉ of 9″ sq. cake 3″ x 3″ x 1″, 3 oz.)	258	12.0	3.	8.	
With boiled white icing:						
Made with butter[1]	¹/₉ of 9″ sq. cake (4 oz.)	299	12.0	7.	5.	
Made with vegetable shortening[2]	¹/₉ of 9″ sq. cake (4 oz.)	299	12.0	3.	9.	
With chocolate icing:						
Made with butter[3]	¹/₁₆ of 10″ layer cake (3.5 oz.)	229	12.7	8.	6.	

(USDA): United States Department of Agriculture
*Prepared as Package Directs
[1]Principal sources of fat: butter, egg & milk.
[2]Principal sources of fat: vegetable shortening, egg & milk.
[3]Principal sources of fat: butter, egg, milk & chocolate.

Food and Description	Measure or Quantity	Sodium (mg.)	Total	Fats in grams — Satu- rated	Unsatu- rated	Choles- terol (mg.)
Made with vegetable shortening[1]	$^1/_{16}$ of 10″ layer cake (3.5 oz.)	229	13.9	4.	9.	
With uncooked white icing:						
Made with butter[2]	$^1/_{16}$ of 10″ layer cake (3.5 oz.)	227	12.7	7.	6.	
Made with vegetable shortening[3]	$^1/_{16}$ of 10″ layer cake (3.5 oz.)	227	11.8	3.	9.	
White, home recipe:						
Without icing:						
Made with butter[2]	$^1/_9$ of 9″ sq. cake (3″ x 3″ x 1″, 3 oz.)	278	13.8	8.	6.	
Made with vegetable shortening[3]	$^1/_9$ of 9″ sq. cake (3″ x 3″ x 1″, 3 oz.)	278	13.8	4.	10.	
With coconut icing:						
Made with butter[2]	$^1/_{16}$ of 10″ layer cake (3.5 oz.)	257	13.3	7.	6.	
Made with vegetable shortening[3]	$^1/_{16}$ of 10″ layer cake (3.5 oz.)	257	13.3	4.	9.	
With uncooked white icing:						
Made with butter[2]	$^1/_{16}$ of 10″ layer cake (3.5 oz.)	234	12.9	7.	6.	
Made with vegetable shortening[3]	$^1/_{16}$ of 10″ layer cake (3.5 oz.)	234	12.9	4.	9.	
Yellow, home recipe:						
Without icing:						
Made with butter[2]	$^1/_9$ of 9″ sq. cake (3″ x 3″ x 1″, 3 oz.)	222	10.9	6.	5.	
Made with vegetable shortening[3]	$^1/_9$ of 9″ sq. cake (3″ x 3″ x 1″, 3 oz.)	222	10.9	3.	8.	
With caramel icing:						
Made with butter[2]	$^1/_{16}$ of 10″ layer cake (3.5 oz.)	226	11.7	6.	6.	
Made with vegetable shortening[3]	$^1/_{16}$ of 10″ layer cake (3.5 oz.)	226	11.7	3.	9.	
With chocolate icing, 2-layer:						
Made with butter[4]	$^1/_{16}$ of 9″ cake (2.6 oz.)	156	9.8	5.	4.	
Made with vegetable shortening[5]	$^1/_{16}$ of 9″ cake (2.6 oz.)	156	9.8	3.	7.	33

(USDA): United States Department of Agriculture
*Prepared as Package Directs
[1]Principal sources of fat: vegetable shortening, egg, milk & chocolate.
[2]Principal sources of fat: butter, egg & milk.
[3]Principal sources of fat: vegetable shortening, egg & milk.
[4]Principal sources of fat: butter, egg, milk & chocolate.
[5]Principal sources of fat: vegetable shortening, egg, milk & chocolate.

Food and Description	Measure or Quantity	Sodium (mg.)	—Fats in grams— Total	Satu- rated	Unsatu- rated	Choles- terol (mg.)

CAKE FROSTING (See **CAKE ICING & CAKE ICING MIX**)

CAKE ICING:

Food and Description	Measure or Quantity	Sodium	Total	Satu-rated	Unsatu-rated	Cholesterol
Butterscotch (Betty Crocker)	$^1/_{12}$ of 16.5-oz. can	157	6.7			
Caramel, home recipe (USDA)[1]	4 oz.	94	7.6	4.	3.	
Cherry (Betty Crocker)	$^1/_{12}$ of 16.5-oz. can	97	6.6			
Chocolate, home recipe (USDA)[2]	1 cup (9.7 oz.)	168	38.2	22.	16.	
Chocolate (Betty Crocker)	$^1/_{12}$ of 16.5-oz. can	113	7.7			
Coconut, home recipe (USDA)[3]	1 cup (5.8 oz.)	196	12.8	12.	Tr.	
Dark Dutch fudge (Betty Crocker)	$^1/_{12}$ of 16.5-oz. can	108	7.0			
Lemon (Betty Crocker)	$^1/_{12}$ of 16.5-oz. can	101	6.6			
Milk chocolate (Betty Crocker)	$^1/_{12}$ of 16.5-oz. can	107	7.0			
Vanilla (Betty Crocker)	$^1/_{12}$ of 16.5-oz. can	96	6.6			
White, boiled, home recipe (USDA)	1 cup (3.3 oz.)	134	0.			
White, uncooked, home recipe (USDA)[1]	4 oz.	56	7.5	4.	3.	

CAKE ICING MIX:

Food and Description	Measure or Quantity	Sodium	Total	Satu-rated	Unsatu-rated	Cholesterol
*Banana (Betty Crocker)	$^1/_{12}$ of cake's icing	46	2.8			
*Butter Brickle (Betty Crocker)	$^1/_{12}$ of cake's icing	66	2.8			
*Caramel (Betty Crocker)	$^1/_{12}$ of cake's icing	75	2.7			
*Caramel apple (Betty Crocker)	$^1/_{12}$ of cake's icing	42	1.9			
*Cherry, creamy (Betty Crocker)	$^1/_{12}$ of cake's icing	69	2.8			
*Cherry fluff (Betty Crocker)	$^1/_{12}$ of cake's icing	29	Tr.			
*Cherry fudge (Betty Crocker)	$^1/_{12}$ of cake's icing	40	3.1			
*Chocolate, fluffy (Betty Crocker)	$^1/_{12}$ of cake's icing	32	2.6			
Chocolate fudge (USDA)[4]	1 oz.	27	2.8	<1.	2.	
*Chocolate fudge, prepared with water & fat (USDA)[5]	8 oz.	354	32.7	14.	19.	
*Chocolate fudge (Betty Crocker)	$^1/_{12}$ of cake's icing	41	3.1			
*Chocolate malt (Betty Crocker)	$^1/_{12}$ of cake's icing	48	3.0			
*Chocolate walnut (Betty Crocker)	$^1/_{12}$ of cake's icing	37	3.6			
*Coconut-pecan (Betty Crocker)	$^1/_{12}$ of cake's icing	42	4.4			
*Coconut, toasted (Betty Crocker)	$^1/_{12}$ of cake's icing	72	3.3			
*Dark chocolate fudge (Betty Crocker)	$^1/_{12}$ of cake's icing	40	3.1			
Fudge, creamy, contains nonfat						

(USDA): United States Department of Agriculture
*Prepared as Package Directs
[1] Principal sources of fat: butter & milk.
[2] Principal sources of fat: chocolate, butter & milk.
[3] Principal source of fat: coconut.
[4] Principal sources of fat: vegetable shortening & cocoa.
[5] Principal sources of fat: vegetable shortening, butter & cocoa.

Food and Description	Measure or Quantity	Sodium (mg.)	— Fats in grams —			Choles- terol (mg.)
			Total	Satu- rated	Unsatu- rated	
dry milk (USDA):						
Dry[1]	1 oz.	75	2.1	<1.	2.	
*Prepared with water (USDA)[2]	1 cup (8.6 oz)	568	15.9			
*Prepared with water & fat (USDA)[3]	8 oz.	728	34.5	16.	19.	
*Fudge nugget (Betty Crocker)	$^1/_{12}$ of cake's icing	33	3.0			
*Lemon, creamy (Betty Crocker)	$^1/_{12}$ of cake's icing	63	2.7			
*Lemon fluff (Betty Crocker)	$^1/_{12}$ of cake's icing	32	Tr.			
*Milk chocolate (Betty Crocker)	$^1/_{12}$ of cake's icing	47	2.2			
*Orange (Betty Crocker)	$^1/_{12}$ of cake's icing	66	2.7			
*Pineapple (Betty Crocker)	$^1/_{12}$ of cake's icing	63	2.7			
*Sour cream, chocolate fudge (Betty Crocker)	$^1/_{12}$ of cake's icing	40	3.1			
*Sour cream, white (Betty Crocker)	$^1/_{12}$ of cake's icing	65	2.0			
*Spice, creamy (Betty Crocker)	$^1/_{12}$ of cake's icing	77	2.8			
*White, creamy (Betty Crocker)	$^1/_{12}$ of cake's icing	66	2.7			
*White, fluffy (Betty Crocker)	$^1/_{12}$ of cake's icing	29	Tr.			
CAKE MIX. Most cake mixes are listed by kind of cake, such as **ANGEL FOOD CAKE MIX, CHOCOLATE CAKE MIX,** etc.						
White:						
(USDA)[4]	1 oz.	106	3.4	<1.	2.	
*Made with egg whites & water, with chocolate icing, 2 layers (USDA)[5]	$^1/_{16}$ of 9″ cake (2.5 oz.)	161	7.6	3.	5.	1
*(Betty Crocker) layer	$^1/_{12}$ of cake	273	4.8			
*(Duncan Hines)	$^1/_{12}$ of cake (2.6 oz.)	314	4.4			
*(Swans Down)	$^1/_{12}$ of cake (2.5 oz.)	350	2.5			
*Sour cream (Betty Crocker) layer	$^1/_{12}$ of cake	274	5.3			
Yellow:						
(USDA)	1 oz.	115	3.7			
*Made with eggs & water, with chocolate icing (USDA)	$^1/_{16}$ of 9″ cake (2.6 oz.)	170	8.5			36

(USDA): United States Department of Agriculture
*Prepared as Package Directs
[1]Contains nonfat dry milk.
[2]Principal sources of fat: vegetable shortening & cocoa.
[3]Principal sources of fat: vegetable shortening, butter & cocoa.
[4]Principal source of fat: vegetable shortening.
[5]Principal sources of fat: vegetable shortening, chocolate, egg & milk.

Food and Description	Measure or Quantity	Sodium (mg.)	— Fats in grams —			Choles-terol (mg.)
			Total	Satu-rated	Unsatu-rated	
*(Betty Crocker) layer	1/12 of cake	273	5.6			
*Butter recipe (Betty Crocker)	1/12 of cake	266	13.7			
*(Duncan Hines)	1/12 of cake (2.7 oz.)	347	6.1			50
*Golden butter (Duncan Hines)	1/12 of cake (3.3 oz.)	345	13.9			
*(Swans Down)	1/12 of cake (2.5 oz.)	294	3.2			48

CANADIAN WHISKY (See
DISTILLED LIQUOR)

CANDIED FRUIT (See
individual kinds)

CANDY. The following values of candies from the U. S. Department of
Agriculture are representative of the
types sold commercially. These values may be useful when individual
brands or sizes are not known:

Almond:						
Chocolate-coated[1]	1 oz.	17	12.4	2.	10.	
Chocolate-coated[1]	1 cup (6.3 oz.)	106	78.7	13.	66.	
Sugar-coated or Jordan[2]	1 oz.	6	5.3	Tr.	5.	
Butterscotch[3]	1 oz.	19	1.0	<1.	Tr.	
Candy Corn	1 oz.	60	.6			
Caramel:						
Plain[4]	1 oz.	64	2.9	1.	2.	
Plain with nuts[5]	1 oz.	58	4.6	2.	3.	
Chocolate[6]	1 oz.	64	2.9	1.	2.	
Chocolate with nuts[7]	1 oz.	58	4.6	2.	3.	
Chocolate-flavored roll[4]	1 oz.	56	2.3	1.	1.	
Chocolate:						
Bittersweet[8]	1 oz.	<1	11.3	6.	5.	
Milk:						
Plain[8]	1 oz.	27	9.2	5.	4.	
With almonds	1 oz.	23	10.1			
With peanuts	1 oz.	19	10.8			
Semisweet[8]	1 oz.	<1	10.1	6.	4.	

(USDA): United States Department of Agriculture
*Prepared as Package Directs
[1]Principal sources of fat: almonds, chocolate & vegetable shortening.
[2]Principal source of fat: almonds.
[3]Principal source of fat: butter.
[4]Principal sources of fat: animal & vegetable shortening.
[5]Principal sources of fat: vegetable shortening & nuts.
[6]Principal sources of fat: animal & vegetable shortening & chocolate.
[7]Principal sources of fat: chocolate, vegetable shortening & nuts.
[8]Principal sources of fat: chocolate & cacao butter.

Food and Description	Measure or Quantity	Sodium (mg.)	—Fats in grams—			Choles-terol (mg.)
			Total	Satu-rated	Unsatu-rated	
Sweet[1]	1 oz.	9	10.0	6.	4.	
Chocolate discs, sugar-coated[2]	1 oz.	20	5.6	3.	2.	
Coconut center, chocolate-coated[2]	1 oz.	56	5.0	3.	2.	
Fondant, plain[3]	1 oz.	60	.6	Tr.	Tr.	
Fondant, chocolate-covered[4]	1 oz.	52	3.0	1.	2.	
Fudge:						
Chocolate fudge[4]	1 oz.	54	3.5	1.	2.	
Chocolate fudge, chocolate-coated[5]	1 oz.	65	4.5	2.	3.	
Chocolate fudge with nuts[6]	1 oz.	48	4.9	2.	3.	
Chocolate fudge with nuts, chocolate-coated[7]	1 oz.	58	5.9	2.	4.	
Vanilla fudge[8]	1 oz.	59	3.1	1.	2.	
Vanilla fudge with nuts[7]	1 oz.	53	4.6	2.	3.	
With peanuts & caramel, chocolate-coated[9]	1 oz.	36	6.5	2.	5.	
Gum drops	1 oz.	10	.2			
Hard	1 oz.	9	.3			
Honeycombed hard candy, with peanut butter, chocolate-covered[10]	1 oz.	46	5.5	2.	4.	
Jelly beans	1 oz.	3	.1			
Marshmallows	1 oz.	11	Tr.			
Mints, uncoated	1 oz.	60	.6			
Nougat & caramel, chocolate-covered[4]	1 oz.	49	3.9	2.	2.	
Peanut bar[11]	1 oz.	3	9.1	2.	7.	
Peanut brittle, no added salt or soda[11]	1 oz.	9	2.9	<1.	2.	
Peanuts, chocolate-covered[12]	1 oz.	17	11.7	3.	9.	
Raisins, chocolate-covered[2]	1 oz.	18	4.8	3.	2.	
Vanilla creams, chocolate-covered[5]	1 oz.	52	4.8	2.	3.	

(USDA): United States Department of Agriculture
*Prepared as Package Directs
[1]Principal sources of fat: chocolate & cacao butter.
[2]Principal sources of fat: chocolate, milk & cacao butter.
[3]Principal source of fat: butter.
[4]Principal sources of fat: animal & vegetable shortening & chocolate.
[5]Principal sources of fat: chocolate & vegetable shortening.
[6]Principal sources of fat: chocolate, animal & vegetable shortening & English walnuts.
[7]Principal sources of fat: animal & vegetable shortening & nuts.
[8]Principal sources of fat: animal & vegetable shortening.
[9]Principal sources of fat: vegetable shortening, chocolate & peanuts.
[10]Principal sources of fat: vegetable shortening, chocolate & peanut butter.
[11]Principal source of fat: peanuts.
[12]Principal sources of fat: peanuts, chocolate & vegetable shortening.

Food and Description	Measure or Quantity	Sodium (mg.)	Fats in grams — Total	Satu- rated	Unsatu- rated	Choles- terol (mg.)
CANDY, COMMERCIAL (See also **CANDY, DIETETIC):**						
Almonds, chocolate-covered:						
Candy-coated (Hershey's)	1 oz.	11	7.9			
(Kraft)	1 piece (3 grams)	1	1.1			
Almond Cluster (Kraft)	1 piece (.4 oz.)	7	4.7			
Almond Toffee Bar (Kraft)	1 oz.	65	8.4			
Brazil nuts, chocolate-covered						
(Kraft)	1 piece (6 grams)	3	2.8			
Bridge Mix:						
Almond (Kraft)	1 piece (4 grams)	2	1.7			
Caramelette (Kraft)	1 piece (3 grams)	6	.5			
Jelly (Kraft)	1 piece (3 grams)	7	.5			
Malted milk ball (Kraft)	1 piece (3 grams)	2	1.0			
Mintette (Kraft)	1 piece (3 grams)	2	.5			
Peanut (Kraft)	1 piece (1 gram)	<1	.6			
Peanut crunch (Kraft)	1 piece (5 grams)	26	1.2			
Raisin (Kraft)	1 piece (1 gram)	<1	.2			
(Nabisco)	1 piece (2 grams)		.3			
Butternut (Hollywood)	1¼-oz. bar		7.7			
Butterscotch Skimmers (Nabisco)	1 piece (6 grams)		.2			
Caramel:						
Caramelette (Kraft)	1 piece (3 grams)	6	.5			
Chocolate (Kraft)	1 piece (8 grams)	21	.7			
Chocolate, bar (Kraft)	1 piece (6 grams)	17	.6			
Coconut (Kraft)	1 piece (8 grams)	18	.9			
Vanilla (Kraft)	1 piece (8 grams)	22	.7			
Vanilla, bar (Kraft)	1 piece (6 grams)	18	.6			
Vanilla, chocolate-covered						
(Kraft)	1 piece (9 grams)	20	1.5			
Vanilla, *Twisteroo* (Kraft)	1 piece (6 grams)	17	.5			
Cashew cluster (Kraft)	1 piece (.4 oz.)	7	4.2			
Cashew crunch, canned (Planters)	1 oz.	55	7.3			0
Charleston Chew:						
10¢ size	1 bar (1⅛ oz.)		4.1			
5¢ size	1 bar (¾ oz.)		2.5			
Bite-size	1 piece (7 grams)		.8			
Cherry, chocolate-covered:						
Dark (Nabisco)	1 piece (.6 oz.)		1.5			
Dark (Nabisco) *Welch's*	1 piece (.6 oz.)		1.5			
Milk (Nabisco)	1 piece (.6 oz.)		1.4			
Milk (Nabisco) *Welch's*	1 piece (.6 oz.)		1.4			

(USDA): United States Department of Agriculture
*Prepared as Package Directs

Food and Description	Measure or Quantity	Sodium (mg.)	—Fats in grams—			Choles-terol (mg.)
			Total	Satu-rated	Unsatu-rated	
Chocolate bar:						
Milk chocolate:						
(Ghirardelli)	1.1-oz. bar	26	9.5			
(Hershey's)	1 oz.	28	9.5	6.	4.	
(Hershey's)	¼-oz. miniature	7	2.4	1.	1.	
Mint chocolate (Ghirardelli)	1.1-oz. bar	26	9.8			
Semisweet (Ghirardelli) *Eagle*	1 sq. (1 oz.)		8.9			
Semisweet (Nestlé's)	1 oz.		8.0			
Special Dark (Hershey's)	1 oz.	4	9.5			
Special Dark (Hershey's)	¼-oz. miniature	1	2.4			
(Nestlé's)	1 oz.		9.5			
Chocolate bar with almonds:						
(Ghirardelli)	1.1-oz. bar	25	10.4			
(Hershey's)	1 oz.	23	10.4			
(Nestlé's)	1 oz.		10.8			
Chocolate block, milk:						
(Ghirardelli)	1 sq. (1 oz.)	23	8.4			
(Hershey's)	1 oz.	16	8.8			
Chocolate Crisp Bar (Ghirardelli)	1-oz. bar	25	9.2			
Chocolate Crisp Bar (Kraft)	1 oz.	73	6.4			
Chocolate Crunch Bar (Nestlé's)	1 oz.		7.9			
Cluster:						
Crispy (Nabisco)	1 piece (.6 oz.)		.8			
Peanut, chocolate-covered (Kraft)	1 piece (.4 oz.)	5	4.3			
Royal Clusters (Nabisco)	1 piece (.6 oz.)		4.7			
Coco-Mello (Nabisco)	1 piece (.7 oz.)		3.7			
Coconut:						
Bar (Nabisco) *Welch's*	1 piece (1.1 oz.)		4.5			
Cream egg (Hershey's)	1 oz.	2	4.5			
Squares (Nabisco)	1 piece (.5 oz.)		1.4			
Eggs (Nabisco) *Chuckles*	1 piece (2 grams)		Tr.			
Fiddle Faddle	1½-oz. packet		3.4			
Frappe (Nabisco) *Welch's*	1 piece (1.1 oz.)		3.9			
Fruit'n Nut chocolate bar						
(Nestlé's)	1 oz.		8.3			
Fudge:						
Bar (Nabisco) *Welch's*	1 piece (1.1 oz.)		6.5			
Bar (Tom Houston)	1 bar (1.5 oz.)	72	7.3			
Fudgies, bar (Kraft)	1 piece (7 grams)	15	.9			
Fudgies, regular (Kraft)	1 piece (8 grams)	19	1.0			
Fudgies, *Twisteroo* (Kraft)	1 piece (6 grams)	15	.8			
Home Style (Nabisco)	1 piece (.7 oz.)		3.5			
Nut, bars or squares (Nabisco)	1 piece (.5 oz.)		3.2			

(USDA): United States Department of Agriculture
*Prepared as Package Directs

Food and Description	Measure or Quantity	Sodium (mg.)	— Fats in grams —			Choles- terol (mg.)
			Total	Satu- rated	Unsatu- rated	
Good & Fruity	1 oz.		<.1			
Good & Plenty	1 oz.		<.1			
Hard candy (F & F) *Sherbit*	1 piece		0.			(0)
Hershey-Ets, candy-coated	1 oz.	10	5.8			
Hollywood	1½-oz. bar		6.1			
Jelly (See also individual flavors and brand names in this section):						
Beans, rings (Nabisco) *Chuckles*	1 piece (.4 oz.)		<.1			
Jujubes (Nabisco) *Chuckles*	1 piece (4 grams)		Tr.			
Kisses, milk chocolate (Hershey's)	1 piece (5 grams)	5	1.5	<1.	<1.	
Krackel Bar (Hershey's)	1 oz.	33	8.3			
Krackel Bar (Hershey's)	¼-oz. miniature	8	2.1			
Licorice (Nabisco) *Chuckles*	1 piece (.4 oz.)		Tr.			
Twist (American Licorice Co):						
Black	1 piece (10 grams)	10	.1			
Red	1 piece (9 grams)	8	.2			
Life Savers (Beech-Nut):						
All but butter rum & butterscotch	1 drop (3 grams)	<1	0.			
Butter rum & butterscotch	1 drop (3 grams)	12	0.			
Mint	1 piece (2 grams)	<1	0.			
Malted Milk Crunch (Nabisco)	1 piece (2 grams)		.5			
Marshmallow:						
Chocolate (Kraft)	1 piece (7 grams)	4	.2			
Coconut (Kraft)	1 piece (.4 oz.)	12	1.1			
Eggs (Nabisco) *Chuckles*	1 piece (10 grams)		<.1			
Flavored, miniature (Kraft)	1 piece (<1 gram)	<1	0.			
Flavored, regular (Kraft)	1 piece (7 grams)	<1	0.			
White, miniature (Kraft)	1 piece (<1 gram)	<1	0.			
White, regular (Kraft)	1 piece (7 grams)	<1	0.			
Mary Jane (Miller):						
1¢ size	1 piece (9 grams)		<.1	Tr.	Tr.	
5¢ size	1 piece (1¼ oz.)		.4	Tr.	Tr.	
Milk Shake (Hollywood)	1¼-oz. bar		4.0			
Mint or peppermint:						
Buttermint (Kraft)	1 piece (2 grams)	4	<.1			
Encore (Kraft)	1 piece (2 grams)	3	0.			
Jamaica Mints (Nabisco)	1 piece (6 grams)		<.1			
Liberty Mints (Nabisco)	1 piece (6 grams)		<.1			
Mini-mint (Kraft)	1 piece (3 grams)	2	.5			
Party (Kraft)	1 piece (2 grams)	4	<.1			

(USDA): United States Department of Agriculture
*Prepared as Package Directs

Food and Description	Measure or Quantity	Sodium (mg.)	— Fats in grams —			Cholesterol (mg.)
			Total	Saturated	Unsaturated	
Pattie, chocolate-covered:						
Junior Mint Pattie (Nabisco)	1 piece (2 grams)		.2			
Peppermint pattie (Nabisco)	1 piece (.5 oz.)		1.4			
Sherbit, pressed mints (F & F)	1 piece		0.			
Thin (Nabisco)	1 piece (.4 oz.)		1.0			
Wafers (Nabisco)	1 piece (2 grams)		.6			
Mr. Goodbar (Hershey's)	1 oz.	16	10.4			
Mr. Goodbar (Hershey's)	¼-oz. miniature	4	2.6			
Nibs (Y & S) cherry	1¾ oz.	100	1.3			0
Nibs (Y & S) licorice	1¾ oz.	208	1.1			0
North Pole (F & F)	1⅜-oz. bar		2.3			
Nougat centers (Nabisco)						
Chuckles	1 piece (4 grams)		Tr.			
Nutty Crunch (Nabisco)	1 piece (½ oz.)		3.2			
$100,000 Bar (Nestlé's)	1 oz.		5.4			
Orange slices (Nabisco) *Chuckles*	1 piece (8 grams)		Tr.			
Payday (Hollywood)	1¼-oz. bar		5.6			
Peanut, chocolate-covered:						
(Hershey's) candy-coated	1 oz.	7	7.1			
(Kraft)	1 piece (2 grams)	1	.9			
(Nabisco)	1 piece (4 grams)		1.7			
(Tom Houston)	1 oz.	17	11.6			
Peanut Brittle:						
(Kraft)	1 oz.	145	5.0			
Coconut (Kraft)	1 oz.	189	4.5			
Jumbo Peanut Block Bar						
(Planters)	1 oz.	55	7.3			0
Peanut Butter Cup (Reese's)	1 oz.	99	8.6			
Peanut Butter Egg (Reese's)	1 oz.	119	8.6			
Peanut Plank (Tom Houston)	1 bar (1.5 oz.)	4	13.5			
Pom Poms (Nabisco)	1 piece (3 grams)		.5			
Raisin, chocolate-covered:						
(Ghirardelli)	1 bar (1.1 oz.)	23	8.4			
(Nabisco)	1 piece (<1 gram)		.2			
Screaming Yellow Zonkers	1 oz.		2.7			
Spearmint leaves:						
(Nabisco) *Chuckles*	1 piece (8 grams)		Tr.			
(Quaker City)	1 oz.		.2			
Spice-flavored sticks & drops						
(Nabisco) *Chuckles*	1 piece (4 grams)		Tr.			
Spice-flavored strings (Nabisco)						
Chuckles	1 piece (5 grams)		Tr.			
Sprigs, sweet chocolate (Hershey's)	1 oz.	6	7.7			

Food and Description	Measure or Quantity	Sodium (mg.)	—Fats in grams— Total	Satu- rated	Unsatu- rated	Choles- terol (mg.)
Stars, chocolate:						
(Kraft)	1 piece (3 grams)		.7			
(Nabisco)	1 piece (3 grams)		.9			
Sugar Babies (Nabisco)	1 piece (2 grams)		<.1			
Sugar Daddy (Nabisco):						
Caramel sucker	1 piece (1.1 oz.)		1.5			
Giant sucker, caramel	1 piece (1 lb.)		19.5			
Junior	1 piece (.4 oz.)		.5			
Junior sucker, choco-flavored	1 piece (.4 oz.)		.6			
Nugget	1 piece (.4 oz.)		.5			
Nugget	1 piece (7 grams)		.3			
Sugar Mama (Nabisco)	1 piece (.8 oz.)		2.6			
Sugar Wafer (F & F)	1¼-oz. pkg.		8.0			
Taffy:						
Chocolate (Kraft)	1 piece (7 grams)	22	.7			
Coffee (Kraft)	1 piece (7 grams)	22	.7			
Rum butter (Kraft)	1 piece (7 grams)	22	.7			
Vanilla (Kraft)	1 piece (7 grams)	22	.7			
Tootsie Roll:						
Regular:						
1¢ size or midgee	1 piece (7 grams)	15	.5	Tr.	Tr.	0
2¢ size	1 piece (.4 oz.)	25	.8	Tr.	Tr.	0
5¢ size	1 piece (¾ oz.)	50	1.6	Tr.	1.	0
10¢ size	1 piece (1½ oz.)	98	3.3	Tr.	3.	0
Twin pak	10¢ size (1¼ oz.)	84	2.7	Tr.	2.	0
Twin pak	15¢ size (2 oz.)	135	4.4	<1.	4.	0
Vending-machine size	1 piece (5 grams)	11	.4	Tr.	Tr.	0
Pop, 2 for 5¢	1 piece (.5 oz.)	8	.3	Tr.	Tr.	0
Pop, 5¢ size	1 piece (1 oz.)	17	.6	Tr.	Tr.	0
Pop-drop	1 piece (5 grams)	3	<.1	Tr.	Tr.	0
Triple Decker bar (Nestlé's)	1 oz.		9.4			
Twizzlers (Y & S):						
Chocolate	1 oz.	64	.9			0
Grape	1 oz.	52	.2			0
Licorice	1¾ oz.	220	.8			0
Strawberry	1¾ oz.	91	.4			0
Strawberry	1-oz. bar	52	.2			0
Variety pack (Nabisco) *Chuckles*	1 piece (.4 oz.)		.1			
Variety pack (Nabisco) *Chuckles*	2-oz. pack		.5			
Walnut Hill (F & F)	1⅜-oz. bar		6.0			
Whirligigs (Nabisco)	1 piece (6 grams)		.5			

(USDA): United States Department of Agriculture
*Prepared as Package Directs

Food and Description	Measure or Quantity	Sodium (mg.)	—Fats in grams—			Choles- terol (mg.)
			Total	Satu- rated	Unsatu- rated	
CANDY DIETETIC:						
Almonds, chocolate-covered						
(Estee)	1 piece (4 grams)	4	1.5	<1.	<1.	<1
Chocolate, assorted:						
Milk (Estee)	1 piece (8 grams)	9	3.7	2.	2.	<1
Slimtreats	1 piece (2 grams)		.7			
Chocolate bar, almonds:						
(Estee)	1 bar (¾ oz.)	28	9.1	5.	4.	1
(Estee)	1 section of 2-oz. bar	3	1.0	<1.	Tr.	Tr.
(Estee)	1 section of 4-oz. bar	19	6.1	3.	3.	<1
Chocolate bar, bittersweet:						
(Estee)	1 bar (¾ oz.)	22	8.5	5.	4.	<1
(Estee)	1 section of 2-oz. bar	2	.9	<1.	Tr.	<1
(Estee)	1 section of 4-oz. bar	15	5.7	3.	2.	<1
Chocolate bar, coconut (Estee)	1 bar (¾ oz.)	30	8.9	6.	3.	1
Chocolate bar, crunch (Estee)	1 bar (⅝ oz.)		6.7	4.	3.	1
Chocolate bar, crunch (Estee)	1 section of 3-oz. bar		4.0	2.	2.	<1
Chocolate bar, fruit-nut (Estee)	1 section of 4-oz. bar	16	5.4	3.	3.	<1
Chocolate bar, milk:						
(Estee)	1 bar (¾ oz.)	31	8.9	5.	4.	1
(Estee)	1 section of 2-oz. bar	3	1.0	<1.	Tr.	<1
(Estee)	1 section of 4-oz. bar	21	6.0	3.	3.	<1
Chocolate bar, peppermint (Estee)	1 bar (¾ oz.)	31	8.9	5.	4.	1
Chocolate bar, white (Estee)	1 section of 4-oz. bar	23	5.1	4.	1.	2
Chocolettes, milk or peppermint						
(Estee)	1 piece (3 grams)	5	1.2	1.	Tr.	<1
Creams, assorted or peppermint						
(Estee)	1 piece (8 grams)	12	3.6	2.	1.	<1
Gum Drops (Estee)	1 piece	<1	Tr.			0
Hard candy:						
Assorted (Estee)	1 piece	<1	Tr.			0
Coffee (Estee)	1 piece	<1	.1			0
Slimtreats	1 piece (3 grams)		0.			
Mint, any flavor (Estee)	1 piece	Tr.	Tr.			0
Nut, chocolate-covered (Estee)	1 piece (8 grams)	9	3.4	2.	2.	<1
Peanut butter cup (Estee)	1 piece (7 grams)	6	3.1	2.	1.	<1
Peanut, chocolate-covered (Estee)	1 piece (1 gram)	1	.5	Tr.	Tr.	Tr.
Petit fours (Estee)	1 piece (8 grams)	8	2.9	1.	2.	<1
Raisin, chocolate-covered (Estee)	1 piece (1 gram)	1	.3	Tr.	Tr.	Tr.
TV mix (Estee)	1 piece (2 grams)	1	.8	Tr.	Tr.	Tr.
CANE SYRUP (USDA)	Any quantity		0.			0

(USDA): United States Department of Agriculture
*Prepared as Package Directs

Food and Description	Measure or Quantity	Sodium (mg.)	—Fats in grams—			Choles-terol (mg.)
			Total	Satu-rated	Unsatu-rated	
CANTALOUPE, fresh:						
Whole	1 lb. (weighed whole)	27	.2			0
Whole, medium (USDA)	5″ dia. melon (1⅔ lbs., weighed with skin & cavity contents)	46	.4			0
Cubed (USDA)	½ cup (2.9 oz.)	10	<.1			0
CAPE GOOSEBERRY (See **GROUND-CHERRY**)						
CAPERS (Crosse & Blackwell)	1 T. (.6 oz.)	306				
CAPICOLA or CAPACOLA SAUSAGE (USDA)	1 oz.		13.0	5.	8.	
CAP'N CRUNCH:						
(Quaker)	¾ cup (1 oz.)	280	3.0			
Crunchberries (Quaker)	¾ cup (1 oz.)	147	2.1			
Peanut butter (Quaker)	¾ cup (1 oz.)	250	3.8			
Vanilly (Quaker)	¾ cup (1 oz.)	141	1.7			
CARAMBOLA, raw (USDA):						
Whole	1 lb. (weighed whole)	9	2.1			0
Flesh only	4 oz.	2	.6			0
CARAMEL CAKE, home recipe (USDA)[1] :						
Without icing[2]	¹/₉ of 9″ sq. cake (3 oz.)	262	14.9	8.	7.	
With caramel icing	3 oz.	214	12.6			
CARAMEL CAKE MIX:						
*(Duncan Hines)	¹/₁₂ of cake (2.7 oz.)	397	6.1			50
*Apple (Betty Crocker) layer	¹/₁₂ of cake	275	6.0			
*Pudding (Betty Crocker)	¹/₆ of cake	328	4.7			
***CARAMEL PUDDING MIX,** nut (Royal)	½ cup (5.1 oz.)	380	5.7			14

(USDA): United States Department of Agriculture
*Prepared as Package Directs
[1]Made with sodium aluminum sulfate-type baking powder.
[2]Principal sources of fat: butter, egg & milk.

Food and Description	Measure or Quantity	Sodium (mg.)	— Fats in grams —			Cholesterol (mg.)
			Total	Saturated	Unsaturated	
CARAWAY SEED (Information supplied by General Mills Laboratory)	1 oz.	13	<.1			(0)
(Spice Islands)	1 tsp.	<1				(0)
CARDAMOM:						
Ground (Spice Islands)	1 tsp.	<1				(0)
Whole (Spice Islands)	1 seed	Tr.				(0)
CARISSA or NATAL PLUM, raw:						
Whole (USDA)	1 lb. (weighed whole)		5.1			0
Flesh only (USDA)	4 oz.		1.5			0
CARNATION INSTANT BREAKFAST:						
Butterscotch[1]	1 pkg. (1.2 oz.)	Tr.	.2	Tr.	Tr.	<1
Chocolate[2]	1 pkg. (1.3 oz.)	Tr.	.9	Tr.	Tr.	Tr.
Chocolate fudge[2]	1 pkg. (1.3 oz.)	Tr.	1.1	<1.	<1.	Tr.
Chocolate malt[3]	1 pkg. (1.2 oz.)	Tr.	1.4	<1.	<1.	2
Chocolat marshmallow[2]	1 pkg. (1.3 oz.)	Tr.	.9	Tr.	Tr.	Tr.
Coffee[4]	1 pkg. (1.3 oz.)	Tr.	.3	Tr.	Tr.	<1
Eggnog[1]	1 pkg. (1.2 oz.)	Tr.	.5	Tr.	Tr.	21
Special Morning, chocolate[2]	1 pkg. (1 oz.)	Tr.	1.3	<1.	<1.	<1
Special Morning, chocolate malt[3]	1 pkg. (1 oz.)	Tr.	1.4	<1.	<1.	2
Special Morning, strawberry[4]	1 pkg. (1 oz.)	Tr.	.4	Tr.	Tr.	<1
Special Morning, vanilla[4]	1 pkg. (1 oz.)	Tr.	.4	Tr.	Tr.	<1
Strawberry[1]	1 pkg. (1.2 oz.)	Tr.	.2	Tr.	Tr.	<1
Vanilla[1]	1 pkg. (1.2 oz.)	Tr.	.2	Tr.	Tr.	<1
Vanilla ice creme[1]	1 pkg. (1.2 oz.)	Tr.	.2	Tr.	Tr.	<1
CAROB FLOUR (See **FLOUR**)						
CAROUSEL WINE (Gold Seal)	3 fl. oz. (3.3 oz.)	3	0.			(0)
CARP, raw (USDA):						
Whole	1 lb. (weighed whole)	68	5.7			
Meat only	4 oz.	57	4.8			

(USDA): United States Department of Agriculture
*Prepared as Package Directs
[1]Principal source of fat: milk.
[2]Principal sources of fat: milk, cocoa & lecithin.
[3]Principal sources of fat: wort solids, milk, cocoa & lecithin.
[4]Principal sources of fat: milk & lecithin.

Food and Description	Measure or Quantity	Sodium (mg.)	Total	Satu- rated	Unsatu- rated	Choles- terol (mg.)
				Fats in grams		

CARROT:
Raw (USDA):

Whole	1 lb. (weighed with full tops)	126	.5			0
Partially trimmed	1 lb. (weighed without tops, with skins)	175	.7			0
Trimmed	5½" x 1" carrot (1.8 oz.)	24	.1			0
Trimmed	25 thin strips (1.8 oz.)	24	.1			0
Chunks	½ cup (2.4 oz.)	32	.1			0
Diced	½ cup (2.5 oz.)	34	.1			0
Grated or shredded	½ cup (1.9 oz.)	26	.1			0
Slices	½ cup (2.3 oz.)	30	.1			0
Strips	½ cup (2 oz.)	27	.1			0

Boiled, without salt (USDA):

Chunks, drained	½ cup (2.9 oz.)	27	.2			0
Diced, drained	½ cup (2.5 oz.)	23	.1			0
Slices, drained	½ cup (2.7 oz.)	25	.2			0

Canned, regular pack:

Diced, solids & liq. (USDA)	½ cup (4.3 oz.)	290	.2			0
Diced, drained solids (USDA)	½ cup (2.8 oz.)	189	.2			0
Drained liq. (USDA)	4 oz.	268	0.			0
Drained solids (Butter Kernel)	½ cup (4.1 oz.)					
Drained solids (Del Monte)	½ cup (2.8 oz.)	1852			0	
Sliced, solids & liq. (Comstock-Greenwood)	½ cup (2.6 oz.)	177	.2			(0)
Solids & liq. (Del Monte)	½ cup (4 oz.)	266	.2			0
Solids & liq. (Stokely-Van Camp)	½ cup (4 oz.)		.2			(0)

Canned, dietetic pack:

Low sodium, solids & liq. (USDA)	4 oz. (by wt.)	44	.1			0
Low sodium, drained solids (USDA)	½ cup (2.8 oz.)	31	<.1			0
Diced, solids & liq. (Blue Boy)	4 oz.	36	.7			(0)
Diced, solids & liq. (Tillie Lewis)	½ cup (4.3 oz.)	40	.1			0
Sliced, solids & liq. (Blue Boy)	4 oz.	43	<.1			(0)
Sliced (S and W), *Nutradiet,* unseasoned	4 oz.	45	.1			(0)
Dehydrated (USDA)	1 oz.	76	.4			0

(USDA): United States Department of Agriculture
*Prepared as Package Directs

Food and Description	Measure or Quantity	Sodium (mg.)	—Fats in grams—			Cholesterol (mg.)
			Total	Satu-rated	Unsatu-rated	
Frozen:						
Nuggets in butter sauce (Green Giant)	⅓ of 10-oz. pkg.	350	2.4			
Sliced, honey glazed (Green Giant)	⅓ of 10-oz. pkg.	217	1.7			
With brown sugar glaze (Birds Eye)	½ cup (3.3 oz.)	500	2.4			0
CASABA MELON, fresh (USDA):						
Whole	1 lb. (weighed whole)	27	Tr.			0
Flesh only	4 oz.	14	Tr.			0
CASHEW NUT:						
Salted:						
(USDA)	1 oz.	57	13.0	2.	11.	0
(USDA)	½ cup (2.5 oz.)	140	32.0	6.	26.	0
(USDA)	5 large or 8 med. (.4 oz.)	21	4.9	<1.	4.	0
(Tom Houston)	15 nuts (1.1 oz.)	60	13.7			(0)
Dry roasted (Flavor House)	1 oz.	36	13.4			(0)
Dry roasted (Planters)	1 oz.	340	13.0	2.	11.	0
Dry roasted (Skippy)	1 oz.	142	13.2	3.	11.	0
Oil roasted, *Freshnut*	1 oz.	140	14.5	3.	11.	0
Oil roasted (Planters)	1 oz.	220	14.1			0
Unsalted:						
(USDA)	1 oz.	4	13.0	2.	11.	0
(USDA)	½ cup (2.5 oz.)	10	32.0	6.	26.	0
(USDA)	5 large or 8 med. (.4 oz.)	2	4.9	<1.	4.	0
CATAWBA WINE:						
(Gold Seal) 13-14% alcohol	3 fl. oz. (3.3 oz.)	3	0.			(0)
(Great Western) pink, 13% alcohol	3 fl. oz.	36	0.			0
CATFISH, freshwater, raw, fillet (USDA)	4 oz.	68	3.5			
CATSUP:						
Regular pack:						
(USDA)	½ cup (5 oz.)	1469	.6			0
(USDA)	1 T. (.6 oz.)	188	<.1			0
(Bama)	1 T. (.6 oz.)	180	.3			(0)

(USDA): United States Department of Agriculture
*Prepared as Package Directs

Food and Description	Measure or Quantity	Sodium (mg.)	—Fats in grams— Total	Satu- rated	Unsatu- rated	Choles- terol (mg.)
(Del Monte)	1 T. (.7 oz.)	263	<.1			0
(Heinz)	1 T.	180	Tr.			(0)
(Hunt's)	½ cup (4.8 oz.)	1620	.3			(0)
(Hunt's)	1 T. (.6 oz.)	202	<.1			(0)
(Nalley's)	1 oz.		.1			(0)
(Stokely-Van Camp)	1 T. (.6 oz.)		<.1			(0)
Dietetic pack:						
Low sodium (USDA)	½ cup (5 oz.)	7-49	.6			0
Low sodium (USDA)	1 T. (.6 oz.)	<1-6	<.1			0
(Tillie Lewis)	1 T. (.5 oz.)	4	Tr.			(0)
CAULIFLOWER:						
Raw (USDA):						
Whole	1 lb. (weighed untrimmed)	23	.4			0
Flowerbuds	1 lb. (weighed trimmed)	59	.9			0
Buds	½ cup (1.8 oz.)	6	.1			0
Slices	½ cup (1.5 oz.)	5	<.1			0
Boiled, without salt, drained (USDA)	½ cup (2.2 oz.)	6	.1			0
Frozen:						
Not thawed (USDA)	10-oz. pkg.	31	.6			0
Boiled, drained (USDA)	½ cup (3.2 oz.)	9	.2			0
(Birds Eye)	⅓ of 10-oz. pkg.	38	.2			0
Au gratin (Stouffer's)	⅓ of 10-oz. pkg.	279	8.0			
Cut, in butter sauce (Green Giant)	⅓ of 10-oz. pkg.	321	2.4			
Hungarian, with sour cream sauce, casserole (Green Giant)	⅓ of 10-oz. pkg.	529	3.3			
In cheese sauce (Green Giant)	⅓ of 10-oz. pkg.	298	2.6			
CAULIFLOWER, SWEET, PICKLED (Smucker's)	1 bud (.5 oz.)	150	Tr.			(0)
CAVIAR, STURGEON (USDA):						
Pressed	1 oz.		4.7			
Whole eggs	1 oz.	624	4.3			>85
Whole eggs	1 T. (.6 oz.)	352	2.4			>48
CELERIAC ROOT, raw (USDA):						
Whole	1 lb. (weighed unpared)	390	1.2			0
Pared	4 oz.	113	.3			0

(USDA): United States Department of Agriculture
*Prepared as Package Directs

Food and Description	Measure or Quantity	Sodium (mg.)	— Fats in grams —			Choles- terol (mg.)
			Total	Satu- rated	Unsatu- rated	
CELERY, all varieties:						
Raw (USDA):						
Whole	1 lb. (weighed untrimmed)	429	.3			0
1 large outer stalk	8″ x 1½″ at root end (1.4 oz.)	50	<.1			0
3 small inner stalks	5″ x ¾″ (1.8 oz.)	63	<.1			0
Diced, chopped or cut in chunks	½ cup (2.1 oz.)	76	<.1			0
Slices	½ cup (1.9 oz.)	67	<.1			0
Boiled without salt, drained:						
Diced or cut in chunks	½ cup (2.7 oz.)	67	<.1			0
Slices	½ cup (3 oz.)	74	<.1			0
CELERY CABBAGE (See **CABBAGE, CHINESE**)						
CELERY SEASONING (French's)	1 tsp. (5 grams)	1430	.1			(0)
CELERY SEED:						
Ground (Spice Islands)	1 tsp.	2				(0)
Whole (Spice Islands)	1 tsp.	4				(0)
CELERY SOUP, Cream of:						
Condensed (USDA)[1]	8 oz. (by wt.)	1805	9.5	2.	7.	
*Prepared with equal volume water (USDA)[1]	1 cup (8.5 oz.)	955	5.0	1.	4.	
*Prepared with equal volume milk (USDA)[1]	1 cup (8.6 oz.)	1039	9.3	2.	7.	
*(Campbell)	1 cup	930	4.4	1.	3.	
*(Heinz)	1 cup (8½ oz.)	1010	6.3			
CEREAL BREAKFAST FOODS (See kind of cereal such as **CORN FLAKES** or brand name such as *KIX*)						
CERVELAT (USDA):						
Dry	1 oz.		10.7			
Soft	1 oz.		6.9			
CHABLIS WINE:						
(Gold Seal) 12% alcohol	3 fl. oz.	3	0.			
(Great Western) 12.5% alcohol	3 fl. oz.	31	0.			0

(USDA): United States Department of Agriculture
*Prepared as Package Directs
[1]Principal sources of fat: corn oil & milk.

Food and Description	Measure or Quantity	Sodium (mg.)	Total	Fats in grams Satu- rated	Unsatu- rated	Choles- terol (mg.)
(Great Western) Diamond, 12.5% alcohol	3 fl. oz.	<1	0.			0
CHAMPAGNE:						
(Gold Seal) brut, or pink, extra dry, 12% alcohol	3 fl. oz. (3.2 oz.)	3	0.			(0)
(Great Western) regular, brut, extra dry, pink or special reserve, 12.5% alcohol	3 fl. oz.	31	0.			0
CHARD, Swiss (USDA):						
Raw, whole	1 lb. (weighed untrimmed)	613	1.3			0
Raw, trimmed	4 oz.	167	.3			0
Boiled without salt, drained	½ cup (3.4 oz.)	83	.2			0
CHARLOTTE RUSSE, with ladyfingers, whipped cream filling, home recipe (USDA)[1]	4 oz.	49	16.6	8.	9.	
CHAYOTE, raw (USDA):						
Whole	1 lb. (weighed unpared)	19	.4			0
Pared	4 oz.	6	.1			0
***CHEDDAR CHEESE SOUP,** canned (Campbell)	1 cup	886	9.4	4.	6.	
CHEERIOS, cereal (General Mills)	1¼ cups (1 oz.)	320	2.0			(0)
CHEESE:						
American or cheddar:						
Natural:						
(USDA)	1 oz.	198	9.1	5.	4.	28
(USDA)	1″ cube (.6 oz.)	119	5.5	3.	2.	17
Diced (USDA)	1 cup (4.6 oz.)	917	42.2	24.	19.	130
Grated or shredded (USDA)	1 cup (3.9 oz.)	777	35.7	20.	16.	110
Grated or shredded (USDA)	1 T. (7 grams)	48	2.2	1.	1.	7
(Kraft)	1 oz.	193	9.1			
Cheddar (Sealtest)	1 oz.	193	9.4			
Grated (Kraft)	1 oz.	836	6.7			

(USDA): United States Department of Agriculture
*Prepared as Package Directs
[1]Principal sources of fat: cream & egg.

Food and Description	Measure or Quantity	Sodium (mg.)	Fats in grams Total	Satu-rated	Unsatu-rated	Choles-terol (mg.)
Shredded (Kraft)	1 oz.	193	9.1			
Sharp cheddar, *Wispride*	1 T. (.5 oz.)		4.4			
Process:						
(USDA) regular	1 oz.	322	8.5	5.	4.	25
(USDA) regular	1″ cube (.6 oz.)	204	5.4	3.	3.	16
(USDA) regular	3½″ x 3⅜″ x ⅛″ slice (1 oz.)	322	8.5	5.	4.	25
(USDA) reduced sodium	1 oz.	184	8.5	5.	4.	25
(USDA) reduced sodium	1″ cube (.6 oz.)	117	5.4	3.	3.	16
(Borden)	¾-oz. slice	247	6.7	5.	2.	31
(Kraft) loaf or slice	1 oz.	554	8.6			
(Sealtest)	1 oz.	369	8.6			
Dried, sharp cheddar (Data from General Mills Laboratory)	1 oz.	521	14.2			
Armenian String (Sierra)	1 oz.		6.2			
Asiago (Frigo)	1 oz.	193	9.1			
Bleu or Blue:						
Natural (USDA)	1 oz.		8.6	5.	4.	24
Natural (USDA)	1″ cube (.6 oz.)		5.2	3.	2.	15
Natural, crumbled (USDA)	1 cup (4.8 oz.)		41.2	23.	18.	117
(Frigo)	1 oz.	511	8.2			
(Kraft) natural	1 oz.	510	8.2			
(Stella)	1 oz.		9.1			
Wispride	2 T. (1 oz.)		8.8			
Bondost (Kraft) natural	1 oz.	147	8.4			
Brick:						
(USDA) natural	1 oz.		8.6	5.	4.	26
(Kraft) natural	1 oz.	204	8.4			
(Kraft) process, slices	1 oz.	445	8.3			
Camembert, domestic:						
Natural (USDA)	1 oz.		7.0	4.	3.	26
Natural (USDA)	2¼″ x 2⅛″ x 1⅛″ wedge (3 to a 4-oz. pkg.)		9.4	5.	4.	35
(Borden)	1 oz.	290	7.1			
(Kraft) natural	1 oz.	295	7.0			
Caraway (Kraft) natural	1 oz.	193	8.9			
Casino Swiss (Kraft) natural	1 oz.	85	7.9			
Chantelle, natural (Kraft)	1 oz.	272	7.4			
Cheddar (See American)						
Cheez-ola, process (Fisher)	1 oz.	454	6.5	<1.	5.	1
Colby, natural:						
(USDA)	1 oz.		9.1			27

(USDA): United States Department of Agriculture
*Prepared as Package Directs

Food and Description	Measure or Quantity	Sodium (mg.)	Fats in grams — Total	Satu- rated	Unsatu- rated	Choles- terol (mg.)
(Kraft)	1 oz.	193	9.0			
Cottage cheese:						
Creamed, unflavored:						
(USDA)	1 oz.	65	1.2	<1.	<1.	5
(USDA)	8-oz. pkg.	520	9.5	5.	5.	43
(USDA)	1 packed cup (8.6 oz.)	561	10.3	5.	5.	47
(USDA)	1 T. (.5 oz.)	34	.6	Tr.	Tr.	3
(Borden)	8-oz. container	519	9.5	4.	5.	34
(Kraft)	1 oz.	141	1.2			
(Sealtest)	1 cup (7.9 oz.)	907	9.2			
California (Breakstone)	8-oz. container	920	9.3			24
California (Breakstone)	1 T. (.6 oz.)	64	.6			2
Light n' Lively (Sealtest)	1 cup (7.9 oz.)	916	2.2			
Lite Line (Borden)	1 cup	970	4.5			
Low fat (Breakstone)	8-oz. container	1168	4.1			11
Low fat (Breakstone)	1 T. (.6 oz.)	82	.3			<1
Low fat, 2% fat (Sealtest)	1 cup (7.9 oz.)	934	4.5			
Tangy small curd (Breakstone)	8-oz. container	920	9.3			24
Tangy small curd (Breakstone)	1 T. (.6 oz.)	64	.6			2
Tiny soft curd (Breakstone)	8-oz. container	920	9.3			24
Tiny soft curd (Breakstone)	1 T. (.6 oz.)	64	.6			2
Creamed, flavored:						
Chive (Breakstone)	8-oz. container	920	9.3			24
Chive (Breakstone)	1 T. (.6 oz.)	64	.6			2
Chive (Sealtest)	1 cup (7.9 oz.)	907	9.2			
Chive-pepper (Sealtest)	1 cup (7.9 oz.)	853	9.0			
Peach, low fat (Breakstone)	8-oz. container	784	4.3			11
Peach, low fat (Breakstone)	1 T. (.6 oz.)	57	.3			<1
Peach-pineapple (Sealtest)	1 cup (7.9 oz.)	728	7.6			
Pineapple, low fat (Break- stone)	8-oz. container	784	4.3			11
Pineapple, low fat (Break- stone)	1 T. (.6 oz.)	56	.3			6
Pineapple (Sealtest)	1 cup (7.9 oz.)	726	7.6			
Spring Garden Salad (Sealtest)	1 cup (7.9 oz.)	903	9.4			
Uncreamed:						
(USDA)	8-oz. pkg.	658	.7			16
(USDA)	1 oz.	82	<.1			2
(USDA)	1 packed cup (7 oz.)	580	.6			14
(Sealtest)	1 cup (7.9 oz.)	13	.7			
Pot, unsalted (Borden)	8-oz. pkg.	45	.7	Tr.	Tr.	3

(USDA): United States Department of Agriculture
*Prepared as Package Directs

Food and Description	Measure or Quantity	Sodium (mg.)	Total	Satu- rated	Unsatu- rated	Choles- terol (mg.)
					Fats in grams	
Pot style (Breakstone)	8-oz. container	920	.7			0
Pot style (Breakstone)	1 T. (.6 oz.)	64	Tr.			0
Skim milk (Breakstone)	8-oz. container	16	.7			0
Skim milk (Breakstone)	1 T. (.6 oz.)	1	Tr.			0
Country Charm, natural (Fisher)	1 oz.		7.4			22
Cream cheese:						
Plain, unwhipped:						
(USDA)	1 oz.	71	10.7	6.	5.	31
(USDA)	3-oz. pkg. (2⅞" x ⅞")	212	32.0	18.	14.	94
(USDA)	8-oz. pkg.	568	85.6	48.	38.	252
(USDA)	½ cup (4.1 oz.)	288	43.4	24.	19.	128
(USDA)	1" cube (.6 oz.)	40	6.0	3.	3.	18
(USDA)	1 T. (.5 oz.)	35	5.3	3.	2.	16
(Breakstone)	1 oz.	25	9.5			113
(Breakstone)	1 T. (.5 oz.)	12	4.8			56
(Kraft)	1 oz.	113	9.7			
(Sealtest)	1 oz.	114	9.5			
Hostess (Kraft)	1 oz.	113	9.5			
Philadelphia (Kraft)	1 oz.	113	9.7			
Philadelphia, imitation (Kraft)	1 oz.	204	3.4			
Plain, whipped (Breakstone):						
Temp-Tee	1 oz.	25	9.5			113
Temp-Tee	1 T. (9 grams)	8	3.1			37
Flavored, unwhipped:						
Chive (Kraft) *Hostess*	1 oz.	170	8.0			
Chive (Kraft) *Philadelphia*	1 oz.	170	8.0			
Olive-pimento (Kraft) *Hostess*	1 oz.	193	8.3			
Pimento (Kraft) *Hostess*	1 oz.	170	8.2			
Pimento (Kraft) *Philadelphia*	1 oz.	170	8.2			
Pineapple (Kraft) *Hostess*	1 oz.	125	7.8			
Roquefort (Kraft) *Hostess*	1 oz.	261	7.5			
Flavored, whipped (Kraft):						
Bacon & horseradish	1 oz.	159	9.4			
Blue	1 oz.	169	9.3			
Catalina	1 oz.	261	9.1			
Chive	1 oz.	170	8.8			
Onion	1 oz.	180	8.8			
Pimento	1 oz.	181	8.6			
Salami	1 oz.	126	8.3			
Smoked salmon	1 oz.	170	8.2			
Edam:						
Natural (USDA)	1 oz.		7.9			29

(USDA): United States Department of Agriculture
*Prepared as Package Directs

Food and Description	Measure or Quantity	Sodium (mg.)	—Fats in grams—			Choles- terol (mg.)
			Total	Satu- rated	Unsatu- rated	
(House of Gold)	1 oz.	204	7.9			
Natural (Kraft)	1 oz.	204	7.9			
Farmer, midget (Breakstone)	1 oz.	111	2.3			6
Farmer, midget (Breakstone)	1 T. (.5 oz.)	53	1.1			3
Fontina, natural (Kraft)	1 oz.	204	9.2			
Fontina (Stella)	1 oz.		9.1			
Frankenmuth, natural (Kraft)	1 oz.	193	9.1			
Gjetost, natural (Kraft)	1 oz.	170	8.2			
Gorgonzola, natural (Kraft)	1 oz.	397	9.1			
Gouda, natural (Kraft)	1 oz.	204	8.5			
Gruyère:						
Natural (Kraft)	1 oz.	91	8.4			
Swiss Knight	1 oz.		8.5			
Jack-dry, natural (Kraft)	1 oz.	204	8.2			
Jack-fresh, natural (Kraft)	1 oz.	181	7.7			
Lager-Kase, natural (Kraft)	1 oz.	215	8.8			
Leyden, natural (Kraft)	1 oz.	204	3.8			
Liederkranz (Borden)	1 oz.	271	7.3			
Limburger:						
Natural (USDA)	1 oz.		7.9	4.	4.	28
Natural (Kraft)	1 oz.	227	7.9			
Monterey Jack (Frigo)	1 oz.	204	8.3			
Monterey Jack, natural (Kraft)	1 oz.	204	8.3			
Mozzarella:						
Low moisture, natural (USDA)	1 oz.		7.5			27
Low moisture, part-skim, natural (USDA)	1 oz.		5.9			18
(Frigo)	1 oz.	227	4.7			
(Sierra)	1 oz.		6.2			
Low moisture, part-skim, natural (Kraft)	1 oz.	227	5.6			
Low moisture part-skim, pizza, natural (Kraft)	1 oz.	227	4.7			
Shredded (Kraft)	1 oz.	227	4.7			
Muenster:						
Natural (USDA)	1 oz.		8.6			25
Natural (Borden)	1 oz.					
Natural (Kraft)	1 oz.	204	8.1			
Process, slices (Kraft)	1 oz.	366	8.3			
Neufchâtel:						
Natural (USDA)	2⅞" x 2" x ⅞" pkg. (3 oz.)		20.4			64
Natural (Kraft) *Calorie-Wise*	1 oz.	113	6.1			

(USDA): United States Department of Agriculture
*Prepared as Package Directs

Food and Description	Measure or Quantity	Sodium (mg.)	— Fats in grams —			Choles- terol (mg.)
			Total	Satu- rated	Unsatu- rated	
Process (Borden)	1 oz.	149	6.5			
Nuworld, natural (Kraft)	1 oz.	397	8.4			
Old English, process, loaf or slices (Kraft)	1 oz.	445	8.6			
Parmesan:						
Natural:						
(USDA)	1 oz.	208	7.4	4.	3.	27
(Frigo)	1 oz.	341	6.7			
(Kraft)	1 oz.	340	6.7			
(Stella)	1 oz.		7.5			
Grated:						
Natural (USDA)	1 oz.	247	9.4			32
Natural (USDA)	1 cup loosely packed (3.7 oz.)	923	35.2			120
Natural (USDA)	1 cup pressed down (4.9 oz.)	1219	46.5			158
Natural (USDA)	1 T. loosely packed (7 grams)	58	2.2			8
Natural (USDA)	1 T. pressed down (9 grams)	78	3.0			10
(Borden)	1 oz.	323	7.1			
(Buitoni)	1 oz.		7.4			
(Frigo)	1 T. (6 grams)	88	1.7			
(Kraft)	1 oz.	408	7.9			
Shredded (Kraft)	1 oz.	363	7.1			
Parmesan & Romano, grated:						
(Borden) natural	1 oz.		8.6			
(Kraft)	1 oz.	411	8.2			
Pepato (Frigo)	1 oz.	341	7.7			
Pimento American, process:						
(USDA)	1 oz.		8.6	4.	4.	
Loaf or slices (Kraft)	1 oz.	445	8.5			
Pinconning, natural (Kraft)	1 oz.	193	9.1			
Pizza:						
(Frigo)	1 oz.	227	3.8			
Low fat, part skim, shredded (Kraft)	1 oz.	284	5.5			
Port du Salut, natural (Kraft)	1 oz.	227	8.0			
Primost, natural (Kraft)	1 oz.	170	8.2			
Provolone:						
Natural (USDA)	1 oz.		7.8			28
(Frigo)	1 oz.	284	7.6			
Natural (Kraft)	1 oz.	284	7.6			

(USDA): United States Department of Agriculture
*Prepared as Package Directs

Food and Description	Measure or Quantity	Sodium (mg.)	—Fats in grams—			Choles-terol (mg.)
			Total	Satu-rated	Unsatu-rated	
Ricotta:						
Natural (USDA)	1 oz.		3.7			14
Part skim, natural (USDA)	1 oz.		2.3			9
(Breakstone)	1 oz.	42	3.1			8
(Breakstone)	1 T. (.6 oz.)	24	1.7			5
Natural (Kraft)	1 oz.	23	3.3			
(Sierra)	1 oz.		3.6			
Romano:						
(Frigo)	1 oz.	341	7.7			
Natural (Stella)	1 oz.		7.8			
Grated:						
(Borden)	1 oz.	488	6.2			
(Buitoni)	1 oz.		8.1			
(Frigo)	1 T. (6 grams)	90	2.0			
(Kraft)	1 oz.	420	9.4			
Shredded (Kraft)	1 oz.	374	8.5			
Romano & Parmesan:						
Plain (Kraft)	1 oz.	417	9.0			
Flavored (Kraft):						
Bacon smoke	1 oz.	562	8.6			
Garlic	1 oz.	401	9.0			
Onion	1 oz.	401	9.0			
Roquefort, natural:						
(USDA)	1 oz.		8.6	4.	4.	
(USDA)	1″ cube (.6 oz.)		5.2	3.	2	
(Kraft)	1 oz.	465	8.8			
Sage, natural (Kraft)	1 oz.	193	9.1			
Sap Sago, natural (Kraft)	1 oz.	510	2.6			
Sardo Romano, natural (Kraft)	1 oz.	397	7.7			
Scamorze:						
(Frigo)	1 oz.	227	4.7			
Natural (Kraft)	1 oz.	215	7.7			
Special cure, process, slices						
(Kraft)	1 oz.	403	8.6			
Supercure, process, slices						
(Kraft)	1 oz.	403	8.6			
Swiss, domestic:						
Natural:						
(USDA)	1 oz.	201	7.9	4.	4.	28
(USDA)	1″ cube (.5 oz.)	106	4.2	2.	2.	15
(USDA)	1¼-oz. slice (7½″ x 4″ x ¹/₁₆″)	248	9.8	5.	5.	35
(Kraft)	1 oz.	85	7.9			

(USDA): United States Department of Agriculture
*Prepared as Package Directs

87

Food and Description	Measure or Quantity	Sodium (mg.)	Total	—Fats in grams— Satu- rated	Unsatu- rated	Choles- terol (mg.)
(Sealtest)	1 oz.	85	8.0			
Process:						
(USDA) regular	1-oz. slice (3½″ x 3⅜″ x ⅛″)	331	7.6	4.	3.	26
(USDA) regular	1″ cube (.6 oz.)	210	4.8	3.	2.	17
(USDA) reduced sodium	1-oz. slice (3½″ x 3⅜ ″ x ⅛″)	193	7.6	4.	3.	26
(USDA) reduced sodium	1″ cube (.6 oz.)	126	4.8	3.	2.	17
(Borden)	1-oz. slice	308	7.0	5.	2.	37
(Borden)	¾-oz. slice	231	5.2	4.	2.	28
(Kraft) slices	1 oz.	445	7.1			
With Muenster (Kraft) loaf	1 oz.	456	7.8			
Washed curd, natural (Kraft)	1 oz.	193	8.6			
CHEESE CAKE, frozen (Mrs. Smith's)	¹/₆ of 8″ cake (4 oz.)	320	11.3			
CHEESE CAKE MIX:						
*(Jell-O)	⅛ of cake including crust (3.3 oz.)	359	11.6			23
*(Royal) *No-Bake*	⅛ of 9″ cake including crust (3.2 oz.)	400	10.8			8
CHEESE DIP (See **DIP**)						
CHEESE FONDUE, home recipe (USDA)	4 oz.	615	20.8	10.	10.	
CHEESE FOOD, process:						
American:						
(USDA)	1-oz. slice (3½″ x 3⅜″ x ⅛″)		6.8	4.	3.	20
(USDA)	1″ cube (.6 oz.)		4.3	2.	2.	13
(USDA)	1 T. (.5 oz.)		3	2.	2.	10
(Borden)	1″ x 1″ x 1 piece (.8 oz.)	356	5.0			34
Grated (Borden)	1 oz.	717	6.7			
Grated, used in *Kraft Dinner*	1 oz.	769	6.7			
Slices (Kraft)	1 oz.	396	6.9			
Cheez'n bacon, slices (Kraft)	¾-oz. slice	343	6.2			
Links (Kraft) *Handi-Snack:*						
Bacon	1 oz.	393	6.9			
Garlic	1 oz.	393	6.8			

(USDA): United States Department of Agriculture
*Prepared as Package Directs

Food and Description	Measure or Quantity	Sodium (mg.)	— Fats in grams —			Choles- terol (mg.)
			Total	Satu- rated	Unsatu- rated	
Jalapeño	1 oz.	404	6.8			
Nippy	1 oz.	393	6.8			
Smokelle	1 oz.	393	7.0			
Swiss	1 oz.	414	6.7			
Loaf:						
Munst-ett (Kraft)	1 oz.	553	8.0			
Pizzalone, loaf (Kraft)	1 oz.	442	6.6			
Super blend (Kraft)	1 oz.	376	6.9			
Pimento, slices (Kraft)	1 oz.	405	6.8			
Salami, slices (Kraft)	1 oz.	390	6.8			
Swiss (Borden)	.7-oz. slice	265	5.0	4.	1.	20
Swiss (Borden)	.8-oz. slice	290	5.5	4.	1.	22
Swiss, slices (Kraft)	1 oz.	396	6.6			

CHEESE PIE:

Food and Description	Measure or Quantity	Sodium (mg.)	Total	Satu- rated	Unsatu- rated	Choles- terol (mg.)
(Tastykake)	4-oz. pie		15.0			
Frozen, pineapple (Mrs. Smith's)	¹/₆ of 8″ pie (4 oz.)	415	11.8			
Frozen, pineapple (Mrs. Smith's)	¹/₈ of 10″ pie (5.4 oz.)	523	14.5			

CHEESE PUFF, frozen (Durkee)   1 piece (.5 oz.)   5.8

CHEESE SOUFFLE:

Food and Description	Measure or Quantity	Sodium (mg.)	Total	Satu- rated	Unsatu- rated	Choles- terol (mg.)
Home recipe (USDA)[1]	4 oz.	413	19.4	10.	9.	189
Home recipe (USDA)[1]	¼ of 7″ soufflé (3.9 oz.)	400	18.8	10.	9.	184
Frozen (Stouffer's)	⅓ of 12-oz. pkg.	606	16.7			

CHEESE SPREAD:

Food and Description	Measure or Quantity	Sodium (mg.)	Total	Satu- rated	Unsatu- rated	Choles- terol (mg.)
American, process:						
(USDA) regular	1 oz. (2¾″ x 2¼″ x ¼″)	461	6.1			18
(USDA) regular	1 T. (.5 oz.)	228	3.0			9
(USDA) regular shredded	1 packed cup (4 oz.)	1836	24.2			72
(USDA) reduced sodium	1 oz. (2¾″ x 2¼″ x ¼″)	323	6.1			18
(USDA) reduced sodium	1 T. (.5 oz.)	159	3.0			9
(USDA) reduced sodium	1 packed cup (4 oz.)	1287	24.2			72
(Borden)	1 oz.	403	6.1	4.	2.	44
(Borden) Vera Sharp	1 oz.		5.7			
(Kraft) *Swankyswig*	1 oz.	490	5.8			
(Nabisco) *Snack Mate*	1 tsp. (5 grams)	65	1.1			
Bacon, process (Kraft):						
Squeez-A-Snak	1 oz.	516	6.9			

(USDA): United States Department of Agriculture
*Prepared as Package Directs
[1]Principal sources of fat: butter, cheese, egg & milk.

Food and Description	Measure or Quantity	Sodium (mg.)	— Fats in grams —			Choles- terol (mg.)
			Total	Satu- rated	Unsatu- rated	
Swankyswig	1 oz.	366	7.6			
Cheddar, process (Nabisco)						
Snack Mate	1 tsp. (5 grams)	62	1.1			
Cheddar, seasoned, process						
(Nabisco) *Snack Mate*	1 tsp. (5 grams)	58	1.1			
Cheez Whiz, process (Kraft)	1 oz.	465	5.8			
Count Down (Fisher)	1 oz.	439	.3	Tr.	Tr.	1
Garlic, process:						
(Kraft) *Squeez-A-Snak*	1 oz.	530	7.0			
(Kraft) *Swankyswig*	1 oz.	366	6.5			
Hickory smoke, process						
(Nabisco) *Snack Mate*	1 tsp. (5 grams)	64	1.1			
Imitation (Fisher) *Chef's*						
Delight	1 oz.	380	1.1			3
Imitation (Fisher)						
Mellow Age	1 oz.	380	1.1			3
Imitation, process (Kraft)						
Calorie-Wise or Tasty-loaf	1 oz.	488	1.7			
Jalapeño, process (Kraft)						
Cheez Whiz	1 oz.	445	5.7			
Limburger natural (Kraft)	1 oz.	431	5.8			
Neufchâtel:						
Bacon & horseradish (Kraft)						
Party Snacks	1 oz.	170	6.8			
Chipped beef (Kraft) *Party*						
Snacks	1 oz.	227	6.0			
Chive (Kraft) *Party Snacks*	1 oz.	170	6.2			
Clam (Kraft) *Party Snacks*	1 oz.	170	5.8			
Olive-pimento (Kraft)						
Swankyswig	1 oz.	204	6.5			
Onion (Kraft) *Party Snacks*	1 oz.	242	6.1			
Pimento (Kraft) *Party Snacks*	1 oz.	159	5.8			
Pimento (Kraft) *Swankyswig*	1 oz.	125	5.8			
Pineapple (Kraft) *Swankyswig*	1 oz.	102	5.8			
Relish (Kraft) *Swankyswig*	1 oz.	136	5.8			
Roka (Kraft) *Swankyswig*	1 oz.	295	7.5			
Old English, process (Kraft)						
Swankyswig	1 oz.	366	8.1			
Onion, French, process (Nabisco)						
Snack Mate	1 tsp. (5 grams)	65	1.1			
Pimento:						
(Kraft) *Cheez Whiz*	1 oz.	465	5.8			
Process (Kraft) *Squeez-A-Snak*	1 oz.	556	7.3			

(USDA): United States Department of Agriculture
*Prepared as Package Directs

Food and Description	Measure or Quantity	Sodium (mg.)	— Fats in grams —			Choles- terol (mg.)
			Total	Satu- rated	Unsatu- rated	
Process (Nabisco) *Snack Mate*	1 tsp. (5 grams)	60	1.1			
(Sealtest)	1 oz.	239	5.8			
Velveeta, process (Kraft)	1 oz.	462	5.8			
Sharp, process (Kraft) *Squeez-A-Snak*	1 oz.	496	7.2			
Sharpie, process (Kraft)	1 oz.	420	7.3			
Smoke, process (Kraft) *Squeez-A-Snak*	1 oz.	519	6.9			
Smokelle, process (Kraft) *Swankyswig*	1 oz.	366	7.2			
Velva Kreme, process (Borden)	1 oz.	106	9.2	3.	6.	
Velveeta, process (Kraft)	1 oz.	462	5.8			

CHEESE STRAW:

Food and Description	Measure or Quantity	Sodium (mg.)	Total	Satu- rated	Unsatu- rated	Choles- terol (mg.)
Made with lard (USDA)[1]	1 oz.	204	8.5	4.	5.	9
Made with lard (USDA)[1]	5″ x ⅜″ x ⅜″ piece (6 grams)	43	1.8	<1.	1.	2
Made with vegetable shortening (USDA)[2]	1 oz.	204	8.5	3.	6.	
Frozen (Durkee)	1 piece (8 grams)		2.2			

CHELOIS WINE (Great Western)

Food and Description	Measure or Quantity	Sodium (mg.)	Total	Satu- rated	Unsatu- rated	Choles- terol (mg.)
12.5% alcohol	3 fl. oz.	37	0.			0

CHERIMOYA, raw (USDA):

Food and Description	Measure or Quantity	Sodium (mg.)	Total	Satu- rated	Unsatu- rated	Choles- terol (mg.)
Whole	1 lb. (weighed with skin &seeds)		1.1			0
Flesh only	4 oz.		.5			0

CHERRI-BERRI, soft drink (Hoffman) 6 fl. oz.

Food and Description	Measure or Quantity	Sodium (mg.)	Total	Satu- rated	Unsatu- rated	Choles- terol (mg.)
		15	0.			(0)

CHERRY:
Sour:
  Fresh (USDA):

Food and Description	Measure or Quantity	Sodium (mg.)	Total	Satu- rated	Unsatu- rated	Choles- terol (mg.)
Whole	1 lb. (weighed with stems)	7	1.1			0
Whole	1 lb. (weighed without stems)	8	1.3			0
Pitted	½ cup (2.7 oz.)	2	.2			0
Canned, syrup pack, pitted (USDA):						
Light syrup	4 oz. (with liq.)	1	.2			0

(USDA): United States Department of Agriculture
*Prepared as Package Directs
[1]Principal sources of fat: lard & milk.
[2]Principal sources of fat: vegetable shortening, milk & lard.

91

Food and Description	Measure or Quantity	Sodium (mg.)	Total	Fats in grams Satu- rated	Unsatu- rated	Choles- terol (mg.)
Heavy syrup	4 oz. (with liq.)	1	.2			0
Heavy syrup	½ cup (4.6 oz.)	1	.3			0
Extra heavy syrup	4 oz. (with liq.)	1	.2			0
Canned, water pack, pitted, solids & liq.:						
(USDA)	½ cup (4.3 oz.)	2	.2			0
(Stokely-Van Camp)	½ cup (4 oz.)		.2			
Frozen, pitted (USDA):						
Sweetened	½ cup (4.6 oz.)	3	.5			0
Unsweetened	4 oz.	2	.5			0
Sweet:						
Fresh (USDA):						
Whole, with stems	1 lb. (weighed with stems)	8	1.2			0
Whole, with stems	½ cup (2.3 oz.)	1	.2			0
Pitted	½ cup (2.9 oz.)	2	.2			0
Canned, syrup pack, with pits, solids & liq.:						
Light syrup (USDA)	4 oz.	1	.2			0
Heavy syrup (USDA)	4 oz.	1	.2			0
Heavy syrup, dark (Del Monte)	½ cup (4.3 oz.)	2	.6			0
Heavy syrup, Royal Anne (Del Monte)	½ cup (4.6 oz.)	2	.5			0
Heavy syrup, light or dark (Stokely-Van Camp)	½ cup (4.2 oz.)		.2			(0)
Extra heavy syrup (USDA)	4 oz.	1	.2			0
Canned, syrup pack, pitted, solids & liq.:						
Light syrup (USDA)	4 oz.	1	.2			0
Heavy syrup (USDA)	4 oz.	1	.2			0
Heavy syrup (USDA)	½ cup (4.2 oz.)	1	.2			0
Heavy syrup (Del Monte)	½ cup (4.3 oz.)	6	1.5			(0)
Extra heavy syrup (USDA)	4 oz.	1	.2			0
Canned, water pack, with pits, solids & liq. (USDA)	4 oz.	1	.2			0
Canned, water or dietetic pack, pitted, solids & liq.:						
(USDA)	4 oz.	1	.2			0
(Blue Boy)	4 oz.	1	Tr.			(0)
Royal Anne: (Diet Delight)	½ cup (4.4 oz.)	6	Tr.			(0)

(USDA): United States Department of Agriculture
*Prepared as Package Directs

Food and Description	Measure or Quantity	Sodium (mg.)	—Fats in grams—			Choles-terol (mg.)
			Total	Satu-rated	Unsatu-rated	
(S and W) *Nutradiet*, low calorie	14 whole cherries (3.5 oz.)	3	Tr.			(0)
(S and W) *Nutradiet*, unsweetened	14 whole cherries (3.5 oz.)	2	Tr.			(0)
(Tillie Lewis)	½ cup (4.5 oz.)	<10	.2			0
Dark (S and W) *Nutradiet*, low calorie	4 oz.	1	.1			(0)
Frozen, quick-thaw (Birds Eye)	½ cup (5 oz.)	3	.3			0

CHERRY, BLACK, SOFT DRINK
(See **CHERRY SOFT DRINK**)

CHERRY CAKE:

*Mix (Duncan Hines)	$^1/_{12}$ of cake (2.6 oz.)	299	5.2			
*Mix, chip, layer (Betty Crocker)	$^1/_{12}$ of cake	278	4.8			
*Mix, upside down (Betty Crocker)	$^1/_9$ of cake	226	10.2			

CHERRY, CANDIED (USDA) 1 oz. .6 0

CHERRY COLA (See **COLA SOFT DRINK**)

CHERRY DRINK (Hi-C) 6 fl. oz. (6.3 oz.) Tr. Tr. 0

CHERRY, MARASCHINO (USDA) 1 oz. (with liq.) <.1 0

CHERRY PIE:

Home recipe, made with lard, 2 crust (USDA)[1]	$^1/_6$ of 9″ pie (5.6 oz.)	480	17.9	6.	12.	
Home recipe, made with vegetable shortening, 2 crust (USDA)[2]	$^1/_6$ of 9″ pie (5.6 oz.)	480	17.9	5.	13.	
(Hostess)	4½-oz. pie	637	10.8			
Cherry-apple (Tastykake)	4-oz. pie		14.7			
Frozen:						
Unbaked (USDA)	5 oz.	287	15.1	4.	11.	
Baked (USDA)	5 oz.	325	17.0	4.	13.	
(Banquet)	5-oz. serving		15.0			
(Morton)	$^1/_6$ of 20-oz. pie	239	10.9			
(Morton)	$^1/_6$ of 24-oz. pie	407	18.4			
(Morton)	$^1/_8$ of 46-oz. pie	299	16.7			

(USDA): United States Department of Agriculture
*Prepared as Package Directs
[1]Principal sources of fat: lard & butter.
[2]Principal sources of fat: vegetable shortening & butter.

Food and Description	Measure or Quantity	Sodium (mg.)	Total	Fats in grams — Satu- rated	Unsatu- rated	Choles- terol (mg.)
(Mrs. Smith's)	¹/₆ of 8″ pie (4.3 oz.)	302	14.2			
(Mrs. Smith's) old fashion	¹/₆ of 9″ pie (4.3 oz.)	470	23.2			
(Mrs. Smith's) golden deluxe	⅛ of 10″ pie (5.6 oz.)	410	18.4			
Tart (Pepperidge Farm)	3-oz. pie tart	196	14.9			
CHERRY PIE FILLING:						
(Comstock)	1 cup (10¾ oz.)	240	.5			
(Lucky Leaf)	8 oz.	104	.4			
CHERRY PRESERVE or JAM:						
Sweetened (Bama)	1 T. (.7 oz.)	2	<.1			
Dietetic or low calorie (S and W)						
Nutradiet	1 T. (.5 oz.)		Tr.			(0)
CHERRY SOFT DRINK:						
Sweetened:						
(Canada Dry) bottle or can	6 fl. oz.	13+	0.			0
(Clicquot Club)	6 fl. oz.	12	0.			0
(Cott)	6 fl. oz.	12	0.			0
(Dr. Brown's) black	6 fl. oz.	14	0.			0
(Fanta)	6 fl. oz.	7	0.			0
(Hoffman) black	6 fl. oz.	14	0.			0
(Key Food) black	6 fl. oz.	14	0.			0
(Kirsch) black	6 fl. oz.	<1	0.			0
(Mission)	6 fl. oz.	12	0.			0
(Nedick's) black	6 fl. oz.	14	0.			0
(Shasta) black	6 fl. oz.	22	0.			0
(Waldbaum) black	6 fl. oz.	14	0.			0
(Yukon Club) black	6 fl. oz.	14	0.			0
Unsweetened or low calorie:						
(Clicquot Club)	6 fl. oz.	46	0.			0
(Cott)	6 fl. oz.	46	0.			0
(Dr. Brown's) black	6 fl. oz.	57	0.			0
(Hoffman) black	6 fl. oz.	57	0.			0
(Key Food) black	6 fl. oz.	57	0.			0
(Mission)	6 fl. oz.	46	0.			0
(No-Cal) black	6 fl. oz.	12	0.			0
(Shasta) black	6 fl. oz.	37	0.			0
(Waldbaum) black	6 fl. oz.	57	0.			0
(Yukon Club) black	6 fl. oz.	71	0.			0
CHERRY SYRUP, dietetic						
(No-Cal) black	1 tsp.	<1	0.			0

(USDA): United States Department of Agriculture
*Prepared as Package Directs

Food and Description	Measure or Quantity	Sodium (mg.)	—Fats in grams—			Choles- terol (mg.)
			Total	Satu- rated	Unsatu- rated	
CHERRY TURNOVER, frozen						
(Pepperidge Farm)	3.3-oz. turnover	259	20.0			
CHERVIL:						
Raw (USDA)	1 oz.		.3			0
Dry (Spice Islands)	1 tsp.	<1				(0)
CHESTNUT (USDA):						
Fresh, in shell	1 lb. (weighed in shell)	22	5.5			0
Fresh, shelled	4 oz.	7	1.7			0
Dried, in shell	1 lb. (weighed in shell)	45	15.3			0
Dried, shelled	4 oz.	14	4.6			0
CHESTNUT FLOUR (See **FLOUR, CHESTNUT**)						
CHEWING GUM:						
Sweetened:						
Beechies	1 tablet (2 grams)	<1	0.			(0)
Beech-Nut	1 stick (3 grams)	<1	0.			(0)
Doublemint	1 stick (3 grams)	<1	Tr.			0
Juicy Fruit	1 stick (3 grams)	<1	Tr.			0
Spearmint (Wrigley's)	1 stick (3 grams)	<1	Tr.			0
Unsweetened or dietetic:						
All flavors (Estee)	1 stick		<.1			(0)
*Care*Free* (Beech-Nut)	1 stick (3 grams)	<1	0.			(0)
(Harvey's)	1 stick	Tr.	0.			0
Peppermint (Amurol)	1 stick					
CHICKEN (See also **CHICKEN, CANNED**) (USDA):						
Broiler, cooked, meat only	4 oz.	75	4.3	1.	3.	99
Capon, raw, ready-to-cook	1 lb. (weighed with bones)		70.2	22.	48.	324
Capon, raw, meat with skin	4 oz.		24.9			92
Fryer:						
Raw:						
Ready-to-cook	1 lb. (weighed with bone)		15.1	5.	10.	310
Meat & skin	1 lb.		23.1	7.	16.	367
Meat only	1 lb.	263	12.2	4.	8.	358

(USDA): United States Department of Agriculture
*Prepared as Package Directs

Food and Description	Measure or Quantity	Sodium (mg.)	—Fats in grams—			Cholesterol (mg.)
			Total	Satu-rated	Unsatu-rated	
Dark meat with skin	1 lb.	304	28.6	9.	20.	399
Light meat with skin	1 lb.	227	17.7	5.	12.	304
Dark meat without skin	1 lb.	304	17.2	5.	12.	399
Light meat without skin	1 lb.	227	6.8	2.	5.	358
Skin only	4 oz.		19.4	6.	14.	
Back	1 lb. (weighed with bone)		23.5	8.	16.	198
Breast	1 lb. (weighed with bone)		8.6	3.	6.	239
Leg or drumstick	1 lb. (weighed with bone)		10.6	3.	8.	239
Neck	1 lb. (weighed with bone)		20.5	7.	14.	177
Rib	1 lb. (weighed with bone)		12.5	4.	8.	187
Thigh	1 lb. (weighed with bone)		19.1	6.	13.	275
Wing	1 lb. (weighed with bone)		16.5	5.	12.	180
Fried. A 2½-pound chicken (weighed with bone before cooking) will give you:						
Back[1]	1 back (2.2 oz.)		8.5	3.	6.	35
Breast[1]	½ breast (3.3 oz.)		4.9	2.	3.	61
Leg or drumstick[1]	1 leg (2 oz.)		3.8	1.	3.	34
Neck[1]	1 neck (2.1 oz.)		7.3	3.	5.	37
Rib[1]	1 rib (.7 oz.)		2.2	<1	2.	12
Thigh[1]	1 thigh (2.3 oz.)		5.7	2.	4.	44
Wing[1]	1 wing (1¾ oz.)		4.3	1.	3.	25
Fryer:						
Fried:						
Meat, skin & giblets[1]	4 oz.	88	13.4	3.	10.	91
Meat & skin[1]	4 oz.	88	13.5	3.	10.	91
Meat only[1]	4 oz.	88	8.8	3.	6.	88
Dark meat with skin[1]	4 oz.	100	15.4	5.	10.	103
Light meat with skin[1]	4 oz.	77	11.2	4.	8.	90
Dark meat without skin[1]	4 oz.	100	10.5	3.	7.	103
Light meat without skin[1]	4 oz.	77	6.9	2.	5.	103
Skin only[1]	1 oz.		8.2	3.	6.	
Hen & cock:						
Raw:						
Ready-to-cook	1 lb. (weighed with bones)		82.1	26.	56.	324

(USDA): United States Department of Agriculture
*Prepared as Package Directs
[1]Principal source of fat: vegetable shortening.

Food and Description	Measure or Quantity	Sodium (mg.)	—Fats in grams— Total	Satu-rated	Unsatu-rated	Choles-terol (mg.)
Meat & skin	1 lb.		85.3	27.	58.	367
Meat only	1 lb.	263	31.8	10.	21.	445
Dark meat without skin	1 lb.	304	34.0	11.	23.	399
Light meat without skin	1 lb.	227	16.8	5.	11.	304
Stewed:						
Meat, skin & giblets	4 oz.		25.2	8.	17.	91
Meat & skin	4 oz.		25.9	8.	18.	99
Meat only	4 oz.	62	10.1	3.	7.	99
Chopped	½ cup (2.5 oz.)	40	6.4	2.	4.	63
Diced	½ cup (2.4 oz.)	37	6.0	2.	4.	58
Ground	½ cup (2 oz.)	31	5.0	2.	3.	49
Roaster:						
Raw:						
Ready-to-cook	1 lb. (weighed with bones)		59.3	19.	40.	324
Meat, skin & giblets	1 lb.		54.0	18.	36.	445
Meat & skin	1 lb.		57.2	19.	38.	445
Meat only	1 lb.	263	20.4	7.	14.	367
Dark meat without skin	1 lb.	304	21.3	7.	14.	399
White meat without skin	1 lb.	227	14.5	5.	10.	304
Roasted:						
Total edible	4 oz.		22.9	7.	16.	99
Meat, skin & giblets	4 oz.		15.9	5.	11.	92
Meat & skin	4 oz.		16.7	5.	11.	99
Meat only	4 oz.	87	7.1	2.	5.	99
Dark meat without skin	4 oz.	100	7.4	2.	5.	99
Light meat without skin	4 oz.	75	5.6	2.	3.	99
CHICKEN A LA KING:						
Home recipe (USDA)	1 cup (8.6 oz.)	760	34.3	12.	22.	186
Canned (College Inn)	5-oz. serving		10.0			
Canned (Richardson & Robbins)	1 cup (7.9 oz.)	1055	14.8			
Canned (Swanson)	1 cup	1000	16.0			
Frozen (Banquet)	5-oz. bag		5.0			
CHICKEN BOUILLON/BROTH, cube or powder (See also **CHICKEN SOUP**):						
(Croyden House)	1 tsp. (5 grams)	680	Tr.	Tr.	0.	
(Herb-Ox)	1 cube (4 grams)	910	.1			
(Herb-Ox) instant	1 packet (5 grams)	940	.1			
(Maggi)	1 cube (4 grams)	747	.2			
(Maggi) instant	1 tsp. (4 grams)	699	.2			

(USDA): United States Department of Agriculture
*Prepared as Package Directs

Food and Description	Measure or Quantity	Sodium (mg.)	— Fats in grams —			Choles- terol (mg.)
			Total	Satu- rated	Unsatu- rated	
(Steero)	1 cube (4 grams)		.2			
(Wyler's) regular	1 cube (4 grams)		.3			
(Wyler's) no salt added	1 cube (4 grams)	2	.4			
(Wyler's) instant	1 tsp.		.3			
CHICKEN CACCIATORE, canned						
(Hormel)	1-lb can		17.7			
CHICKEN, CANNED:						
Boned:						
(USDA)	4 oz.		13.3	5.	9.	
(USDA)	½ cup (3 oz.)		9.9	3.	7.	
(Lynden Farms) solids & liq.	5-oz. jar	446	13.3	4.	9.	
(Lynden Farms) with broth	11-oz. jar	980	29.5	10.	20.	
(Lynden Farms) with broth	29-oz. can	2137	174.6	59.	115.	
(Swanson) with broth	5-oz. can	705	11.0			
Fat, rendered, with onion						
(Lynden Farms)	¼ of 12.5-oz. jar	4	85.1	28.	57.	
Whole (Lynden Farms) solids & liq.	¼ of 52-oz. can	1100	95.9	33.	63.	
CHICKEN, CREAMED, frozen						
(Stouffer's)	11½-oz. pkg.	1015	44.4			
CHICKEN DINNER:						
Canned:						
Dumpling (College Inn)	5-oz. serving (½ can)		11.0			
Noodle (Heinz)	8½-oz. can	1123	7.6			
Noodle (Lynden Farms)	14-oz. jar	993	23.8	10.	14.	
Noodle with vegetables						
(Lynden Farms)	15-oz. can	2121	22.5	8.	15.	
Frozen:						
(Weight Watchers)	10-oz. luncheon		2.8			
Boneless chicken (Swanson)						
Hungry Man	19-oz. dinner	1995	33.6			
Chicken & dumplings:						
Buffet (Banquet)	2-lb. pkg.		57.1			
(Morton)	12-oz. dinner	1506	17.4			
(Morton) 3-course	1-lb 5-oz. dinner	1220	32.9			
Chicken livers & onion						
(Weight Watchers)	11½-oz. luncheon		2.0			
Creole (Weight Watchers)	12-oz. luncheon	748	6.8			

(USDA): United States Department of Agriculture
*Prepared as Package Directs

Food and Description	Measure or Quantity	Sodium (mg.)	—Fats in grams—			Choles- terol (mg.)
			Total	Satu- rated	Unsatu- rated	
Fried:						
With mashed potato, carrots peas, corn & beans (USDA)[1]	12 oz.	1170	28.9	10.	19.	
(Banquet):						
Meat compartment	5 oz.		21.4			
Corn compartment	2.5 oz.		1.9			
Potato compartment	3.5 oz.		1.7			
Complete dinner	11-oz. dinner		25.0			
(Morton)	11-oz. dinner	1153	25.0			
(Morton) 3-course	1-lb. 1-oz. dinner	1531	49.7			
(Swanson)	11½-oz. dinner	1174	29.3	9.	21.	
(Swanson) 3-course	15-oz. dinner	1431	27.3			
With shoestring potato						
(Swanson)	25-oz. pkg.	1190	107.9			

CHICKEN & DUMPLINGS (See CHICKEN DINNER)

CHICKEN, FREEZE DRY

(Wilson) *Campsite:*						
Dry	2 oz.	284	5.8	2.	4.	142
*Reconstituted	4 oz.	217	4.6	1.	3.	114

CHICKEN FRICASSEE:

Home recipe (USDA)[2]	1 cup (8.5 oz.)	370	22.3	7.	15.	96
Canned (Lynden Farms)	14.5-oz. can	1607	29.5	10.	19.	
Canned (Richardson & Robbins)	1 cup (7.9 oz.)	1028	12.8			

CHICKEN, FRIED, frozen:

(Banquet) whole chicken	2 lbs.		127.8			
(Banquet) ½ chicken	14 oz.		55.9			
(Swanson) halves	2 pieces (17¼ oz.)	1086	69.6			
(Swanson) quarters	4 pieces (17¼ oz.)	1087	68.4			
(Swanson)	16-oz. pkg.	1176	73.5			
(Swanson)	32-oz. pkg.	2322	145.2			
With whipped potato (Swanson)	7-oz. pkg.	895	23.1			

CHICKEN GIBLETS:

Capon, raw	2 oz.		8.3			
Fryer, raw	2 oz.		1.8			

(USDA): United States Department of Agriculture
*Prepared as Package Directs
[1]Principal sources of fat: vegetable shortening, chicken & butter.
[2]Principal source of fat: chicken.

Food and Description	Measure or Quantity	Sodium (mg.)	—Fats in grams— Total	Saturated	Unsaturated	Cholesterol (mg.)
Fryer, fried, from a 2½-lb. chicken	1 heart, gizzard & liver (2.1 oz.)		6.7			
Hen & cock, raw	2 oz.		6.6			
Roaster, raw	2 oz.		2.7			
CHICKEN GIZZARD (USDA):						
Raw	2 oz.	37	1.5			82
Simmered	2 oz.	32	1.9			111
CHICKEN LIVER PUFF, frozen (Durkee)	1 piece (.5 oz.)		4.7			
CHICKEN & NOODLES:						
Home recipe (USDA)[1]	1 cup (8.5 oz.)	600	18.5	5.	14.	96
Canned (College Inn)	5-oz. serving		7.0			
Frozen (Banquet) buffet	2-lb. pkg.		21.2			
Frozen, escalloped (Stouffer's)	11½-oz. pkg.	1199	37.9			
CHICKEN PIE:						
Baked, home recipe (USDA)[2]	4¼" pie (8 oz.)	581	30.6	11.	19.	70
Baked, home recipe (USDA)[2]	⅓ of 9" pie (8.2 oz.)	594	31.3	12.	20.	72
Frozen:						
Commercial, unheated (USDA)[3]	8-oz. pie	933	26.1	7.	19.	29
(Banquet)	8-oz. pie		21.1			
(Banquet)	2-lb. 4-oz. pie		62.3			
(Morton)	8-oz. pie	1048	25.4			
(Stouffer's)	10-oz. pkg.	1507	50.6			
(Swanson)	8-oz. pie	981	24.4			
(Swanson) deep dish	16-oz. pie	2090	38.8			
CHICKEN, POTTED (USDA)	1 oz.		5.4			
CHICKEN PUFF, frozen (Durkee)	1 piece (.5 oz.)		4.8			
CHICKEN RAVIOLI, in sauce (Lynden Farms)	14.5-oz. can	2318	10.5	3.	8.	
CHICKEN SOUP, canned:						
(Campbell) *Chunky*	1 cup	960	5.4			
*Barley (Manischewitz)	1 cup		2.1			

(USDA): United States Department of Agriculture
*Prepared as Package Directs
[1]Principal sources of fat: chicken & egg.
[2]Principal sources of fat: vegetable shortening, cream, chicken & butter.
[3]Principal sources of fat: vegetable shortening, chicken, cream & corn oil.

Food and Description	Measure or Quantity	Sodium (mg.)	—Fats in grams—			Choles-terol (mg.)
			Total	Satu-rated	Unsatu-rated	
Broth:						
*(Campbell)	1 cup	750	1.8	Tr.	2.	
(Lynden Farms)	1 cup (8 oz.)	848	0.			
(Richardson & Robbins)	1 cup (8.1 oz.)	1152	1.1			
(Swanson)	1 cup	1056	1.6			
*Dietetic (Claybourne)	8 oz. (by wt.)	27	.9			
*Diet (Slim-ette)	8 oz. (by wt.)	36	.3			
With rice (Richardson & Robbins)	1 cup (8.1 oz.)	1136	1.4			
Consommé:						
Condensed (USDA)	8 oz. (by wt.)	1367	.2			
*Prepared with equal volume water (USDA)	1 cup (8.5 oz.)	722	Tr.			
Cream of:						
Condensed (USDA)	8 oz. (by wt.)	1836	10.9	2.	9.	
*Prepared with equal volume milk (USDA)	1 cup (8.6 oz.)	1054	10.3	2.	8.	
*Prepared with equal volume water (USDA)	1 cup (8 oz.)	970	5.8	Tr.	6.	
*(Campbell)	1 cup	885	5.0	1.	4.	
*(Heinz)	1 cup (8.5 oz.)	962	5.3			
(Heinz) *Great American*	1 cup (8.5 oz.)	1080	5.4			
*& Dumplings (Campbell)	1 cup	972	5.5	1.	4.	
Gumbo:						
Condensed (USDA)	8 oz. (by wt.)	1798	3.0			
*Prepared with equal volume water (USDA)	1 cup (8.5 oz.)	950	1.4			
*(Campbell)	1 cup	919	1.3	\Tr.	1.	
Creole (Heinz) *Great American*	1 cup (8¾ oz.)	1031	2.0			
*& Kasha (Manischewitz)	1 cup		1.4			
& Noodle:						
Condensed (USDA)	8 oz. (by wt.)	1852	3.6			
*Prepared with equal volume water (USDA)	1 cup (8.8 oz.)	1020	2.0			
*(Campbell)	1 cup	930	1.7	Tr.	1.	6
*Noodle-O's (Campbell)	1 cup	855	1.9	1.	1.	
*(Heinz)	1 cup (8.5 oz.)	1091	2.5			
*(Manischewitz)	1 cup		1.2			
Dietetic (Tillie Lewis)	1 cup (8 oz.)	40	1.4			
With dumplings (Heinz) *Great American*	1 cup (8.5 oz.)	1123	3.4			
*With stars (Campbell)	1 cup	1050	1.5	Tr.	1.	
*With stars (Heinz)	1 cup (8.5 oz.)	1058	2.3			

(USDA): United States Department of Agriculture
*Prepared as Package Directs

Food and Description	Measure or Quantity	Sodium (mg.)	Total	— Fats in grams — Satu- rated	Unsatu- rated	Choles- terol (mg.)
& Rice:						
Condensed (USDA)	8 oz. (by wt.)	1734	2.3			
*Prepared with equal volume water (USDA)	1 cup (8.5 oz.)	917	1.2			
*(Campbell)	1 cup	748	1.5	Tr.	1.	
*(Heinz)	1 cup (8.5 oz.)	996	2.7			
*(Manischewitz)	1 cup		1.1			
With mushrooms (Heinz) *Great American*	1 cup (8.5 oz.)	1123	3.7			
Vegetable:						
Condensed (USDA)	8 oz. (by wt.)	1916	4.5			
*Prepared with equal volume water (USDA)	1 cup (8.6 oz.)	1034	2.4			
*(Campbell)	1 cup	912	2.0	Tr.	2.	
*(Heinz)	1 cup (8.5 oz.)	1070	3.6			
*(Manischewitz)	1 cup		1.5			

CHICKEN SOUP MIX:

Food and Description	Measure or Quantity	Sodium (mg.)	Total	Satu- rated	Unsatu- rated	Choles- terol (mg.)
Cream of (Lipton) *Cup-a-Soup*	1 pkg. (.8 oz.)	859	5.1	4.	<1.	4
& Noodle:						
(USDA)[1]	2-oz. pkg.	2438	5.7	2.	4.	
*(USDA)	1 cup (8.1 oz.)	554	1.4			
*(Lipton)	1 cup	888	1.8	<1.	1.	16
(Lipton) *Cup-a-Soup*	1 pkg. (.4 oz.)	931	.8	Tr.	<1.	16
*(Wyler's)	1 cup		1.3			
Ring-O-Noodle (Lipton)	1 cup	827	1.2	Tr.	<1.	24
*With diced chicken (Lipton)	1 cup	971	2.2	<1.	2.	21
With meat (Lipton) *Cup-a-Soup*	1 pkg. (.4 oz.)	989	.9	Tr.	<1.	12
& Rice:						
(USDA)	1 oz.	1237	1.9	1.	1.	
*(USDA)	1 cup (8 oz.)	591	.9			
*(Lipton)	1 cup (8 oz.)	948	2.1			
*(Wyler's)	1 cup		1.2			
*Vegetable (Lipton)	1 cup	1160	2.2	<1.	2.	7

CHICKEN SPREAD:

Food and Description	Measure or Quantity	Sodium (mg.)	Total	Satu- rated	Unsatu- rated	Choles- terol (mg.)
(Swanson)	5-oz. can	830	21.0			
(Underwood)	4¾-oz. can	1014	21.9			
(Underwood)	1 T. (.5 oz.)	104	2.3			

CHICKEN STEW:

Food and Description	Measure or Quantity	Sodium (mg.)	Total	Satu- rated	Unsatu- rated	Choles- terol (mg.)
Canned:						
(B & M)	1 cup (7.9 oz.)	936	2.7			

(USDA): United States Department of Agriculture
*Prepared as Package Directs
[1]Principal sources of fat: vegetable shortening & egg.

Food and Description	Measure or Quantity	Sodium (mg.)	—Fats in grams— Total	Satu- rated	Unsatu- rated	Choles- terol (mg.)
(Swanson)	1 cup	1018	5.9			
With dumplings (Heinz)	8.5-oz. can	1174	8.2			
Freeze dry, canned (Wilson) *Campsite:;*						
Dry	4½-oz. can	3375	19.8	7.	13.	
*Reconstituted	1 lb.	2608	15.4	5.	10.	
CHICKEN STOCK BASE						
(French's)	1 tsp. (3 grams)	480	.2			
CHICKEN TAMALE PIE,						
canned (Lynden Farms)	½ tamale pie with sauce (3.8 oz.)	554	8.3	2.	6.	
CHICK PEA or GARBANZO (USDA):						
Dry	1 lb.	118	21.8	2.	20.	0
Dry	1 cup (7.1 oz.)	52	9.6	Tr.	10.	0
CHICORY GREENS, raw (USDA):						
Untrimmed	½ lb. (weighed untrimmed)		.6			0
Trimmed	4 oz.		.3			0
CHICORY, WITLOOF, Belgian or French endive, raw, bleached head (USDA):						
Untrimmed	½ lb. (weighed untrimmed)	14	.2			0
Trimmed, cut	½ cup (.9 oz.)	2	<.1			0
CHILI or CHILI CON CARNE:						
Canned, with beans:						
(USDA)	1 cup (8.8 oz.)	1328	15.2	8.	8.	
(Armour Star)	15½-oz. can		33.8			
(Chef Boy-Ar-Dee)	¼ of 30-oz. can	912	15.5			
(Heinz)	8¾-oz. can	1332	18.4			
(Hormel)	7½ oz.	933	17.9	6.	9.	32
(Nalley's) mild or hot	8 oz.		17.9			
(Swanson)	1 cup	1041	13.2			
(Van Camp)	1 cup (8 oz.)		14.0			
(Wilson)	½ of 15½-oz. can	1167	16.5	8.	9.	34
Canned without beans:						
(USDA) not less than 60% meat nor more than 8% cereal & seasonings[1]	1 cup (9 oz.)		37.7	18.	20.	

(USDA): United States Department of Agriculture
*Prepared as Package Directs
[1]Principal source of fat: beef.

Food and Description	Measure or Quantity	Sodium (mg.)	Total	—Fats in grams— Satu- rated	Unsatu- rated	Choles- terol (mg.)
(Armour Star)	15½-oz. can		66.3			
(Chef Boy-Ar-Dee)	½ of 15¼-oz. can	927	22.5			
(Hormel)	7½ oz.	878	25.6			
(Nalley's)	8 oz.		20.4			
(Van Camp)	1 cup (8.1 oz.)		34.0			
(Wilson)	½ of 15½-oz. can	1204	35.4	18.	18.	60
Frozen, with beans (Banquet)	8-oz. bag		18.4			
CHILI BEEF SOUP:						
*(Campbell)	1 cup	975	4.2	2.	2.	
*(Heinz)	1 cup (8¾ oz.)	1130	5.5			
(Heinz) *Great American*	1 cup (8¾ oz.)	1146	5.9			
CHILI CON CARNE MIX (Durkee):						
*With meat & beans	2½ cups					
	(2¼-oz. pkg.)	2347	38.0			
*Without meat & beans	1¼ cups					
	(2¼-oz. pkg.)	2116	2.2			
CHILI CON CARNE SPREAD						
(Oscar Mayer):						
With beans	1 oz.	243	4.0			
Without beans	1 oz.	453	5.7			
***CHILI DOG SAUCE MIX**						
(McCormick)	1 serving (.9 oz.)	97	.2			
CHILI PEQUIN (Spice Islands)	1 pod	Tr.				(0)
CHILI POWDER:						
With added seasoning (USDA)	1 T. (.5 oz.)	236	1.9			
(Chili Products)	½ oz.		1.6			
CHILI SAUCE:						
(USDA)	½ cup (4.4 oz.)	1659	.4			
(USDA)	1 T. (.5 oz.)	201	<.1			
(USDA) low sodium	½ cup (4.4 oz.)	6–43	.4			
(USDA) low sodium	1 T. (.5 oz.)	<1–5	<.1			
(Del Monte)	1 T. (.5 oz.)	280	<.1			
(Heinz)	1 T.	191	Tr.			
(Hunt's)	½ cup (4.8 oz.)	1549	.3			
(Hunt's)	1 T. (.6 oz.)	194	<.1			
(Ortega)	¼ cup (2.1 oz.)	364	Tr.			
(Stokely-Van Camp)	1 T. (.5 oz.)		Tr.			

(USDA): United States Department of Agriculture
*Prepared as Package Directs

Food and Description	Measure or Quantity	Sodium (mg.)	— Fats in grams —			Choles- terol (mg.)
			Total	Satu- rated	Unsatu- rated	

CHILI SEASONING MIX:

Chili-O (French's)	1¾-oz. pkg.	3800	1.2			
*(Kraft)	1 oz.	49	2.3			
(Lawry's)	1.6-oz. pkg.		2.5			
*(Wyler's)	6 fl. oz.		.8			
Powder (Spice Islands)	1 tsp.	1				

CHINESE DATE (See **JUJUBE**)

CHINESE DINNER, frozen:

Beef chop suey (Chun King)	11-oz. dinner		10.0			
(Banquet) dinner:						
Meat compartment	7 oz.		5.0			
Rice compartment	4 oz.		2.6			
Complete dinner	11-oz. dinner		7.6			
Chicken chow mein (Chun King)	11-oz. dinner		12.0			
Egg foo young (Chun King)	11-oz. dinner		10.0			
Shrimp chow mein (Chun King)	11-oz. dinner		10.0			
(Swanson)	11-oz. dinner	1606	12.6			

CHINESE VEGETABLES
(See **VEGETABLE, MIXED**)

CHIPS (See **CRACKERS** for corn chips and **POTATO CHIPS**)

CHITTERLINGS, canned (Hormel)	1-lb. 2-oz. can		76.5			
CHIVES, raw (USDA)	1 oz.		<.1			0

CHOCOLATE, BAKING:

Bitter or unsweetened:

(USDA)	1 oz.	1	15.0	9.	6.	0
Grated (USDA)	½ cup (2.3 oz.)	3	35.0	20.	15.	0
(Baker's)	1-oz. sq.	4	14.3			0
Pre-melted, *Choco-Bake*	1-oz. packet	Tr.	13.5			
(Hershey's)	1 oz.	11	15.7			
Sweetened:						
Bittersweet (USDA)	1 oz.	<1	11.3	7.	5.	0
Chips, milk (Hershey's)	1 oz.	28	9.5			
Chips, semisweet (Baker's)	¼ cup (1.5 oz.)	9	10.5			0
Chips, semisweet (Ghirardelli)	⅓ cup (2 oz.)		16.3			
Chips, semisweet (Hershey's)	1 oz.	4	9.2			

(USDA): United States Department of Agriculture
*Prepared as Package Directs

Food and Description	Measure or Quantity	Sodium (mg.)	Total	Satu-rated	Unsatu-rated	Choles-terol (mg.)
German's, sweet (Baker's)	4½ sq. (1 oz.)	7	9.3			0
Morsels, milk (Nestlé's)	1 oz.	<1	8.3			
Morsels, semisweet (Nestlé's)	6-oz. pkg.	4	47.6			
Morsels, semisweet (Nestlé's)	1 oz.	<1	8.0			
Semisweet, small pieces (USDA)	½ cup (3 oz.)	2	30.3	17.	13.	
Semisweet (Baker's)	1-oz. sq.	1	8.9			0

CHOCOLATE CAKE:
Home recipe (USDA):
Without icing:

Food and Description	Measure or Quantity	Sodium (mg.)	Total	Satu-rated	Unsatu-rated	Choles-terol (mg.)
Made with butter[1]	3 oz.	250	14.6	8.	7.	
Made with vegetable shortening[2]	3 oz.	250	14.6	4.	10.	
With chocolate icing, 2-layer	¹/₁₆ of 10″ cake (4.2 oz.)	282	19.7			52
With chocolate icing, 2-layer	¹/₁₆ of 9″ cake (2.6 oz.)	176	12.3			32
With uncooked white icing	¹/₁₆ of 10″ cake (4.2 oz.)	281	17.5			
Fudge, frozen (Pepperidge Farm)	¹/₆ of cake (3 oz.)	299	15.0			
German chocolate, frozen (Morton)	2.2-oz. serving	243	12.6			
Golden, frozen (Pepperidge Farm)	¹/₆ of cake (3 oz.)	252	15.6			

CHOCOLATE CAKE MIX (See also **FUDGE CAKE MIX**):

Food and Description	Measure or Quantity	Sodium (mg.)	Total	Satu-rated	Unsatu-rated	Choles-terol (mg.)
Chocolate malt (USDA)[3]	1 oz.	156	3.0	1.	2.	
*Chocolate malt,[2] uncooked white icing (USDA)[4]	4 oz.	361	9.9			
*Chocolate malt layer (Betty Crocker)	¹/₁₂ of cake	264	5.8			
*Chocolate pudding (Betty Crocker)	¹/₆ of cake	338	4.9			
*Deep chocolate (Duncan Hines)	¹/₁₂ of cake (2.7 oz.)	483	6.1			50
*German chocolate layer (Betty Crocker)	¹/₁₂ of cake	286	5.8			
*German chocolate (Swans Down)	¹/₁₂ of cake (2.5 oz.)	376	3.6			
*Milk chocolate layer (Betty Crocker)	¹/₁₂ of cake	281	6.0			
*Swiss chocolate (Duncan Hines)	¹/₁₂ of cake (2.7 oz.)	405	6.1			50

(USDA): United States Department of Agriculture
*Prepared as Package Directs
[1]Principal sources of fat: butter, chocolate, egg & milk.
[2]Principal sources of fat: vegetable shortening, chocolate, egg & milk.
[3]Principal sources of fat: vegetable shortening, chocolate & milk.
[4]Prepared with eggs & water.

Food and Description	Measure or Quantity	Sodium (mg.)	— Fats in grams —			Choles- terol (mg.)
			Total	Satu- rated	Unsatu- rated	

CHOCOLATE CANDY (See **CANDY**)

CHOCOLATE DRINK:
Canned (Borden) 9½-fl.-oz. can 47 6.7
Mix:
  Hot (USDA)[1] 1 oz. 108 3.0 2. 1.
  Hot (USDA)[1] 1 cup (4.9 oz.) 531 14.7 8. 6.
  Dutch, instant (Borden) 2 heaping tsps. (¾ oz.) 5 .7
  Instant (Ghirardelli) 1 T. (.4 oz.) 2 .5
  Quik (Nestlé's) 2 heaping tsp. (.6 oz.) 32 .3

CHOCOLATE, GROUND
(Ghirardelli) ¼ cup (1.3 oz.) 46 3.8

CHOCOLATE, HOT, home
recipe (USDA) 1 cup (8.8 oz.) 120 12.5 8. 5. 31

CHOCOLATE ICE CREAM (See
also individual brands):
(Borden) 9.5% fat ¼ pt. (2.3 oz.) 31 6.3
(Prestige) French ¼ pt. (2.6 oz.) 36 10.9
(Sealtest) ¼ pt. (2.3 oz.) 42 6.4

CHOCOLATE PIE:
Chiffon, home recipe (USDA):
  Made with lard[2] ¹/₆ of 9″ pie (4.9 oz.) 353 21.4 8. 13.
  Made with vegetable shortening[3] ¹/₆ of 9″ pie (4.9 oz.) 353 21.4 6. 16.
Meringue, home recipe (USDA):
  Made with lard[2] ¹/₆ of 9″ pie (4.9 oz.) 358 16.8 6. 11.
  Made with vegetable shortening[3] ¹/₆ of 9″ pie (4.9 oz.) 358 16.8 4. 13.
Nut (Tastykake) 4½-oz. pie 18.4
Frozen:
  Cream:
    (Banquet) 2½-oz. serving 8.7
    (Mrs. Smith's) ¹/₆ of 8″ pie (2.3 oz.) 150 13.2
  Tart (Pepperidge Farm) 1 pie tart (3 oz.) 211 18.0
  Velvet nut (Kraft) ¹/₆ of 16¾-oz. pie 77 18.6

(USDA): United States Department of Agriculture
*Prepared as Package Directs
[1]Principal sources of fat: chocolate & milk.
[2]Principal sources of fat: lard & butter.
[3]Principal sources of fat: vegetable shortening & butter.

Food and Description	Measure or Quantity	Sodium (mg.)	Fats in grams — Total	Satu- rated	Unsatu- rated	Choles- terol (mg.)
CHOCOLATE PIE FILLING (See **CHOCOLATE PUDDING MIX**)						
CHOCOLATE PUDDING,						
sweetened:						
Home recipe with starch base (USDA)[1]	½ cup (4.6 oz.)	73	6.1	4.	2.	
Chilled:						
Dark chocolate (Breakstone)	5-oz. container	195	13.2			0
Light chocolate (Breakstone)	5-oz. container	145	13.5			0
(Sealtest)	4 oz.	128	3.5			
Canned:						
(Betty Crocker)	½ cup	239	5.2			
(Hunt's)[2]	5-oz. can	146	12.9	2.	10.	
(Thank You)	½ cup (4.5 oz.)		5.1			
Fudge (Betty Crocker)	½ cup	175	5.2			
Fudge (Del Monte)	5-oz. can	233	5.8			
Fudge (Hunt's)[2]	5-oz. can	150	12.9	3.	10.	
Milk chocolate (Del Monte)	5-oz. can	251	5.7			
CHOCOLATE PUDDING or PIE FILLING MIX:						
Sweetened:						
Regular:						
Dry (USDA)[3]	1 oz.	127	6.0	3.	3.	
*Prepared with milk (USDA)[1]	½ cup (4.6 oz.)	168	3.9	3.	1.	16
*(Jell-O)	½ cup (5.2 oz.)	167	5.1			13
*(Royal)	½ cup (5.1 oz.)	140	5.5			14
*(Royal) *Dark 'N' Sweet*	½ cup (5.1 oz.)	140	5.9			14
*Fudge (Jell-O)	½ cup (5.2 oz.)	167	5.1			13
*Milk chocolate (Jell-O)	½ cup (5.2 oz.)	167	5.1			13
Instant:						
Dry (USDA)	1 oz.	115	.5			
*Prepared with milk, without cooking (USDA)[1]	4 oz.	141	2.8	1.	2.	
*(Jell-O)	½ cup (5.4 oz.)	486	5.2			13
*(Royal)	½ cup (5.1 oz.)	320	5.5			14
*(Royal) *Dark 'N' Sweet*	½ cup (5.1 oz.)	320	5.5			14
*Fudge (Jell-O)	½ cup (5.4 oz.)	486	5.2			13
*Low calories or dietetic (D-Zerta)	½ cup (4.6 oz.)	72	4.8			13

(USDA): United States Department of Agriculture
*Prepared as Package Directs
[1]Principal sources of fat: milk & chocolate.
[2]Principal source of fat: soybean oil.
[3]Principal source of fat: chocolate.

Food and Description	Measure or Quantity	Sodium (mg.)	— Fats in grams —			Choles- terol (mg.)
			Total	Satu- rated	Unsatu- rated	
CHOCOLATE RENNET MIX:						
Powder:						
Dry (Junket)	1 oz.	26	.7			
*(Junket)	4 oz.	61	4.1			
Tablet:						
Dry (Junket)	1 tablet (<1 gram)	197	Tr.			
*& sugar (Junket)	4 oz.	98	3.9			
CHOCOLATE SOFT DRINK:						
Sweetened:						
(Clicquot Club) cream	6 fl. oz.	11	0.			0
(Cott) cream	6 fl. oz.	11	0.			0
(Hoffman) *Cocoa Cooler*	6 fl. oz.	14	0.			0
(Hoffman) cream	6 fl. oz.	14	0.			0
(Mission) cream	6 fl. oz.	11	0.			0
(Yukon Club) cream	6 fl. oz.	14	0.			0
Low calorie:						
(Clicquot Club)	6 fl. oz.	45	0.			0
(Cott)	6 fl. oz.	45	0.			0
(Hoffman)	6 fl. oz.	39	0.			0
(Mission)	6 fl. oz.	45	0.			0
(No-Cal)	6 fl. oz.	12	0.			0
(Shasta)	6 fl. oz.	37	0.			0
CHOCOLATE SYRUP:						
Sweetened:						
Fudge (USDA)[1]	1 fl. oz. (1.3 oz.)	34	5.2	3.	2.	
Fudge (USDA)[1]	1 T. (.7 oz.)	17	2.6	1.	1.	
Thin type (USDA)	1 fl. oz. (1.3 oz.)	20	.8	Tr.	Tr.	
Thin type (USDA)	1 T. (.7 oz.)	10	.4	Tr.	Tr.	
(Hershey's)	1 T. (1 oz.)	14	.3			
(Smucker's)	1 T. (.6 oz.)	8	.2	Tr.	Tr.	
Low calorie:						
(Slim-ette) *Chocotop*	1 T. (.5 oz.)		.2			
(Tillie Lewis)	1 T. (.5 oz.)		.2			
CHOCO-NUT SUNDAE CONE						
(Sealtest)	2½ fl. oz. (2.1 oz.)	57	9.9			
CHOP SUEY:						
Home recipe, with meat (USDA)[2]	1 cup (8.8 oz.)	1052	17.0	7.	10.	64

(USDA): United States Department of Agriculture
*Prepared as Package Directs
[1]Principal sources of fat: chocolate, animal & vegetable shortening & milk.
[2]Principal sources of fat: butter, beef & pork.

Food and Description	Measure or Quantity	Sodium (mg.)	Total	Satu-rated	Unsatu-rated	Choles-terol (mg.)
				— Fats in grams —		
Canned:						
With meat (USDA)[1]	1 cup (8.8 oz.)	1378	8.0	2.	6.	7
Chicken (Hung's)	8 oz.		5.6			
Meatless (Hung's)	8 oz.		6.0			
Mix:						
(Durkee)	1½-oz. pkg.	828	2.1			
*With meat & vegetables						
(Durkee)	3½ cups (1½-oz. pkg.)	1164	64.8			
Frozen:						
Beef (Banquet) cooking bag	7-oz. bag		4.2			
Beef (Banquet) buffet	2-lb. pkg.		19.3			
Beef dinner (Banquet):						
Meat compartment	7 oz.		5.1			
Rice compartment	4 oz.		2.8			
Complete dinner	11-oz. dinner		7.9			

CHOP SUEY VEGETABLES (See
VEGETABLE, MIXED)

CHOW CHOW:

Sour (USDA)	1 oz.	379	.4			0
Sweet (USDA)	1 oz.	149	.3			0
(Crosse & Blackwell)	1 T. (.6 oz.)	254	0.			(0)

CHOW MEIN (See also
CHINESE DINNER):

Home recipe, chicken, without						
noodles (USDA)[2]	4 oz.	325	4.5	1.	3.	
Canned:						
Beef (Chun King) *Divider-Pak*	7-oz. serving (¼ can)		5.0			
Chicken:						
(USDA) without noodles	4 oz.	329	.1			
(Chun King) *Divider-Pak*	7-oz. serving (¼ can)		3.0			
(Hung's)	8 oz.		4.6			
Meatless (Hung's)	8 oz.		5.0			
Pork (Chun King) *Divider-Pak*	7-oz. serving (¼ can)		9.0			
Shrimp (Chun King) *Divider-Pak*	7-oz. serving (¼ can)		1.0			
Frozen:						
Beef (Chun King)	7½-oz. serving					
	(½ pkg.)		3.0			
Chicken:						
(Banquet) cooking bag	7-oz. bag		3.4			

(USDA): United States Department of Agriculture
*Prepared as Package Directs
[1]Principal sources of fat: pork, beef & corn oil.
[2]Principal sources of fat: chicken, corn oil & soybeans.

Food and Description	Measure or Quantity	Sodium (mg.)	— Fats in grams — Total	Satu- rated	Unsatu- rated	Choles- terol (mg.)
(Banquet) buffet	2-lb. pkg.		15.8			
(Chun King)	7½-oz serving					
	(½ pkg.)		4.0			
With rice (Swanson)	8½-oz. pkg.	1080	4.7			
Shrimp (Chun King)	7½-oz. serving					
	(½ pkg.)		2.0			

CHOW MEIN NOODLES (See **NOODLES, CHOW MEIN**)

CHOW MEIN VEGETABLES (See **VEGETABLES, MIXED**)

CHUB, raw (USDA):

Whole	1 lb. (weighed whole)		13.2			
Meat only	4 oz.		10.0			

CHUTNEY, *Major Grey's* (Crosse & Blackwell)

Blackwell)	1 T. (.8 oz.)	294	0.			

CIDER (See **APPLE CIDER**)

CINNAMON:

Ground (Information supplied by General Mills Laboratory)	1 oz.		1.0			(0)
Stick (Spice Islands)	1 stick	<1				(0)
With sugar (French's)	1 tsp. (4 grams)		Tr.			(0)

CINNAMON STICKS, frozen (Aunt Jemima)

Jemima)	3 pieces (1¾ oz.)	283	5.2			

CISCO (See **LAKE HERRING**)

CITRON, CANDIED (USDA)	1 oz.	82	<.1			0
CITRUS COOLER (Hi-C)	6 fl. oz. (6.3 oz.)	Tr.	Tr.			0

CITRUS SOFT DRINK, low calorie:

Flair, sugar-free	6 fl. oz.	52+	0.			0
(No-Cal)	6 fl. oz.	11	0.			0
CLACKERS, cereal (General Mills)	1 cup (1 oz.)	360	2.3			(0)

(USDA): United States Department of Agriculture
*Prepared as Package Directs

Food and Description	Measure or Quantity	Sodium (mg.)	— Fats in grams —			Cholesterol (mg.)
			Total	Saturated	Unsaturated	

CLAM:

Food and Description	Measure or Quantity	Sodium (mg.)	Total	Saturated	Unsaturated	Cholesterol (mg.)
Raw, all kinds, meat & liq. (USDA)	4 oz.		1.0			
Raw, all kinds, meat only (USDA)	4 med. clams (3 oz.)	102	1.4			42
Raw, hard or round (USDA):						
Meat & liq.	1 lb. (weighed in shell)		.6			
Meat only	1 cup (7 round chowders, 8 oz.)	465	2.0			114
Raw, soft (USDA):						
Meat & liq.	1 lb. (weighed in shell)		2.6			
Meat only	1 cup (19 large, 8 oz.)	82	4.3			114
Canned, all kinds:						
Solids & liq. (USDA)	4 oz.		.8			
Meat only (USDA)	½ cup (2.8 oz.)		2.0			50
Chopped (Snow)	4 oz.		.2			
Creamed, with mushrooms (Snow)	4 oz.		6.6			
Minced (Snow)	4 oz.		.3			
Frozen, fried (Mrs. Paul's)	4 oz.		23.6			

CLAM CAKE, frozen, thins

Food and Description	Measure or Quantity	Sodium (mg.)	Total	Saturated	Unsaturated	Cholesterol (mg.)
(Mrs. Paul's)	10-oz. pkg.		23.5			

CLAM CHOWDER:

Food and Description	Measure or Quantity	Sodium (mg.)	Total	Saturated	Unsaturated	Cholesterol (mg.)
Manhattan, canned:						
Condensed (USDA)	8 oz. (by wt.)	1739	4.8			
*Prepared with equal volume water (USDA)	1 cup (8.6 oz.)	938	2.4			
*(Campbell)	1 cup	878	2.4	Tr.	2.	
(Campbell) *Chunky*	1 cup	1055	3.2			
(Crosse & Blackwell)	½ can (6½ oz.)	935	1.5			
*(Doxsee)	1 cup (8.6 oz.)	926	2.7			22
(Heinz) *Great American*	1 cup (8½ oz.)	1293	4.5			
(Snow)	8 oz.	631	4.0	<1.	3.	
New England:						
Canned:						
*(Campbell)	1 cup	930	6.7			
(Crosse & Blackwell)	½ can (6½ oz.)	935	3.5			
*(Doxsee)	1 cup (8.6 oz.)	982	7.3			52
(Snow)	8 oz.	1368	4.1	<1.	3.	
Frozen:						
Condensed (USDA)	8 oz. (by wt.)	1975	14.5			

(USDA): United States Department of Agriculture
*Prepared as Package Directs

Food and Description	Measure or Quantity	Sodium (mg.)	—Fats in grams— Total	Satu- rated	Unsatu- rated	Choles- terol (mg.)
*Prepared with equal volume water (USDA)	1 cup (8.5 oz.)	1044	7.7			
*Prepared with equal volume milk (USDA)	1 cup (8.6 oz.)	1129	12.2			
CLAM FRITTERS, home recipe,[1] (USDA)	1 fritter (2″ x 1¾″, 1.4 oz.)		6.0			52
CLAM JUICE/LIQUOR, canned:						
(USDA)	1 cup (8.3 oz.)		.2			
(Snow)	8 oz.		.2			
CLAM STEW, New England (Snow)	8 oz.		10.3			
CLAM STICKS, frozen (Mrs. Paul's)	4 oz.		7.7			
CLAMATO COCKTAIL (Mott's)	4 oz.		.3			
CLARET WINE (Gold Seal)						
12% alcohol	3 fl. oz. (3.1 oz.)	3	0.			(0)
CLOVE:						
Ground (Spice Islands)	1 tsp.	<1				(0)
Whole (Spice Islands)	1 clove	<1				(0)
CLUB SODA SOFT DRINK:						
Regular:						
(Dr. Brown's)	6 fl. oz.	29	0.			0
(Fanta)	6 fl. oz.	39	0.			0
(Hoffman)	6 fl. oz.	29	0.			0
(Key food)	6 fl. oz.	29	0.			0
(Nedick's)	6 fl. oz.	29	0.			0
(Schweppes)	6 fl. oz.	26	0.			0
(Shasta)	6 fl. oz.	98	0.			0
(Waldbaum)	6 fl. oz.	29	0.			0
(Yukon Club)	6 fl. oz.	29	0.			0
Dietetic:						
(Dr. Brown's)	6 fl. oz.	2	0.			0
(Hoffman)	6 fl. oz.	2	0.			0
(Key Food)	6 fl. oz.	2	0.			0
(Waldbaum)	6 fl. oz.	2	0.			0

(USDA): United States Department of Agriculture
*Prepared as Package Directs
[1]Prepared with flour, baking powder, butter & eggs.

Food and Description	Measure or Quantity	Sodium (mg.)	— Fats in grams —			Cholesterol (mg.)
			Total	Saturated	Unsaturated	
COCOA, dry:						
Plain (USDA):						
Low fat	½ cup (1.5 oz.)	3	3.4	2.	2.	0
Low fat	1 T. (5 grams)	<1	.4	Tr.	Tr.	0
Medium-low fat	½ cup (1.5 oz.)	3	5.5	3.	2.	0
Medium-low fat	1 T. (5 grams)	<1	.7	Tr.	Tr.	0
Medium-high fat	½ cup (1.5 oz.)	3	8.2	5.	3.	0
Medium-high fat	1 T. (5 grams)	<1	1.0	<1.	Tr.	0
High-fat	½ cup (1.5 oz.)	3	10.2	6.	5.	0
High-fat	1 T. (5 grams)	<1	1.3	<1.	<1.	0
Processed with alkali (USDA):						
Medium-low fat	½ cup (1.5 oz.)	308	5.5	3.	2.	0
Medium-low fat	1 T. (5 grams)	39	.7	Tr.	Tr.	0
Medium-high fat	½ cup (1.5 oz.)	308	8.2	5.	3.	0
Medium-high fat	1 T. (5 grams)	39	1.0	<1.	Tr.	0
High-fat	½ cup (1.5 oz.)	308	10.2	6.	5.	0
High-fat	1 T. (5 grams)	39	1.3	<1.	<1.	0
Unsweetened (Droste)	1 T. (7 grams)		1.7			(0)
(Hershey's)	½ cup (1.5 oz.)	32	7.1			(0)
(Hershey's)	1-oz. packet	21	4.7			(0)
(Hershey's)	1 T. (5 grams)	4	.9			(0)
COCOA, HOME RECIPE (USDA)	1 cup (8.8 oz.)	128	11.5	8.	4.	35
COCOA KRISPIES, cereal (Kellogg's)	1 cup (1 oz.)	167	.7			(0)
COCOA MIX:						
With nonfat dry milk (USDA)[1]	1 oz.	149	.8	<1.	Tr.	
Without nonfat dry milk (USDA)[2]	1 oz.	76	.6	Tr.	Tr.	0
(Kraft)	1 oz.	73	.5			
*(Kraft)	1 cup	91	.7			
(Nestlé's) *EverReady*	3 heaping tsp. (.8 oz.)	84	1.			
Hot (Hershey's)	1 oz.	105	2.4			
Hot (Nestlé's)	1-oz. pkg.	139	.3			
Instant:						
(Hershey's)	1 oz.	60	.6			
(Swiss Miss)	1 oz.		.4			
Chocolate marshmallow (Carnation)[3]	1 pkg. (1 oz.)	122	1.6	1.	Tr.	<1
Milk chocolate (Carnation)	1 pkg. (1 oz.)	163	2.6	2.	Tr.	<1
Rich chocolate (Carnation)[3]	1 pkg. (1 oz.)	121	1.6	1.	Tr.	<1

(USDA): United States Department of Agriculture
*Prepared as Package Directs
[1]Principal sources of fat: cocoa & milk.
[2]Principal source of fat: cocoa.
[3]Principal sources of fat: milk, cocoa & modified coconut oil.

Food and Description	Measure or Quantity	Sodium (mg.)	—Fats in grams—			Choles-terol (mg.)
			Total	Satu-rated	Unsatu-rated	
COCOA PEBBLES, cereal (Post)	⅞ cup (1 oz.)	125	.1			0
COCOA PUFFS, cereal (General Mills)	1 cup (1 oz.)	184	.5			(0)
COCONUT:						
Fresh (USDA):						
Whole	1 lb. (weighed in shell)	54	83.3	72.	11.	0
Meat only	4 oz.	26	40.0	34.	6.	0
Meat only	2″ x 2″ x ½″ piece (1.6 oz.)	10	15.9	14.	2.	0
Grated or shredded	1 firmly packed cup (4.6 oz.)	30	45.9	39.	7.	0
Grated	1 lightly packed cup (2.9 oz.)	18	28.2	24.	4.	0
Cream, liq. expressed from grated coconut	4 oz.	5	36.5	32.	5.	0
Milk, liq. expressed from mixture of grated coconut & water	4 oz.		28.2	25.	3.	0
Water, liq. from coconut	1 cup (8.5 oz.)	60	.5			0
Dried, canned or packaged:						
Sweetened, shredded (USDA)	½ lightly packed cup (1.6 oz.)		18.0	16.	2.	0
Unsweetened (USDA)	½ lightly packed cup (1.6 oz.)		30.0	26.	4.	0
Angel Flake (Baker's)	½ cup (1.3 oz.)	86	11.8			0
Cookie (Baker's)	½ cup (2 oz.)	136	18.6			0
Crunchies (Baker's)	½ cup (2.1 oz.)	144	30.2			0
Premium shred (Baker's)	½ cup (1.5 oz.)	100	13.8			0
Southern-style (Baker's)	½ cup (1.5 oz.)		13.4			0
COCONUT CAKE, frozen (Pepperidge Farm)	⅙ of cake (3 oz.)	252	15.1			
*COCONUT CAKE MIX (Duncan Hines)	¹⁄₁₂ of cake (2.7 oz.)	377	6.1			50
COCONUT PIE:						
Cream:						
(Tastykake)	4-oz. pie		27.5			
Frozen:						
(Banquet)	2½-oz. serving		11.6			

(USDA): United States Department of Agriculture
*Prepared as Package Directs

Food and Description	Measure or Quantity	Sodium (mg.)	Total	Fats in grams Satu- rated	Unsatu- rated	Choles- terol (mg.)
(Morton)	¼ of 14.4-oz. pie	189	15.4			
(Mrs. Smith's)	¹/₆ of 8″ pie (2.3 oz.)	132	12.3			
Tart (Pepperidge Farm)	3-oz. pie tart	206	20.1			
Custard:						
Home recipe (USDA)	¹/₆ of 9″ pie (5.4 oz.)	375	19.0			
Frozen:						
Baked (USDA)	5 oz.	358	17.0			
Unbaked (USDA)	5 oz.	338	12.1			
(Banquet)	5-oz. serving		12.0			
(Morton)	¹/₆ of 20-oz. pie	219	9.1			
(Morton)	⅛ of 46-oz. pie	394	30.0			
(Mrs. Smith's)	¹/₆ of 8″ pie (4 oz.)	338	13.2			
(Mrs. Smith's)	⅛ of 10″ pie (5.4 oz.)	438	17.0			

COCONUT PIE FILLING MIX (See also **COCONUT PUDDING MIX**):

Custard & pie crust, dry (USDA)[1]	1 oz.	178	5.7	2.	3.	
*Custard made with egg yolk & milk (USDA)[2]	5 oz. (including crust)	334	11.2	4.	7.	

COCONUT PUDDING MIX:

*Cream, regular (Jell-O)	½ cup (5.2 oz.)	205	6.0			13
*Cream, instant (Jell-O)	½ cup (5.3 oz.)	313	6.1			13
*Toasted, instant (Royal)	½ cup (5.1 oz.)	385	6.1			14

COCO WHEATS, cereal | 2 T. (.6 oz.) | 1 | .3 | | | (0)

COD:

Raw, whole (USDA)	1 lb. (weighed whole)	98	.4			70
Raw, meat only (USDA)	4 oz.	79	.3			57
Raw, meat rinsed in brine (USDA)	4 oz.	289	.3			57
Broiled (USDA)	4 oz.	125	6.0			
Canned (USDA)	4 oz.		.3			
Dehydrated, lightly salted (USDA)	4 oz.	9185	3.2			
Dried, salted (USDA)	4 oz.		.8			93

(USDA): United States Department of Agriculture
*Prepared as Package Directs
[1]Principal sources of fat: vegetable shortening & coconut.
[2]Principal sources of fat: vegetable shortening, coconut, milk & egg.

| Food and Description | Measure or Quantity | Sodium (mg.) | — Fats in grams — | | | Choles- terol (mg.) |
			Total	Satu- rated	Unsatu- rated	
Dried, salted (USDA)	5½" x 1½" x ½" (2.8 oz.)		.6			66
Frozen (Gorton)	⅓ of 1-lb. pkg.	107	.5			

CODFISH CAKE (See **FISH CAKE**)

COFFEE:
  Regular:
  *(Chase & Sanborn) | ¾ cup | 1 | Tr. | | | 0
  Max Pax | ¾ cup | 1 | Tr. | | | 0
  *(Maxwell House) | ¾ cup | Tr. | Tr. | | | 0
  *(Yuban) | ¾ cup | Tr. | Tr. | | | 0

COFFEE:						
Regular:						
*(Chase & Sanborn)	¾ cup	1	Tr.			0
Max Pax	¾ cup	1	Tr.			0
*(Maxwell House)	¾ cup	Tr.	Tr.			0
*(Yuban)	¾ cup	Tr.	Tr.			0
Instant:						
Dry (USDA)	1 oz.	20	Tr.			0
Dry (USDA)	1 rounded tsp. (2 grams)	2	Tr.			0
*(USDA)	1 cup (8.4 oz.)	2	Tr.			0
(Borden)	1 rounded tsp. (2 grams)		<.1			(0)
*(Chase & Sanborn)	¾ cup	1	Tr.			0
Kava (Borden)	1 tsp. (1 gram)	<1	<.1			(0)
*(Maxwell House)	¾ cup	Tr.	Tr.			0
*(Yuban)	¾ cup	Tr.	Tr.			0
Decaffeinated:						
Decaf	1 tsp. (2 grams)	Tr.	0.			(0)
Sanka regular	¾ cup	Tr.	Tr.			0
Sanka instant	¾ cup	Tr.	Tr.			0
Siesta	¾ cup	1	Tr.			0
Freeze-dried:						
Maxim	¾ cup	Tr.	Tr.			0
Sanka	¾ cup	Tr.	Tr.			0
COFFEE CAKE:						
Butterfly (Mrs. Smith's)	1 piece (2¾ oz.)	175	14.2			
Cherry (Mrs. Smith's)	1 piece (2¾ oz.)	180	14.2			
Cinnamon-raisin (Mrs. Smith's)	1 piece (2¾ oz.)	145	11.3			
Cinnamon twist (Pepperidge Farm)	⅙ cake (1.8 oz.)	208	6.6			
Danish, apple, frozen (Morton)	1 cake (13.5 oz.)	1243	47.4			
Danish pastry, without fruit or nuts:						
Individual round (USDA)[1]	1 piece (2.3 oz.)	238	15.3	5.	11.	
Packaged ring (USDA)[1]	12-oz. cake	1244	79.9	24.	56.	

(USDA): United States Department of Agriculture
*Prepared as Package Directs
[1]Principal sources of fat: vegetable shortening, butter & egg.

Food and Description	Measure or Quantity	Sodium (mg.)	Total	— Fats in grams — Satu-rated	Unsatu-rated	Choles-terol (mg.)
Danish pecan twist, frozen						
(Morton)	12-oz. cake	1539	80.6			
Melt-A-Way, frozen (Morton)	13-oz. cake	1625	79.8			
Meltaway (Mrs. Smith's)	1 piece (2¾ oz.)	185	26.9			
Pecan roll (Mrs. Smith's)	1 piece (2¾ oz.)	135	19.8			
COFFEE CAKE MIX:						
Dry (USDA)[1]	1 oz.	174	3.1	<1	3.	
*Prepared with egg & milk (USDA)[2]	2 oz.	244	5.4	1.	4.	
*(Aunt Jemima)	⅛ of cake (1.8 oz.)	211	5.8			
COFFEE SOFT DRINK:						
Sweetened (Hoffman)	6 fl. oz.	14	0.			0
Low calorie (Hoffman)	6 fl. oz.	32	0.			0
Low calorie (No-Cal)	6 fl. oz.	20	0.			0
***COFFEE SOUTHERN,** liqueur,*						
55 proof	1 fl. oz.	Tr.	0.			0
COLA SOFT DRINK:						
Sweetened:						
(Canada Dry) Jamaica	6 fl. oz.	0+	0.			0
(Clicquot Club)	6 fl. oz.	11	0.			0
Coca-Cola	6 fl. oz.	Tr.+	0.			0
(Cott)	6 fl. oz.	11	0.			0
(Dr. Brown's)	6 fl. oz.	3	0.			0
(Hoffman)	6 fl. oz.	3	0.			0
(Key Food)	6 fl. oz.	3	0.			0
(Kirsch)	6 fl. oz.	<1	0.			0
(Mission)	6 fl. oz.	11	0.			0
(Nedick's)	6 fl. oz.	3	0.			0
Pepsi-Cola	6 fl. oz.	Tr.+	0.			0
RC with a twist (Royal Crown)	6 fl. oz. (6.5 oz.)	3+	0.			0
(Royal Crown)	6 fl. oz. (6.5 oz.)	3+	0.			0
(Shasta)	6 fl. oz.	10	0.			0
(Waldbaum)	6 fl. oz.	3	0.			0
(Yukon Club)	6 fl. oz.	3	0.			0
Cherry (Key Food)	6 fl. oz.	14	0.			0
Cherry (Shasta)	6 fl. oz.	10	0.			0
Low calorie:						
(Canada Dry)sugar-free	6 fl. oz.	8+	0.			0

(USDA): United States Department of Agriculture
*Prepared as Package Directs
[1]Principal source of fat: vegetable shortening.
[2]Principal sources of fat: vegetable shortening, egg & milk.

Food and Description	Measure or Quantity	Sodium (mg.)	Fats in grams Total	Satu- rated	Unsatu- rated	Choles- terol (mg.)
(Clicquot Club)	6 fl. oz.	38	0.			0
(Cott)	6 fl. oz.	38	0.			0
Diet Pepsi-Cola, sugar-free	6 fl. oz.	31+	0.			0
Diet Rite, sugar-free	6 fl. oz. (6.4 oz.)	29+	0.			0
(Dr. Brown's)	6 fl. oz.	40	0.			0
(Hoffman)	6 fl. oz.	40	0.			0
(Key Food)	6 fl. oz.	40	0.			0
(Mission)	6 fl. oz.	38	0.			0
(No-Cal)	6 fl. oz.	12	0.			0
(Shasta)	6 fl. oz.	37	0.			0
RC Cola, sugar-free	6 fl. oz.	29+	0.			0
Tab	6 fl. oz.	13	0.			0
(Waldbaum)	6 fl. oz.	40	0.			0
(Yukon Club)	6 fl. oz.	57	0.			0
Cherry (Shasta)	6 fl. oz.	37	0.			0
COLA SYRUP, low calorie (No-Cal)	1 tsp. (5 grams)	<1	0.			0
COLESLAW, not drained (USDA):						
Prepared with commercial French dressing[1]	4 oz.	304	8.3	1.	7.	
Prepared with homemade French dressing, using corn oil[2]	4 oz.	149	13.9	1.	13.	
Prepared with homemade French dressing, using cottonseed oil[3]	4 oz.	149	13.9	3.	10.	
Prepared with mayonnaise[2]	4 oz.	136	15.9	2.	14.	
Prepared with mayonnaise-type salad dressing[2]	4 oz.	141	9.0	1.	8.	
COLLARDS:						
Raw (USDA):						
Leaves, including stems	1 lb.	195	3.2			0
Leaves only	½ lb.		1.2			0
Boiled without salt, drained (USDA):						
Leaves, cooked in large amount of water	½ cup (3.4 oz.)		.7			0
Leaves & stems, cooked in small amount of water	4 oz.	28	.7			0
Leaves, cooked in small amount water	½ cup (3.4 oz.)		.7			0

(USDA): United States Department of Agriculture
*Prepared as Package Directs
[1]Principal sources of fat: soybean oil, cottonseed oil & corn oil.
[2]Principal source of fat: corn oil.
[3]Principal source of fat: cottonseed oil.

Food and Description	Measure or Quantity	Sodium (mg.)	Total	—Fats in grams— Satu- rated	Unsatu- rated	Choles- terol (mg.)
Frozen:						
Not thawed (USDA)	10-oz. pkg.	51	1.1			0
Boiled, chopped, drained (USDA)	½ cup (3 oz.)	14	.3			0
Chopped (Birds Eye)	⅓ of pkg. (3.3 oz.)	17	.4			0
COLLINS MIX (Bar-Tender's)	1 serving (⅝ oz.)	32	.2			(0)
CONCENTRATE, cereal (Kellogg's)	⅓ cup (1 oz.)	82	.1			
CONCORD WINE:						
(Gold Seal) 13–14% alcohol	3 fl. oz. (3.3 oz.)	3	0.			(0)
(Pleasant Valley) red, 12.5% alcohol	3 fl. oz.	23	0.			0
CONSOMME, canned, dietetic pack (Slim-ette)	8 oz. (by wt.)	5	Tr.			
CONSOMME MADRILENE, canned, clear or red (Crosse & Blackwell)	½ can (6½ oz.)		2.4			
COOKIE, COMMERCIAL: The following are listed by type or brand name:						
Almond crescent (Nabisco)	1 piece (7 grams)	23	1.4			
Almond toast, Mandel (Stella D'oro)	1 piece (.5 oz.)		.7			
Angelica Goodies (Stella D'oro)	1 piece (.8 oz.)		3.9			
Anginetti (Stella D'oro)	1 piece (5 grams)		1.7			
Animal Cracker:						
(USDA)	1 oz.	86	2.7			
(Nabisco) *Barnum's*	1 piece (3 grams)	12	.3			
(Sunshine) regular	1 piece (2 grams)		.3			
(Sunshine) iced	1 piece (5 grams)		1.2			
Anisette sponge (Stella D'oro)	1 piece (.5 oz.)		.8			
Anisette toast (Stella D'oro)	1 piece (.4 oz.)		.5			
Applesauce (Sunshine) regular or iced	1 piece (.6 oz.)		3.8			
Arrowroot (Sunshine)	1 piece (4 grams)	12	.4			
Assortment:						
(USDA)	1 oz.	103	5.7			
(Stella D'oro) *Lady Stella*	1 piece (8 grams)		1.6			
(Sunshine) *Lady Joan*	1 piece (9 grams)		1.9			
(Sunshine) *Lady Joan,* iced	1 piece (.4 oz.)		2.4			

(USDA): United States Department of Agriculture
*Prepared as Package Directs

Food and Description	Measure or Quantity	Sodium (mg.)	Total	Fats in grams Satu- rated	Unsatu- rated	Choles- terol (mg.)
Aunt Sally, iced (Sunshine)	1 piece (.8 oz.)		1.6			
Bana-Bee (Nabisco)	6 pieces					
	(1¾-oz. pkg.)	210	12.4			
Big Treat (Sunshine)	1 piece (1.3 oz.)		5.0			
Bordeaux (Pepperidge Farm)	1 piece (8 grams)	23	1.6			
Breakfast Treats (Stella D'oro)	1 piece (.8 oz.)		3.7			
Brown edge wafers (Nabisco)	1 piece (6 grams)	20	1.2			
Brownie:						
(Hostess) 2 to pkg.	1 piece (.9 oz.)	41	4.0			
(Tastykake)	1 pkg. (2¼ oz.)		10.3			
Chocolate nut (Pepperidge Farm)	1 piece (.4 oz.)	22	3.5			
Peanut butter (Tastykake)	1 pkg. (1¾ oz.)		10.2			
Pecan fudge (Keebler)	1 piece (.9 oz.)	47	5.7			
Frozen, with nuts & chocolate icing (USDA)[1]	1 oz.	57	5.8	1.	4.	
Brussels (Pepperidge Farm)	1 piece (8 grams)	21	2.4			
Butter:						
Thin, rich (USDA)[2]	1 oz.	119	4.8	3.	2.	
(Nabisco)	1 piece (5 grams)	17	.9			
(Sunshine)	1 piece (5 grams)		.9			
Buttercup (Keebler)	1 piece (5 grams)	30	1.0			
Butterscotch Fudgies (Tastykake)	1 pkg. (1¾ oz.)		10.4			
Capri (Pepperidge Farm)	1 piece (.6 oz.)	38	4.6			
Cardiff (Pepperidge Farm)	1 piece (4 grams)	8	.8			
Cherry Coolers (Sunshine)	1 piece (6 grams)		1.1			
Chinese almond (Stella D'oro)	1 piece (1.2 oz.)		9.2			
Chocolate & chocolate-covered:						
(USDA)[3]	1 oz.	39	4.5	1.	3.	
Como (Stella D'oro)	1 piece (1.1 oz.)		8.9			
Creme (Wise)	1 piece (7 grams)	28	1.3			
Peanut bars (Nabisco) *Ideal*	1 piece (.6 oz.)	66	5.4			
Pinwheels (Nabisco)	1 piece (1.1 oz.)	30	5.8			
Snaps (Nabisco)	1 piece (4 grams)	14	.7			
Snaps (Sunshine)	1 piece (3 grams)		.5			
Wafers (Nabisco) *Famous*	1 piece (6 grams)	29	.7			
Chocolate chip:						
(USDA)[3]	1 oz.	114	6.0	1.	5.	
(Keebler) old fashioned	1 piece (.6 oz.)	60	3.7			
(Nabisco)	1 piece (7 grams)	19	1.6			
(Nabisco) *Chips Ahoy*	1 piece (.4 oz.)	31	2.1			

(USDA): United States Department of Agriculture
*Prepared as Package Directs
[1]Principal sources of fat: vegetable shortening, egg, milk, nuts & chocolate.
[2]Principal sources of fat: butter, egg & milk.
[3]Principal sources of fat: vegetable shortening, egg & milk.

Food and Description	Measure or Quantity	Sodium (mg.)	Total	Fats in grams — Satu- rated	Unsatu- rated	Choles- terol (mg.)
(Nabisco) *Family Favorites*	1 piece (7 grams)	19	.9			
(Nabisco) Snaps	1 piece (4 grams)	16	.7			
(Pepperidge Farm)	1 piece (.4 oz.)	23	2.9			
(Sunshine) *Chip-A-Roos*	1 piece (.4 oz.)		2.9			
(Tastykake) *Choc-O-Chip*	4 pieces (1¾-oz. pkg.)		14.4			
Cinnamon:						
Crisp (Keebler)	1 piece (4 grams)	27	.6			
Spice, vanilla sandwich						
(Nabisco) *Crinkles*	6 pieces (1⅝-oz. pkg.)	202	9.6			
Sugar (Pepperidge Farm)	1 piece (.4 oz.)	31	2.4			
Toast (Sunshine)	1 piece (3 grams)		.3			
Coconut:						
Bar (USDA)[1]	1 oz.	42	6.9	2.	5.	
Bar (Nabisco)	1 piece (9 grams)	34	.2			
Bar (Sunshine)	1 piece (.4 oz.)		2.3			
(Nabisco) *Family Favorites*	1 piece (3 grams)	11	.7			
Chocolate chip (Nabisco)	1 piece (.5 oz.)	49	4.1			
Chocolate chip (Sunshine)	1 piece (.6 oz.)		4.3			
Chocolate drop (Keebler)	1 piece (.5 oz.)	41	4.2			
Coconut Kiss (Tastykake)	4 pieces (1¾-oz. pkg.)		19.2			
Commodore (Keebler)	1 piece (.5 oz.)	74	2.3			
Como Delight (Stella D'oro)	1 piece (1.1 oz.)		7.9			
Cowboys and Indians (Nabisco)	1 piece (2 grams)	9	.2			
Cream Lunch (Sunshine)	1 piece (.4 oz.)		1.4			
Creme Wafer Stick (Nabisco)	1 piece (9 grams)	9	2.8			
Cup Custard (Sunshine):						
Chocolate	1 piece (.5 oz.)		3.3			
Vanilla	1 piece (.5 oz.)		3.3			
Devil's Food Cake (Nab)	2 pieces (1¼-oz. pkg.)	83	2.2			
Devil's Food Cake (Nabisco)	1 piece (.5 oz.)	31	.8			
Dixie Vanilla (Sunshine)	1 piece (.5 oz.)		1.8			
Dresden (Pepperidge Farm)	1 piece (.6 oz.)	33	4.6			
Egg Jumbo (Stella D'oro)	1 piece (.4 oz.)		.7			
Fig bar:						
(USDA)	1 oz.	71	1.6	Tr.	1.	
(Keebler)	1 piece (.7 oz.)	84	1.2			
(Nab) *Fig Newtons*	1 piece (1-oz. pkg.)	97	2.2			
(Nab) *Fig Newtons*	2 pieces (2-oz. pkg.)	193	4.3			

(USDA): United States Department of Agriculture
*Prepared as Package Directs
[1]Principal sources of fat: butter, egg & coconut.

Food and Description	Measure or Quantity	Sodium (mg.)	— Fats in grams —			Choles- terol (mg.)
			Total	Satu- rated	Unsatu- rated	
(Nabisco) *Fig Newtons*	1 piece (.6 oz.)	53	1.2			
(Sunshine)	1 piece (.4 oz.)		.8			
Fruit, iced (Nabisco)	1 piece (.6 oz.)	77	1.5			
Fudge:						
(Sunshine)	1 piece (.5 oz.)		3.7			
Chip (Pepperidge Farm)	1 piece (.4 oz.)	33	2.5			
Fudge Stripes (Keebler)	1 piece (.4 oz.)	36	2.7			
Gingersnap:						
(USDA)[1]	1 oz.	162	2.5	Tr.	2.	
(USDA) crumbs[1]	1 cup (4.1 oz.)	657	10.2	2.	8.	
(Keebler)	1 piece (6 grams)	88	.6			
(Nabisco) old fashion	1 piece (7 grams)	41	.7			
(Sunshine)	1 piece (6 grams)		.6			
Zu Zu (Nabisco)	1 piece (4 grams)	17	.4			
Golden Bars (Stella D'oro)	1 piece (1 oz.)		5.9			
Golden Fruit (Sunshine)	1 piece (.7 oz.)		.6			
Graham Cracker (See **CRACKERS, GRAHAM**)						
Hermit bar, frosted (Tastykake)	1 pkg. (2 oz.)		7.2			
Home Plate (Keebler)	1 piece (.5 oz.)	58	1.6			
Hostest With The Mostest (Stella D'oro)	1 piece (8 grams)		1.8			
Hydrox (Sunshine):						
Regular or mint	1 piece (.4 oz.)		2.2			
Vanilla	1 piece (.4 oz.)		2.3			
Jan Hagel (Keebler)	1 piece (10 grams)	51	1.6			
Keebies (Keebler)	1 piece (.4 oz.)	44	2.3			
Ladyfingers (USDA)[1]	.4-oz. ladyfinger (3¼" x 1⅜" x 1⅛")	8	.9	Tr.	<1.	40
Lemon:						
(Sunshine)	1 piece (.5 oz.)		3.7			
Jumble rings (Nabisco)	1 piece (.5 oz.)	46	2.3			
Lemon Coolers (Sunshine)	1 piece (6 grams)		1.1			
Nut crunch (Pepperidge Farm)	1 piece (.4 oz.)	23	3.2			
Snaps (Nabisco)	1 piece (4 grams)	12	.4			
Lido (Pepperidge Farm)	1 piece (.6 oz.)	32	5.3			
Lisbon (Pepperidge Farm)	1 piece (5 grams)	13	1.5			
Macaroon:						
(USDA)[2]	1 oz.	10	6.6	5.	2.	
Almond (Tastykake)	2-oz. pkg. (2 pieces)		20.3			
Coconut (Nabisco) *Bake Shop*	1 piece (.7 oz.)	17	4.0			

(USDA): United States Department of Agriculture
*Prepared as Package Directs
[1]Principal sources of fat: vegetable shortening, egg & milk.
[2]Principal sources of fat: coconut & almonds.

Food and Description	Measure or Quantity	Sodium (mg.)	— Fats in grams —			Cholesterol (mg.)
			Total	Saturated	Unsaturated	
Sandwich (Nabisco)	1 piece (.5 oz.)	29	3.4			
Margherite, chocolate (Stella D'oro)	1 piece (.6 oz.)		3.0			
Margherite, vanilla (Stella D'oro)	1 piece (.6 oz.)		3.0			
Marquisette (Pepperidge Farm)	1 piece (8 grams)	18	2.6			
Marshmallow:						
(USDA)	1 oz.	59	3.7			
Fancy Crests (Nabisco)	1 piece (.5 oz.)	29	1.0			
Mallowmars (Nabisco)	1 piece (.5 oz.)	19	2.5			
Mallo Puff (Sunshine)	1 piece (.6 oz.)		1.6			
Minarets (Nabisco)	1 piece (10 grams)	14	2.4			
Puffs (Nabisco)	1 piece (.7 oz.)	25	4.4			
Sandwich (Nabisco)	1 piece (8 grams)	22	.8			
Twirls (Nabisco)	1 piece (1.1 oz.)	32	4.6			
Milano (Pepperidge Farm)	1 piece (.4 oz.)	21	3.5			
Milano, mint (Pepperidge Farm)	1 piece (.5 oz.)	21	4.4			
Mint sandwich (Nabisco) *Mystic*	1 piece (.6 oz.)	46	4.6			
Molasses (USDA)[1]	1 oz.	109	3.0	<1.	2.	
Molasses & Spice (Sunshine)	1 piece (.6 oz.)		1.8			
Naples (Pepperidge Farm)	1 piece (6 grams)	10	1.9			
Nassau (Pepperidge Farm)	1 piece (.6 oz.)	51	4.9			
Oatmeal:						
(Keebler) old fashioned	1 piece (.6 oz.)	76	3.0			
(Nabisco)	1 piece (.6 oz.)	65	3.1			
(Nabisco) *Family Favorites*	1 piece (5 grams)	20	.9			
(Sunshine)	1 piece (.4 oz.)	150	2.3			
Iced (Sunshine)	1 piece (.5 oz.)		2.2			
Irish (Pepperidge Farm)	1 piece (.4 oz.)	41	2.2			
Peanut butter (Sunshine)	1 piece (.6 oz.)		3.6			
Raisin (USDA)[1]	1 oz.	46	4.4	1.	3.	
Raisin (Nabisco) *Bake Shop*	1 piece (.6 oz.)	81	3.0			
Raisin (Pepperidge Farm)	1 piece (.4 oz.)	54	2.6			
Raisin bar (Tastykake)	1 pkg. (2¼ oz.)		10.0			
Old Country Treats (Stella D'oro)	1 piece (.5 oz.)		2.9			
Orleans (Pepperidge Farm)	1 piece (6 grams)	7	1.7			
Peach-apricot pastry (Stella D'oro)	1 piece (.8 oz.)		3.7			
Peanut & peanut butter:						
(USDA)[1]	1 oz.	49	5.4	1.	4.	
Bars, cocoa-covered (Nabisco)						
Crowns	1 piece (.6 oz.)	89	5.1			
Caramel logs (Nabisco) *Heydays*	1 piece (.8 oz.)	35	6.6			
Creme patties (Nab)	3 pieces (½-oz. pkg.)	38	3.8			
Creme patties (Nab)	6 pieces (1-oz. pkg.)	77	7.6			

(USDA): United States Department of Agriculture
*Prepared as Package Directs
[1]Principal sources of fat: vegetable shortening, egg & milk.

Food and Description	Measure or Quantity	Sodium (mg.)	— Fats in grams —			Choles- terol (mg.)
			Total	Satu- rated	Unsatu- rated	
Creme patties (Nabisco)	1 piece (7 grams)	18	1.8			
Creme patties, cocoa-covered						
(Nabisco) *Fancy*	1 piece (.4 oz.)	25	3.3			
Patties (Sunshine)	1 piece (7 grams)		1.4			
Sandwich (Nabisco) *Nutter*						
Butter	1 piece (.5 oz.)	57	3.1			
Pecan Sandies (Keebler)	1 piece (.6 oz.)	52	5.1			
Penguins (Keebler)	1 piece (.8 oz.)	66	5.7			
Pirouette (Pepperidge Farm):						
Chocolate laced	1 piece (7 grams)	11	2.1			
Lemon or original	1 piece (7 grams)	12	2.0			
Pitter Patter (Keebler)	1 piece (.6 oz.)	117	3.8			
Pizzelle, Carolines (Stella D'oro)	1 piece (.4 oz.)		2.0			
Raisin:						
(USDA)[1]	1 oz.	15	1.5	Tr.	1.	
Fruit biscuit (Nabisco)	1 piece (.5 oz.)	19	.6			
Rich 'n Chips (Keebler)	1 piece (.5 oz.)	44	3.8			
Rochelle (Pepperidge Farm)	1 piece (.6 oz.)	38	4.5			
Sandwich, creme:						
(USDA)[1]	1 oz.	137	6.4	2.	5.	
Cameo (Nabisco)	1 piece (.5 oz.)	49	2.6			
Chocolate chip (Nabisco)	1 piece (.5 oz.)	36	3.8			
Chocolate fudge:						
(Keebler)	1 piece (.7 oz.)	97	4.7			
Assorted (Nabisco) *Cookie*						
Break	1 piece (.4 oz.)	33	2.5			
Chocolate (Nabisco) *Cookie*						
Break	1 piece (.4 oz.)	32	2.5			
Orbit (Sunshine)	1 piece (.4 oz.)		2.4			
Oreo (Nab)	4 pieces (1-oz. pkg.)	136	6.0			
Oreo (Nab)	6 pieces (1⅝-oz. pkg.)	222	9.7			
Oreo (Nab)	6 pieces (2⅛-oz. pkg.)	290	12.7			
Oreo (Nabisco)	1 piece (.4 oz.)	49	2.2			
Oreo & Swiss (Nab)	6 pieces (1⅝-oz. pkg.)	159	10.5			
Oreo & Swiss (Nab)	6 pieces (2¼-oz. pkg.)	220	14.5			
Oreo & Swiss, assortment						
(Nabisco)	1 piece (.4 oz.)	35	2.3			
Pride (Nabisco)	1 piece (.4 oz.)	31	2.6			

(USDA): United States Department of Agriculture
*Prepared as Package Directs
[1]Principal sources of fat: vegetable shortening, egg & milk.

Food and Description	Measure or Quantity	Sodium (mg.)	Total	Fats in grams Satu-rated	Unsatu-rated	Choles-terol (mg.)
Social Tea (Nabisco)	1 piece (.4 oz.)	34	2.3			
Swiss (Nab)	4 pieces (1-oz. pkg.)	59	7.2			
Swiss (Nab)	6 pieces (1¾-oz. pkg.)	104	12.6			
Swiss (Nabisco)	1 piece (.4 oz.)	22	2.6			
(Tom Houston)	1 piece (.5 oz.)	149	3.6			
Vanilla (Keebler)	1 piece (.6 oz.)	78	3.8			
Vanilla (Nabisco)	1 piece (.4 oz.)	34	2.4			
Vienna Finger (Sunshine)	1 piece (.5 oz.)		2.9			
Sesame, Regina (Stella D'oro)	1 piece (.4 oz.)		2.3			
Shortbread or shortcake:						
(USDA)	1 oz.	17	6.5	2.	5.	
(USDA)	1¾"-square (8 grams)	5	1.8	Tr.	1.	
(Nabisco) *Dandy*	1 piece (.4 oz.)	29	1.5			
(Pepperidge Farm)	1 piece (.5 oz.)	39	3.9			
Lorna Doone (Nab)	4 pieces (1-oz. pkg.)	146	6.1			
Lorna Doone (Nab)	6 pieces (1½-oz. pkg.)	219	9.1			
Lorna Doone (Nabisco)	1 piece (8 grams)	39	1.6			
Pecan (Nabisco)	1 piece (.5 oz.)	44	4.6			
Scotties (Sunshine)	1 piece (8 grams)		1.8			
Striped (Nabisco)	1 piece (10 grams)	13	2.3			
Vanilla (Tastykake)	6 pieces (2¼-oz. pkg.)		18.6			
Social Tea Biscuit (Nabisco)	1 piece (5 grams)	18	.6			
Spiced wafers (Nabisco)	1 piece (10 grams)	58	1.1			
Sprinkles (Sunshine)	1 piece (.6 oz.)	1.5				
Sugar cookie:						
(Keebler) old fashioned	1 piece (.6 oz.)	55	2.8			
(Pepperidge Farm)	1 piece (.4 oz.)	30	2.4			
(Sunshine)	1 piece (.6 oz.)		3.7			
Brown (Nabisco) *Family Favorite*	1 piece (5 grams)	11	1.3			
Brown (Pepperidge Farm)	1 piece (.4 oz.)	24	2.2			
Rings (Nabisco)	1 piece (.5 oz.)	47	2.5			
Sugar wafer:						
(USDA)[1]	1 oz.	54	5.5	1.	4.	
(Nab) *Biscos*	3 pieces (⅞-oz. pkg.)	32	6.1			
(Nabisco) *Biscos*	1 piece (4 grams)	5	.9			
(Sunshine)	1 piece (9 grams)	36	1.8			
Krisp Kreem (Keebler)	1 piece (6 grams)	14	1.8			
Lemon (Sunshine)	1 piece (9 grams)		1.9			

(USDA): United States Department of Agriculture
*Prepared as Package Directs
[1]Principal sources of fat: vegetable shortening, egg & milk.

Food and Description	Measure or Quantity	Sodium (mg.)	—Fats in grams—			Choles- terol (mg.)
			Total	Satu- rated	Unsatu- rated	
Swedish Kreme (Keebler)	1 piece (5.7 oz.)	81	5.2			
Tahiti (Pepperidge Farm)	1 piece (.5 oz.)	17	5.4			
Toy (Sunshine)	1 piece (3 grams)		.4			
Vanilla creme (Wise)	1 piece (7 grams)	23	1.4			
Vanilla snap (Nabisco)	1 piece (3 grams)	10	.3			
Vanilla wafer:						
(USDA)[1]	1 oz.	71	4.6	1.	3.	
(Keebler)	1 piece (4 grams)	18	.9			
(Nabisco) *Nilla*	1 piece (4 grams)	12	.7			
(Sunshine) small	1 piece (3 grams)		.6			
Venice (Pepperidge Farm)	1 piece (.4 oz.)	13	3.3			
Waffle creme (Nabisco) *Biscos*	1 piece (8 grams)	10	2.1			
Yum Yums (Sunshine)	1 piece (.5 oz.)		3.2			
COOKIE, DIETETIC:						
Almond chocolate wafer (Estee)	1 piece	2	1.7			
Angel puffs (Stella D'oro)	1 piece (3 grams)	2	1.0			
Apple pastry (Stella D'oro)	1 piece (.8 oz.)	42	3.9			
Assorted (Estee)	1 piece (6 grams)	3	1.4	Tr.	1.	
Assorted filled wafers (Estee)	1 piece (5 grams)	3	1.3	Tr.	1.	
Banana wafers (Estee)	1 piece	12	8.5			
Beljuin Treats (Estee)	1 piece	10	1.7			
Chocolate chip (Estee)	1 piece (6 grams)	11	1.0	Tr.	<1.	
Chocolate Holland filled wafer (Estee)	1 piece (3 grams)	1	1.2	Tr.	<1.	1
Chocolate & vanilla wafer (Estee)	1 piece (4 grams)	3	1.3	Tr.	1.	
Fig pastry (Stella D'oro)	1 piece (.9 oz.)	20	3.7			
Fruit flavored wafer (Estee)	1 piece (4 grams)	1	1.2	Tr.	<1.	1
Have-A-Heart (Stella D'oro)	1 piece (.7 oz.)	22	5.1			
Holland bittersweet wafer (Estee)	1 piece	3	8.3			
Holland milk chocolate wafer (Estee)	1 piece (.8 oz.)	3	8.3	3.	6.	8
Kichel (Stella D'oro)	1 piece (1 gram)	2	.5			
Monties (Estee)	1 piece	10	2.2			
Oatmeal raisin (Estee)	1 piece (7 grams)	2	1.4	Tr.	1.	2
Pastry stick (Estee)	1 piece (7 grams)	3	2.1	2.	Tr.	
Peach-apricot pastry (Stella D'oro)	1 piece (.8 oz.)	14	4.2			
Prune pastry (Stella D'oro)	1 piece (.8 oz.)	15	3.4			
Royal Nuggets (Stella D'oro)	1 piece (<1 gram)	1	.1			
Sandwich, chocolate (Estee)	1 piece (8 grams)	3	2.1	<1.	2.	
Sandwich, Duplex (Estee)	1 piece (9 grams)	3	2.0	<1.	1.	

(USDA): United States Department of Agriculture
*Prepared as Package Directs
[1]Principal sources of fat: vegetable shortening, egg & milk.

Food and Description	Measure or Quantity	Sodium (mg.)	— Fats in grams —			Choles- terol (mg.)
			Total	Satu- rated	Unsatu- rated	
Sandwich, lemon (Estee)	1 piece (.5 oz.)	20	3.0	2.	1.	<1
Vanilla filled wafer (Estee)	1 piece (4 grams)	3	1.3	Tr.	1.	
Vanilla Holland filled wafer (Estee)	1 piece (3 grams)	1	1.2	Tr.	<1.	1
Vanilla & strawberry wafer (Estee)	1 piece	3	1.3			
Wafer cake (Estee)	1 piece (9 grams)	4	3.3	1.	2.	3

COOKIE DOUGH, refrigerated:

Food and Description	Measure or Quantity	Sodium (mg.)	Total	Satu- rated	Unsatu- rated	Choles- terol (mg.)
Unbaked, plain (USDA)[1]	1 oz.	141	6.4	1.	5.	
Baked, plain (USDA)[1]	1 oz.	155	7.1	2.	6.	

COOKIE, HOME RECIPE:

Food and Description	Measure or Quantity	Sodium (mg.)	Total	Satu- rated	Unsatu- rated	Choles- terol (mg.)
Brownie with nuts (USDA):						
Made with butter[2]	1 oz.	71	8.5	3.	6.	24
Made with butter[2]	.7-oz. piece (1¾" x 1¾" x ⅞")	50	6.0	2.	4.	17
Made with vegetable shortening[3]	1 oz.	71	8.9	2.	7.	
Chocolate Chip (USDA):						
Made with butter[4]	1 oz.	99	8.0	4.	4.	
Made with vegetable shortening[5]	1 oz.	99	8.5	2.	6.	
Sugar, soft, thick (USDA):						
Made with butter[6]	1 oz.	90	4.3	2.	2.	
Made with vegetable shortening[7]	1 oz.	90	4.8	1.	4.	

COOKIE MIX:

Food and Description	Measure or Quantity	Sodium (mg.)	Total	Satu- rated	Unsatu- rated	Choles- terol (mg.)
Plain, dry (USDA)[8]	1 oz.	100	6.9	2.	5.	
*Plain, prepared with egg & water (USDA)[1]	1 oz.	98	6.9	2.	5.	
*Plain, prepared with milk (USDA)[9]	1 oz.	98	6.7	2.	5.	
Brownie:						
Dry, with egg (USDA)[10]	1 oz.	85	3.4	<1.	3.	
Dry, without egg (USDA)[11]	1 oz.	55	4.6	1.	4.	

(USDA): United States Department of Agriculture
*Prepared as Package Directs
[1] Principal sources of fat: vegetable shortening & egg.
[2] Principal sources of fat: pecans, butter, chocolate & egg.
[3] Principal sources of fat: pecans, vegetable shortening, chocolate & egg.
[4] Principal sources of fat: butter, chocolate, walnuts & egg.
[5] Principal sources of fat: vegetable shortening, chocolate, walnuts & egg.
[6] Principal sources of fat: butter, egg & milk.
[7] Principal sources of fat: vegetable shortening, egg & milk.
[8] Principal source of fat: vegetable shortening.
[9] Principal sources of fat: vegetable shortening & milk.
[10] Principal sources of fat: vegetable shortening, cocoa & egg.
[11] Principal sources of fat: vegetable shortening & cocoa.

Food and Description	Measure or Quantity	Sodium (mg.)	—Fats in grams— Total	Satu- rated	Unsatu- rated	Choles- terol (mg.)
*Dry, with egg, prepared with water & nuts (USDA)[1]	1 oz.	62	5.3	<1.		4.
*Dry, without egg, prepared with egg, water & nuts (USDA)[1]	1 oz.	47	5.7	1.		5.
*Butterscotch (Betty Crocker)	1½″ sq.	68	2.1			
*"Cake like," family size (Duncan Hines)	¹/₂₄ of pan (1.2 oz.)	103	6.9			
*"Cake like," regular size (Duncan Hines)	¹/₁₆ of pan (1.2 oz.)	106	7.1			
*Fudge (Betty Crocker)	1½″ sq.	36	2.2			
*Fudge, supreme (Betty Crocker)	1½″ sq.	28	2.2			
*Fudge, chewy, family size (Duncan Hines)	¹/₂₄ of pan (1.1 oz.)	96	6.4	1.		5.
*Fudge, chewy, regular size (Duncan Hines)	¹/₁₆ of pan (1.2 oz.)	99	6.6	2.		5.
*German chocolate (Betty Crocker)	1½″ sq.	38	2.4			
*Walnut (Betty Crocker)	1½″ sq.	35	2.9			
Chocolate mint (Nestlé's)	1 oz.	111	5.3			
*Date bar (Betty Crocker)	2″ x 1″ bar	35	2.5			
*Macaroon, coconut (Betty Crocker)	1 macaroon (1¾″)	11	3.6			
Lemon (Nestlé's)	1 oz.	111	5.2			
Sugar (Nestlé's)	1 oz.	111	5.2			
Toll House (Nestlé's)	1 oz.	111	5.2			
Toll House, with morsels, prepared with egg (Nestlé's)	1 piece (.4 oz.)	34	2.4			
Toll House, without morsels, prepared without egg (Nestlé's)	1 piece (8 grams)	31	1.8			
*Vienna Dream bar (Betty Crocker)	1 bar (2″ x 1⅓″)	65	4.9			

COOKING FATS (See **FATS**)

COOL 'N CREAMY (Birds Eye)	½ cup (4.4 oz.)	119	6.1			0

CORIANDER, whole or ground

(Spice Islands)	1 tsp.	<1				(0)

CORN:

Fresh, white or yellow (USDA):

Raw, untrimmed, on cob	1 lb. (weighed in husk)	Tr.	1.6			0

(USDA): United States Department of Agriculture
*Prepared as Package Directs
[1]Principal sources of fat: walnuts, vegetable shortening, cocoa & egg.

Food and Description	Measure or Quantity	Sodium (mg.)	—Fats in grams— Total	Satu- rated	Unsatu- rated	Choles- terol (mg.)
Raw, trimmed, on cob	1 lb. (husk removed)	Tr.	2.5			0
Raw, kernels	4 oz.	Tr.	1.1			0
Boiled without salt, kernels, cut from cob, drained	1 cup (5.9 oz.)	Tr.	1.7			0
Boiled without salt, whole	1 ear (5" x 1¾") (4.9 oz.)	Tr.	.8			0
Canned, regular pack:						
Golden or yellow, whole kernel:						
Solids & liq., vacuum pack (USDA)	½ cup (3.7 oz.)	250	.5			0
Solids & liq., wet pack (USDA)	½ cup (4.5 oz.)	302	.8			0
Drained solids, wet pack (USDA)	½ cup (3 oz.)	203	.7			0
Drained liq., wet pack (USDA)	4 oz.	268	Tr.			0
Solids & liq., vacuum pack (Del Monte)	½ cup (3.7 oz.)	220	.5			0
Drained solids, wet pack (Del Monte) Family Style	½ cup (3 oz.)	202	.8			0
Vacuum pack, *Niblets*	⅓ of 12-oz. can	318	.6			(0)
Solids & liq. (Green Giant)	½ of 8.5-oz. can	368	.6			(0)
Shoe peg (Le Sueur)	¼ of 17-oz. can	366	.6			(0)
Solids & liq., wet pack (Stokely-Van Camp)	½ cup (4.5 oz.)		.8			(0)
With peppers, solids & liq. (Del Monte)	½ cup (3.7 oz.)	156	.3			0
With peppers, vacuum pack, *Mexicorn*	⅓ of 12-oz. can	352	.6			(0)
White, whole kernel:						
Solids & liq., wet pack (USDA)	½ cup (4.5 oz.)	302	.8			0
Drained solids, wet pack (USDA)	½ cup (2.8 oz.)	189	.6			0
Drained liq., wet pack (USDA)	4 oz.	268	Tr.			0
Vacuum pack (Green Giant)	⅓ of 12-oz. can	255	.6			(0)
Canned, white or yellow, dietetic pack:						
Solids & liq., wet pack (USDA)	4 oz.	2	.6			0
Drained solids (USDA)	4 oz.	2	.8			0
Drained liq. (USDA)	4 oz. (by wt.)	2	Tr.			0
Solids & liq. (Blue Boy)	4 oz.	3	.6			(0)

(USDA): United States Department of Agriculture
*Prepared as Package Directs

Food and Description	Measure or Quantity	Sodium (mg.)	—Fats in grams—			Choles- terol (mg.)
			Total	Satu- rated	Unsatu- rated	
Solids & liq. (Diet Delight)	½ cup (4.4 oz.)	5	.5			(0)
Solids & liq. (S and W) *Nutradiet,* unseasoned	4 oz.	7	.4			(0)
Solids & liq. (Tillie Lewis)	½ cup	<10	.6			0
Canned, cream style, white or yellow, regular pack:						
Solids & liq. (USDA)	½ cup (4.4 oz.)	295	.8			0
Golden, solids & liq. (Del Monte)	½ cup (4.4 oz.)	656	.4			(0)
Golden, solids & liq. (Green Giant)	½ of 8.5-oz. can	349	.6			(0)
Solids & liq. (Stokely-Van Camp)	½ cup (4.1 oz.)		.7			(0)
Canned, cream style, dietetic pack:						
Solids & liq. (USDA)	4 oz.	2	1.2			0
Solids & liq. (Blue Boy)	4 oz.	3	1.2			(0)
Solids & liq. (S and W) *Nutradiet*	4 oz.	2	.5			(0)
Frozen:						
On the cob:						
Not thawed (USDA)	4 oz.	1	1.1			0
Boiled, drained (USDA)	4 oz.	1	1.1			0
(Birds Eye)	1 ear (3.5 oz.)	1	1.0			0
Niblets Ears	1 ear (4.9 oz.)	21	.8			(0)
Kernel, cut off cob:						
Not thawed (USDA)	4 oz.	1	.6			0
Boiled, drained (USDA)	½ cup (3.2 oz.)	<1	.5			0
(Birds Eye)	½ cup (3.3 oz.)	33	.5			0
Sweet white (Birds Eye)	½ cup (3.3 oz.)	1	.5			0
Cream style (Green Giant)	⅓ of 10-oz. pkg.	198	.4			(0)
In butter sauce:						
& peppers (Green Giant)	⅓ of 10-oz. pkg.	312	2.8			
Yellow, *Niblets*	⅓ of 10-oz. pkg.	383	2.4			
White (Green Giant)	⅓ of 10-oz. pkg.	346	2.8			
Scalloped casserole (Green Giant)	⅓ of 10-oz. pkg.	619	5.7			
With peas & tomatoes (Birds Eye)	⅓ of 10-oz. pkg.	447	.4			0

CORNBREAD:

Food and Description	Measure or Quantity	Sodium (mg.)	Total	Satu- rated	Unsatu- rated	Choles- terol
Corn pone, home recipe, prepared with white, whole-ground corn- meal (USDA)[1]	4 oz.	449	6.0	2.	4.	

(USDA): United States Department of Agriculture
*Prepared as Package Directs
[1]Principal sources of fat: lard & egg.

Food and Description	Measure or Quantity	Sodium (mg.)	—Fats in grams—			Choles- terol (mg.)
			Total	Satu- rated	Unsatu- rated	
Corn sticks, frozen (Aunt Jemima)	3 pieces (1¾ oz.)	360	5.1			
Johnnycake, home recipe, prepared with yellow, degermed cornmeal (USDA)[1]	4 oz.	782	5.9	2.	4.	
Southern-style, home recipe, prepared with degermed corn-meal (USDA)[1]	2½″ x 2½″ x 1⅝″ piece (2.9 oz.)	491	5.0	1.	4.	58
Southern-style, home recipe, prepared with whole-ground cornmeal (USDA)[1]	4 oz.	712	8.2	2.	6.	
Spoonbread, home recipe, prepared with white, whole-ground cornmeal (USDA)[2]	4 oz.	547	12.9	5.	8.	
CORNBREAD MIX:						
Dry (USDA)[3]	1 oz.	328	3.6	<1.	3.	
*Prepared with egg & milk:						
(USDA)[4]	4 oz.	844	9.5	3.	6.	78
(USDA)[4]	2⅜″ muffin (1.4 oz.)	298	3.4	1.	2.	28
(USDA)[4]	2½″ x 2½″ x 1⅜″ piece (1.9 oz.)	409	4.6	2.	3.	38
*(Aunt Jemima)	⅙ of cornbread (2.4 oz.)	575	7.6			
*(Dromedary)	2″ x 2″ piece (1.4 oz.)	294	4.6			
CORN CHEX, cereal, dry	1¼ cups (1 oz.)	304	.1			(0)
CORN CHIPS (See **CRACKERS**)						
CORN CHOWDER, New England (Snow)	8 oz.		6.3			
CORNED BEEF:						
Uncooked, boneless, medium fat (USDA)	1 lb.	5897	113.4	54.	59.	
Cooked, boneless, medium fat (USDA)	4 oz.	1973	34.5	17.	17.	

(USDA): United States Department of Agriculture
*Prepared as Package Directs
[1]Principal sources of fat: lard & egg.
[2]Principal sources of fat: lard, egg & milk.
[3]Principal sources of fat: vegetable shortening & egg.
[4]Principal sources of fat: vegetable shortening, milk & egg.

Food and Description	Measure or Quantity	Sodium (mg.)	Total	—Fats in grams— Satu-rated	Unsatu-rated	Choles-terol (mg.)
Canned:						
Lean (USDA)	4 oz.		9.1	5.	5.	
Medium fat (USDA)	4 oz.		13.6	7.	7.	
Fat (USDA)	4 oz.		20.4	10.	10.	
(Armour Star)	4 oz. (from 12-oz. can)		22.0			
(Hormel) *Dinty Moore*	4 oz.	1191	15.2	6.	7.	79
Brisket (Wilson) *Tender Made*	4 oz.	1435	10.9	5.	5.	71
Packaged (Oscar Mayer)	5-gram slice (16 to 3 oz.)	74	.2			
CORNED BEEF HASH, canned:						
With potato (USDA)[1]	4 oz.	612	12.8	6.	7.	
(Armour Star)	15½-oz. can		60.1			
(Hormel)	7½ oz.	1480	26.2			
(Nalley's)	4 oz.		11.3			
(Van Camp)	½ cup (4.1 oz.)		12.4			
(Wilson)	15½-oz. can	3625	58.0	26.	32.	105
CORNED BEEF HASH DINNER, frozen (Banquet):						
Meat compartment	5.5 oz.		9.9			
Apple compartment	2.8 oz.		.1			
Peas compartment	1.9 oz.		.7			
Complete dinner	10.2-oz. dinner		10.7			
CORNED BEEF SPREAD:						
(Underwood)	1 T. (.5 oz.)	121	2.1			
(Underwood)	4½-oz. can	1117	19.5			
CORN FLAKES, cereal:						
Whole (USDA)	1 cup (1 oz.)	291	.1			0
Crushed (USDA)	1 cup (2.5 oz.)	704	.3			0
Frosted (USDA)	1 cup (1.4 oz.)	310	<.1			0
Country (General Mills)	1¼ cups (1 oz.)	303	.3			(0)
(Kellogg's)	1⅓ cups (1 oz.)	268	.1			(0)
(Ralston)	1 cup (1 oz.)	290	<.1			(0)
(Van Brode)	1 oz.		<.1			(0)
CORN FRITTER:						
Home recipe (USDA)[2]	4 oz.	541	24.4	6.	19.	
Frozen (Mrs. Paul's)	12-oz. pkg.		31.8			

(USDA): United States Department of Agriculture
*Prepared as Package Directs
[1]Principal source of fat: beef.
[2]Principal sources of fat: vegetable shortening, egg, milk & butter.

Food and Description	Measure or Quantity	Sodium (mg.)	— Fats in grams — Total	Satu- rated	Unsatu- rated	Choles- terol (mg.)
CORN GRITS (See **HOMINY**)						
CORNMEAL MIX:						
Bolted (Aunt Jemima/Quaker)	¼ cup (1 oz.)	370	.8			(0)
Degermed (Aunt Jemima/Quaker)	¼ cup (1 oz.)	370	.3			(0)
CORNMEAL, WHITE or YELLOW:						
Dry (USDA):						
Bolted	1 cup (4.3 oz.)	1	4.1	Tr.	4.	0
Degermed	1 cup (4.3 oz.)	1	1.7			0
Self-rising, degermed	1 cup (5 oz.)	1946	1.6			0
Self-rising, whole-ground	1 cup (5 oz.)	1946	4.1	Tr.	4.	0
Whole-ground, unbolted	1 cup (4.3 oz.)	1	4.8	Tr.	5.	0
Cooked:						
*Bolted (Aunt Jemina/Quaker)	⅔ cup	<1	.7			(0)
Degermed (USDA)	1 cup (8.5 oz.)	264	.5			0
*Degermed (Albers)	1 cup		.5			(0)
*Degermed (Aunt Jemima/Quaker)	⅔ cup	<1	.3			(0)
CORN PUDDING, home recipe (USDA)[1]	1 cup (8.6 oz.)	1068	11.5	5.	7.	103
CORN SALAD, raw (USDA):						
Untrimmed	1 lb. (weighed untrimmed)		1.7			0
Trimmed	4 oz.		.5			0
CORN SOUFFLE, frozen (Stouffer's)	12-oz. pkg.	1674	22.0			
CORNSTARCH:						
(USDA)	1 cup (4.5 oz.)	Tr.	Tr.			0
(USDA)	1 T. (8 grams)	Tr.	Tr.			0
(Argo)	1 T. (10 grams)	Tr.	<.1			0
(Duryea's)	1 T. (10 grams)	Tr.	<.1			0
(Kingsford's)	1 T. (10 grams)	Tr.	<.1			0
CORNSTARCH PUDDING (See **VANILLA PUDDING**)						
CORN STICK (See **CORNBREAD**)						

(USDA): United States Department of Agriculture
*Prepared as Package Directs
[1]Principal sources of fat: milk, vegetable shortening & egg.

Food and Description	Measure or Quantity	Sodium (mg.)	— Fats in grams —			Choles- terol (mg.)
			Total	Satu- rated	Unsatu- rated	
CORN SYRUP, light & dark blend:						
(USDA)	1 cup (11.5 oz.)	221	0.			0
(USDA)	1 T. (.7 oz.)	14	0.			0
CORN TOTAL, cereal (General Mills)	1¼ cups (1 oz.)	316	.2			(0)
COTTAGE PUDDING, home recipe (USDA)[1] :						
Without sauce[2]	2 oz.	170	6.4	4.	3.	
With chocolate sauce	2 oz.	132	5.0			
With strawberry sauce	2 oz.	132	5.0			
COUGH DROP:						
(Beech-Nut)	1 drop (2 grams)	<1	0.			(0)
(Estee)	1 drop	<1	Tr.			(0)
(Pine Bros.)	1 drop (3 grams)	<1	0			(0)
COUNT CHOCULA, cereal (General Mills)	1 cup (1 oz.)	152	.8			(0)
COUNTRY-STYLE SAUSAGE, smoked links (USDA)[3]	1 oz.		8.8	3.	6.	
COWPEA, including black-eyed peas (USDA):						
Immature seeds:						
Raw, whole	1 lb. (weighed in pods)	5	2.0			0
Raw, shelled	½ cup (2.5 oz.)	1	.6			0
Boiled without salt, drained	½ cup (2.9 oz.)	<1	1.7			0
Canned, solids & liq.	4 oz.	268	.3			0
Frozen (See **BLACK-EYED PEA,** frozen)						
Young pods with seeds:						
Raw, whole	1 lb. (weighed untrimmed)	17	1.2			0
Boiled without salt, drained	4 oz.	3	.3			0
Mature seeds, dry:						
Raw	½ cup (3 oz.)	29	1.3			0
Boiled without salt, drained	½ cup (4.4 oz.)	10	.4			0

(USDA): United States Department of Agriculture
*Prepared as Package Directs
[1]Made with sodium aluminum sulfate-type baking powder.
[2]Principal sources of fat: butter, egg & milk.
[3]Principal source of fat: pork.

Food and Description	Measure or Quantity	Sodium (mg.)	—Fats in grams— Total	Satu- rated	Unsatu- rated	Choles- terol (mg.)
CRAB, all species:						
Fresh (USDA):						
Steamed, whole	1 lb. (weighed in shell)		4.1			218
Steamed, meat only	1 cup (4.4 oz.)		2.4			125
Canned:						
Drained solids (USDA)	1 packed cup (5.6 oz.)	1600	4.0			162
(Del Monte) Alaska King	7½-oz. can	1178	.6			
Frozen (Wakefield's) Alaska King, thawed & drained	4 oz.	1	1.1			113
CRAB APPLE, fresh (USDA):						
Whole	1 lb. (weighed whole)	4	1.3			0
Flesh only	4 oz.	1	.3			0
CRAB CAKE, frozen, thins (Mrs. Paul's)	10-oz. pkg.		30.9			
CRAB, DEVILED:						
Home recipe (USDA)[1]	1 cup (8.5 oz.)	2081	22.6			245
Frozen (Mrs. Paul's)	4 oz.		10.8			
Frozen, miniature (Mrs. Paul's)	4 oz.		12.1			
CRAB IMPERIAL, home recipe (USDA)[2]	1 cup (7.8 oz.)	1602	16.7			308
CRAB NEWBURG, frozen (Stouffer's)	12-oz. pkg.	1158	45.5			
CRAB SOUP (Crosse & Blackwell)	½ can (6½ oz.)		.9			
CRACKER, PUFFS and CHIPS:						
American Harvest (Nabisco)	1 piece (3 grams)	36	.8			
Arrowroot biscuit (Nabisco)	1 piece (5 grams)	11	.8			
Bacon-flavored thins (Nabisco)	1 piece (2 grams)	32	.6			
Bacon Nips	1 oz.	700	9.4	3.	6.	0
Bacon rinds (Wonder)	1 oz.	220	7.8			
Bacon toast (Keebler)	1 piece (3 grams)	28	.7			
Bakon Tasters (Old London)	½-oz. bag	237	2.1			

(USDA): United States Department of Agriculture
*Prepared as Package Directs
[1]Prepared with bread cubes, butter, parsley, eggs, lemon juice & catsup.
[2]Prepared with butter, flour, milk, onion, green pepper, eggs & lemon juice.

Food and Description	Measure or Quantity	Sodium (mg.)	Total	—Fats in grams— Satu-rated	Unsatu-rated	Choles-terol (mg.)
Bugles (General Mills)	15 pieces (½ oz.)	138	5.3			
Butter (USDA)[1]	1 oz.	310	5.0	2.	3.	
Butter thins (Nabisco)	1 piece (3 grams)	16	.5			
Cheese flavored (See also individual brand names in this grouping):						
(USDA)	1 oz.	295	6.0	2.	4.	
Cheese'n Bacon, sandwich (Nab)	6 pieces (1¼-oz. pkg.)	307	9.6			
Cheese'n Cracker (Kraft)	4 crackers & ¾-oz. cheese	68	1.2			
Cheese *Nips* (Nab)	24 pieces (⅞-oz. pkg.)	422	4.2			
Cheese *Nips* (Nabisco)	1 piece (1 gram)	19	.2			
Cheese'n Rye, sandwich (Nab)	6 pieces (1¼-oz. pkg.)	490	11.8			
Cheese Pixies (Wise)	1-oz. bag	334	11.7			
Chee.Tos, baked	1 oz.	470	9.9	3.	7.	0
Chee.Tos, fried	1 oz.	290	9.8	3.	6.	0
Cheez Doodles (Old London)	1⅛-oz. bag	244	9.6			
Cheez-Its (Sunshine)	1 piece (1 gram)		.3			
Cheez Waffles (Old London)	1 piece (2 grams)	50	.6			
Che-zo (Keebler)	1 piece (<1 gram)	9	.2			
Ritz (Nabisco)	1 piece (3 grams)	35	.9			
Sandwich (Nab)	6 pieces (1¼-oz. pkg.)	426	10.6			
Shapies, dip delights (Nabisco)	1 piece (2 grams)	24	.6			
Shapies, shells (Nabisco)	1 piece (2 grams)	26	.6			
Thins (Pepperidge Farm)	2 pieces (5 grams)	50	.5			
Thins, dietetic (Estee)	1 piece		.2			
Tid-Bit (Nab)	32 pieces (1⅛-oz. pkg.)	524	6.8			
Tid-Bit (Nabisco)	1 piece (<1 gram)	15	.2			
Toast (Keebler)	1 piece (3 grams)	31	.8			
Twists (Nalley)	1 oz.		10.2			
Twists (Wonder)	1 oz.	333	9.6			
Cheese & peanut butter sandwich:						
(USDA)[2]	1 oz.	281	6.8	2.	5.	
(Nab) *O-So-Gud*	4 pieces (1-oz. pkg.)	358	6.8			
(Nab) squares	4 pieces (1-oz. pkg.)	332	7.1			
(Nab) squares	6 pieces (1½-oz. pkg.)	498	10.7			

(USDA): United States Department of Agriculture
*Prepared as Package Directs
[1]Principal sources of fat: vegetable shortening & butter.
[2]Principal sources of fat: vegetable shortening, cheese & peanut butter.

Food and Description	Measure or Quantity	Sodium (mg.)	Total	Fats in grams — Saturated	Unsaturated	Cholesterol (mg.)
(Nab) squares	6 pieces (1¾-oz. pkg.)	581	12.5			
(Nab) variety pack	6 pieces (1½-oz. pkg.)	520	10.5			
(Nab) variety pack	6 pieces (1¾-oz. pkg.)	606	12.3			
Chicken in a Biskit (Nabisco)	1 piece (2 grams)	19	.5			
Chippers (Nabisco)	1 piece (3 grams)	48	.7			
Chipsters (Nabisco)	1 piece (<1 gram)	8	.1			
Clam-flavored crisps (Snow)	1 oz.		8.6			
Club (Keebler)	1 piece (3 grams)	44	.7			
Corn Capers (Wonder)	1 oz.	220	10.0			
Corn cheeze (Tom Houston)	10 pieces (5 grams)	3	2.2			
Corn chips:						
Cornetts	1 oz.		8.5			0
Fritos, regular	1 oz.	160	10.5	3.	8.	0
Fritos, barbecued	1 oz.	190	10.1	2.	8.	0
Korkers (Nabisco)	1 piece (2 grams)	11	.5			
(Old London)	1¾-oz. bag	921	16.3			
(Wise)	1¾-oz. bag	229	17.7			
(Wonder)	1 oz.	220	10.7			
Barbecue (Wise)	1¾-oz. bag	278	17.6			
Corn Diggers (Nabisco)	1 piece (<1 gram)	10	.2			
Crown Pilot (Nabisco)	1 piece (.6 oz.)	64	1.9			
Dipsy Doodles (Old London)	1¾-oz. bag	366	18.3			
Doo Dads (Nabisco)	1 piece (<1 gram)	7	.1			
Escort (Nabisco)	1 piece (4 grams)	37	.9			
Flings, cheese-flavored curls (Nabisco)	1 piece (2 grams)	28	.8			
Flings, Swiss'n ham (Nabisco)	1 piece (2 grams)	16	.7			
Goldfish (Pepperidge Farm):						
Cheddar cheese	10 pieces (6 grams)	87	1.3			
Lightly salted	10 pieces (6 grams)	84	1.2			
Parmesan cheese	10 pieces (6 grams)	87	1.2			
Pizza	10 pieces (6 grams)	77	1.4			
Pretzel	10 pieces (7 grams)	195	.6			
Onion	10 pieces (6 grams)	81	1.2			
Sesame garlic	10 pieces (6 grams)	84	1.3			
Graham:						
(USDA)[1]	2½" sq. (7 grams)	47	.7	Tr.	<1.	
(Nabisco)	1 piece (7 grams)	44	.7			

(USDA): United States Department of Agriculture
*Prepared as Package Directs
[1]Principal source of fat: vegetable shortening.

Food and Description	Measure or Quantity	Sodium (mg.)	Total	Fats in grams — Satu-rated	Unsatu-rated	Choles-terol (mg.)
Chocolate or cocoa-covered:						
(USDA)[1]	1 oz.	115	6.7	2.	5.	
(Keebler) Deluxe	1 piece (9 grams)	27	2.0			
(Nabisco)	1 piece (.4 oz.)	34	2.7			
(Nabisco) *Fancy*	1 piece (.5 oz.)	41	3.3			
(Nabisco) *Pantry*	1 piece (.4 oz.)	41	2.8			
Sweet-Tooth (Sunshine)	1 piece (.4 oz.)		2.2			
Sugar-honey coated (USDA)[2]	1 oz.	143	3.2	<1.	2.	
Sugar-honey coated (Nabisco)						
Honey Maid	1 piece (7 grams)	52	.7			
Hi-Ho (Sunshine)	1 piece (4 grams)		1.0			
Hot Potatas (Old London)	⅝-oz. bag	276	3.4			
Matzo (See **MATZO**)						
Melba toast (See **MELBA**)						
Milk lunch (Nabisco) *Royal Lunch*	1 piece (.4 oz.)	66	2.2			
Munchos	1 oz.	230	10.6	4.	7.	0
Onion flavored:						
Crisps (Snow)	1 oz.		9.2			
French (Nabisco)	1 piece (2 grams)	31	.5			
Funyuns (Frito-Lay)	1 oz.	230	5.7	<1.	5.	0
Meal Mates (Nabisco)	1 piece (4 grams)	51	.5			
Onyums (General Mills)	30 pieces (.5 oz.)	111	5.4			
Rings (Old London)	½-oz. bag	170	2.8			
Rings (Wise)	½-oz. bag	133	2.2			
Rings (Wonder)	1 oz.	440	6.0			
Thins (Pepperidge Farm)	1 piece (3 grams)	20	.2			
Toast (Keebler)	1 piece (3 grams)	29	.7			
Oyster:						
(USDA)[2]	10 pieces (.4 oz.)	110	1.3	Tr.	1.	
(USDA)[2]	1 cup (1 oz.)	312	3.7	<1.	3.	
(Keebler)	1 piece (<1 gram)	4	<.1			
Dandy (Nabisco)	1 piece (<1 gram)	11	<.1			
Mini (Sunshine)	1 piece (<1 gram)		.1			
Oysterettes (Nabisco)	1 piece (<1 gram)	12	<.1			
Peanut butter 'n cheez crackers						
(Kraft)	4 crackers & ¾ oz. peanut butter	264	13.4			
Peanut butter sandwich:						
Adora (Nab)	6 pieces (1½-oz. pkg.)	498	8.6			
Cheese crackers (Wise)	1 piece (6 grams)	68	1.7			

(USDA): United States Department of Agriculture
*Prepared as Package Directs
[1]Principal sources of fat: vegetable shortening, chocolate or cocoa.
[2]Principal source of fat: vegetable shortening.

Food and Description	Measure or Quantity	Sodium (mg.)	Total	Satu- rated	Unsatu- rated	Choles- terol (mg.)
				— Fats in grams —		
Malted milk (Nab)	4 pieces (1-oz. pkg.)	188	6.7			
Malted milk (Nab)	6 pieces (1⅜-oz. pkg.)	259	9.2			
Toasted crackers (Wise)	1 piece (6 grams)	52	1.5			
Pizza Spins (General Mills)	32 pieces (½ oz.)	218	3.8			
Pizza Wheels (Wise)	¾-oz. bag	272	2.4			
Potato crisps (General Mills)	16 pieces (½ oz.)	171	5.1			
Ritz, plain (Nabisco)	1 piece (3 grams)	32	.8			
Rye thins (Pepperidge Farm)	1 piece (3 grams)	13	.2			
Rye toast (Keebler)	1 piece (4 grams)	39	.8			
Rye wafers, whole grain (USDA)	1 piece (1⅞" x 3½," 6 grams)	11	.2			
Rye wafers (Nabisco) *Meal Mates*	1 piece (4 grams)	59	.4			
Ry-Krisp:						
Seasoned	1 whole cracker (7 grams)	96	.6			
Traditional	1 whole cracker (6 grams)	75	<.1			
Saltine:						
(USDA)[1]	4 crackers (.4 oz.)	121	1.3	Tr.	1.	
Krispy (Sunshine) salted tops	1 piece (3 grams)	50	.2			
Krispy (Sunshine) unsalted tops	1 piece (3 grams)	21	.3			
Premium (Nab)	8 pieces (¾-oz. pkg.)	263	2.4			
Premium (Nabisco)	1 piece (3 grams)	35	.3			
Zesta (Keebler)	1 section (3 grams)	34	.3			
Sea toast (Keebler)	1 piece (.5 oz.)	112	1.4			
Sesame:						
(Sunshine) *La Lanne*	1 piece (3 grams)		.7			
Buttery flavored (Nabisco)	1 piece (3 grams)	35	.8			
Wafer (Keebler)	1 piece (3 grams)	32	.8			
Wafer (Nabisco), *Meal Mates*	1 piece (5 grams)	65	.7			
Sip 'n Chips (Nabisco)	1 piece (2 grams)	29	.5			
Sociables (Nabisco)	1 piece (2 grams)	30	.4			
Soda:						
(USDA)[1]	1 oz.	312	3.7	<1.	3.	
(USDA)[1]	2½" sq. (6 grams)	60	.7	Tr.	<1.	
(Nabisco) *Premium*, unsalted tops	1 piece (3 grams)	23	.3			
(Sunshine)	1 piece (4 grams)		.5			
Soya (Sunshine) *La Lanne*	1 piece (3 grams)		.9			
Star Lites (Wise)	1 cup (.5 oz.)	197	2.4			
Swedish rye wafer (Keebler)	1 piece (5 grams)	54	.3			
Taco corn chips (Old London)	1¼-oz. bag	205	7.0			

(USDA): United States Department of Agriculture
*Prepared as Package Directs
[1] Principal source of fat: vegetable shortening.

Food and Description	Measure or Quantity	Sodium (mg.)	—Fats in grams—			Choles-terol (mg.)
			Total	Satu-rated	Unsatu-rated	
Taco tortilla chips (Wonder)	1 oz.	241	8.1			
Tortilla Chips, *Doritos,* regular	1 oz.	130	6.2	1.	5.	0
Tortilla chips, *Doritos,* taco flavor	1 oz.	240	6.1	1.	5.	0
Tortilla chips (Old London)	1½-oz. bag	300	9.8			
Tortilla chips (Wonder)	1 oz.	165	7.9			
Town House	1 piece (3 grams)	42	1.0			
Triangle Thins (Nabisco)	1 piece (2 grams)	24	.3			
Triscuit (Nabisco)	1 piece (4 grams)	30	.8			
Twigs, sesame & cheese (Nabisco)	1 piece (3 grams)	32	.7			
Uneeda Biscuit (Nabisco) unsalted tops	1 piece (5 grams)	35	.6			
Wafer-ets (Hol-Grain):						
Rice, salted	1 piece (3 grams)	2	<.1			
Rice, unsalted	1 piece (3 grams)	<1	<.1			
Wheat, salted	1 piece (2 grams)	5	<.1			
Wheat, unsalted	1 piece (2 grams)	<1	<.1			
Waldorf, low salt (Keebler)	1 piece (3 grams)	<1	.4			
Waverly wafer (Nabisco)	1 piece (4 grams)	46	.8			
Wheat chips (General Mills)	12 pieces (.5 oz.)	139	4.5			
Wheat thins (Nabisco)	1 piece (2 grams)	23	.4			
Wheat toast (Keebler)	1 piece (3 grams)	25	.7			
Whistles (General Mills)	17 pieces (.5 oz.)	265	3.8			
White thins (Pepperidge Farm)	1 piece (3 grams)	22	.2			
Whole-wheat (USDA)[1]	1 oz.	155	3.9	1.	3.	
Whole-wheat, natural (Froumine)	1 piece (.4 oz.)	1	1.2			
CRACKER CRUMBS:						
Graham (USDA)[1]	1 cup (3 oz.)	576	8.1	2.	6.	
Graham (Keebler)	3 oz.	460	9.9			
Graham (Nabisco)	1½ cups (4.6 oz. or 9" pie shell)	823	14.0			
Graham (Sunshine)	3 oz.		7.5			
CRACKER JACK (See **POPCORN**)						
CRACKER MEAL:						
(USDA)[1]	3 oz.	935	11.1	3.	8.	
(USDA)	1 T. (.4 oz.)	110	1.3	Tr.	1.	
(Keebler):						
Fine, medium or coarse	3 oz.	5	.9			
Zesty	3 oz.	1047	9.5			

(USDA): United States Department of Agriculture
*Prepared as Package Directs
[1]Principal source of fat: vegetable shortening.

Food and Description	Measure or Quantity	Sodium (mg.)	— Fats in grams — Total	Satu- rated	Unsatu- rated	Choles- terol (mg.)
(Sunshine)	3 oz.		.9			
Salted (Nabisco)	1 cup (3 oz.)	1022	1.4			
Unsalted (Nabisco)	1 cup (3 oz.)	51	1.4			

CRACKER PIE CRUST MIX
(See **PIECRUST MIX**)

CRANAPPLE DRINK (Ocean Spray):

Regular	½ cup (4.5 oz.)	3	.3			0
Low calorie	½ cup (4.2 oz.)	4	.1			0
*Frozen	½ cup (4.4 oz.)	2	Tr.			0

CRANBERRY:
Fresh:

Untrimmed (USDA)	1 lb. (weighed with stems)	9	3.0			0
Stems removed (USDA)	1 cup (4 oz.)	2	.8			0
(Ocean Spray)	1 oz.	<1	.4			0
Dehydrated (USDA)	1 oz.	5	1.9			

CRANBERRY JUICE COCKTAIL:

(USDA) approx. 33% cranberry juice	½ cup (4.4 oz.)	1	.1			0
Regular (Ocean Spray)	½ cup (4.4 oz.)	2	<.1			0
Low calorie (Ocean Spray)	½ cup (4.4 oz.)	5	<.1			0
Frozen (Ocean Spray)	½ cup (4.4 oz.)	1	Tr.			

CRANORANGE JUICE DRINK,

frozen (Ocean Spray)	½ cup (4.4 oz.)	1	.2			

CRANBERRY-ORANGE RELISH:

Uncooked (USDA)	4 oz.	1	.5			0
(Ocean Spray)	4 oz.	12	.4			0

CRANBERRY PIE (Tastykake)

CRANBERRY PIE (Tastykake)	4-oz. pie		14.6			

CRANBERRY SAUCE:
Home recipe, sweetened,

unstrained (USDA)	4 oz.	1	.3			0

Canned:

Sweetened, strained (USDA)	½ cup (4.8 oz.)	1	.3			0
Jellied (Ocean Spray)	4 oz.		1.5			0
Whole berry (Ocean Spray)	4 oz.		.4			0

(USDA): United States Department of Agriculture
*Prepared as Package Directs

Food and Description	Measure or Quantity	Sodium (mg.)	Fats in grams — Total	Satu- rated	Unsatu- rated	Choles- terol (mg.)
CRANBREAKER MIX						
(Bar-Tender's)	1 serving (⅝ oz.)	Tr.	.2			(0)
CRANPRUNE (Ocean Spray)	½ cup (4.4 oz.)	3	.1			0
CRAPPIE, white, raw, meat only						
(USDA)	4 oz.		.9			
CRAYFISH, freshwater (USDA):						
Raw, in shell	1 lb. (weighed in shell)		.3			
Raw, meat only	4 oz.		.6			
CREAM:						
Half & half:						
(USDA)	1 cup (8.5 oz.)	111	28.3	15.	14.	104
(USDA)	1 T. (.5 oz.)	7	1.8	<1.	<1.	6
10.5% fat (Sealtest)	1 cup (8.5 oz.)	104	25.2			
12% fat (Sealtest)	1 cup (8.5 oz.)	102	28.8			
Light, table, or coffee:						
(USDA)	1 cup (8.5 oz.)	103	49.4	26.	23.	158
(USDA)	1 T. (.5 oz.)	6	3.1	2.	1.	10
16% fat (Sealtest)	1 T. (.5 oz.)	6	2.4			
18% fat (Sealtest)	1 T. (.5 oz.)	6	2.7			
25% fat (Sealtest)	1 T. (.5 oz.)	6	3.8			
Light whipping:						
(USDA)	1 cup (8.4 oz.)	86	74.8	41.	34.	
(USDA)	1 T. (.5 oz.)	5	4.7	3.	2.	
30% fat (Sealtest)	1 T. (.5 oz.)	5	4.5			
Whipped topping, pressurized:						
(USDA)	1 cup (2.1 oz.)		14.	8.	6.	51
(USDA)	1 T. (3 grams)		<1.	Tr.	Tr.	3
Heavy whipping:						
Unwhipped (USDA)	1 cup (8.4 oz. or 2 cups whipped)	76	89.5	50.	40.	317
Unwhipped (USDA)	1 T. (.5 oz.)	5	5.6	3.	2.	20
36% fat (Sealtest)	1 T. (.5 oz.)	5	5.4			
Sour:						
(USDA)	1 cup (8.1 oz.)	99	47.4	25.	22.	152
(USDA)	1 T. (.4 oz.)	5	2.5	1.	1.	8
(Borden)	1 cup (8.6 oz.)	96	43.2			
(Borden)	1 T. (.5 oz.)	6	2.7			
(Breakstone)	8-oz. container	112	41.5			112

(USDA): United States Department of Agriculture
*Prepared as Package Directs

Food and Description	Measure or Quantity	Sodium (mg.)	—Fats in grams— Total	Satu- rated	Unsatu- rated	Choles- terol (mg.)
(Breakstone)	1 T. (.5 oz.)	8	2.8			8
˙(Sealtest)	1 T. (.5 oz.)	6	2.7			
Half & half (Sealtest)	1 T. (.5 oz.)	6	1.8			
Imitation:						
(Sealtest) non-dairy	1 T. (.5 oz.)	16	2.7			
Sour Treat (Delite)	1 T. (.5 oz.)	6	2.2			
Zest (Borden) 13.5% vegetable fat	1 T.	16	2.0			
Sour cream, dried (Data from General Mills)	1 oz.	87	16.2			
Sour dairy dressing (Sealtest)	1 T. (.5 oz.)	7	1.8			
Sour dressing or sour cream, made with nonfat dry milk:						
(USDA)	1 cup (8.3 oz.)		38.	35.	3.	
(USDA)	1 T. (.4 oz.)		2.	2.	Tr.	
Sour dressing, cultured (Breakstone)	1 T. (.5 oz.)	9	2.4			0
CREAMIES (Tastykake):						
Banana cake	1⅞-oz. pkg.		10.1			
Chocolate	1⅞-oz. pkg.		17.7			
Koffee Kake	1⅞-oz. pkg.		10.3			
Vanilla	1⅞-oz. pkg.		16.5			
CREAM PUFF, home recipe, with custard filling (USDA)[1]	3½" x 2" (4.6 oz.)	108	18.1	5.	13.	187
CREAM OF RICE, cereal, no salt added	4 oz.	<1	<.1			
CREAMSICLE (Popsicle Industries)	2½-fl. oz. bar (1.9 oz.)	14	2.6			
CREAM or CREME SOFT DRINK:						
Sweetened:						
(Canada Dry) vanilla	6 fl. oz.	13+	0.			0
(Clicquot Club)	6 fl. oz.	16	0.			0
(Cott)	6 fl. oz.	16	0.			0
(Dr. Brown's)	6 fl. oz.	14	0.			0
(Fanta)	6 fl. oz.	6	0.			0
(Hoffman)	6 fl. oz.	14	0.			0
(Key Food)	6 fl. oz.	14	0.			0

(USDA): United States Department of Agriculture
*Prepared as Package Directs
[1]Principal sources of fat: vegetable shortening, egg & milk.

Food and Description	Measure or Quantity	Sodium (mg.)	Total	—Fats in grams— Satu- rated	Unsatu- rated	Choles- terol (mg.)
(Kirsch)	6 fl. oz.	<1	0.			0
(Mission)	6 fl. oz.	16	0.			0
(Nedick's)	6 fl. oz.	14	0.			0
(Shasta)	6 fl. oz.	20	0.			0
(Waldbaum)	6 fl. oz.	14	0.			0
(Yukon Club)	6 fl. oz.	14	0.			0
Low calorie:						
(Clicquot Club)	6 fl. oz.	45	0.			0
(Cott)	6 fl. oz.	45	0.			0
(Dr. Brown's)	6 fl. oz.	35	0.			0
(Hoffman)	6 fl. oz.	35	0.			0
(Key Food)	6 fl. oz.	35	0.			0
(Mission)	6 fl. oz.	45	0.			0
(No-Cal)	6 fl. oz.	12	0.			0
(Shasta)	6 fl. oz.	39	0.			0
(Waldbaum)	6 fl. oz.	35	0.			0
(Yukon Club)	6 fl. oz.	69	0.			0

CREAM SUBSTITUTE:

Food and Description	Measure or Quantity	Sodium (mg.)	Total	Satu- rated	Unsatu- rated	Choles- terol (mg.)
Liquid, frozen (USDA)	1 cup (8.6 oz.)		27.0	25.	2.	
Liquid, frozen (USDA)	1 tsp. (5 grams)		.6	<1.	Tr.	
Powdered (USDA)	1 cup (3.3 oz.)		33.0	31.	2.	
Powdered (USDA)	1 tsp. (2 grams)		1.0	<1.	Tr.	
Coffee-mate (Carnation)[1]	1 tsp. (2 grams)	3	.7	<1.	Tr.	Tr.
Coffee-mate (Carnation)[1]	1 packet (3 grams)	4	1.1	1.	Tr.	Tr.
Coffee Rich	1 tsp. (5 grams)	2	.6			0
Coffee Twin (Sealtest)	½ fl. oz. (.5 oz.)	6	1.2			
Cremora (Borden)	1 tsp. (2 grams)	2	.7			
Perx	1 tsp. (5 grams)	<1	.6	<1.	0.	0
Poly Perx	1 oz.	4	2.8	Tr.	2.	0

CREAM OF TARTAR

Food and Description	Measure or Quantity	Sodium (mg.)	Total	Satu- rated	Unsatu- rated	Choles- terol (mg.)
(Spice Islands)	1 tsp.	8				(0)

CREAM OF WHEAT, cereal:

Food and Description	Measure or Quantity	Sodium (mg.)	Total	Satu- rated	Unsatu- rated	Choles- terol (mg.)
Instant, dry	1 oz. (¾ cup cooked)	2	.3			
Mix 'n Eat:						
Regular, dry	3½ T. (1 oz.)	220	.4			
Baked apple & cinnamon, dry	3¾ T. (1¼ oz.)	228	.7			
Maple & brown sugar, dry	3¾ T. (1¼ oz.)	206	.5			
Quick, dry	1 oz. (¾ cup cooked)	78	.3			
Regular, dry	1 oz. (¾ cup cooked)	<1	.3			

(USDA): United States Department of Agriculture
*Prepared as Package Directs
[1]Principal source of fat: modified coconut oil.

Food and Description	Measure or Quantity	Sodium (mg.)	— Fats in grams —			Choles- terol (mg.)
			Total	Satu- rated	Unsatu- rated	
CRESS, GARDEN (USDA):						
Raw, whole	1 lb. (weighed untrimmed)	45	2.3			0
Boiled without salt, in small amount of water, short time, drained (USDA)	1 cup (6.3 oz.)	14	1.1			0
Boiled without salt, large amount of water, long time, drained (USDA)	1 cup (6.3 oz.)	14	1.1			0
CRISP RICE, cereal (Van Brode)	1 oz.		<.1			(0)
CRISPY CRITTERS, cereal (Post)	1 cup (1 oz.)	170	1.1			0
CROAKER (USDA):						
Atlantic:						
Raw, whole	1 lb. (weighed whole)	134	3.4			
Raw, meat only	4 oz.	99	2.5			
Baked	4 oz.	136	3.6			
White, raw, meat only	4 oz.		.9			
Yellowfin, raw, meat only	4 oz.		.9			
CRULLER (See **DOUGHNUT**)						
CUCUMBER, fresh (USDA):						
Eaten with skin	½ lb. (weighed with skin)	13	.2			0
Eaten without skin	½ lb. (weighed with skin)	10	.2			0
Not pared, 10-oz. cucumber	7½" x 2" pared cucumber (7.3 oz.)	12	.2			0
Pared	6 slices (2" x ⅛", 1.8 oz.)	3	<.1			0
Pared & diced	½ cup (2.5 oz.)	4	<.1			0
CUMIN SEED, whole or ground (Spice Islands)	1 tsp.	<1				(0)
CUPCAKE:						
Home recipe (USDA)[1]:						
Made with butter, without icing[2]	2¾" cupcake (1.4 oz.)	120	5.1	3.	2.	

(USDA): United States Department of Agriculture
*Prepared as Package Directs
[1]Made with sodium aluminum sulfate-type baking powder.
[2]Principal sources of fat: butter, egg & milk.

Food and Description	Measure or Quantity	Sodium (mg.)	Fats in grams Total	Satu- rated	Unsatu- rated	Choles- terol (mg.)
Made with vegetable shortening, without icing[1]	2¾" cupcake (1.4 oz.)	120	5.6	2.	4.	
With chocolate icing	2¾" cupcake (1.8 oz.)	114	7.0			
With boiled white icing	2¾" cupcake (1.8 oz.)	131	5.2			
With uncooked white icing	2¾" cupcake (1.8 oz.)	114	5.9			
Commercial:						
Chocolate (Tastykake)	1 cupcake (1 oz.)		5.7			
Chocolate, chocolate-creme filled (Tastykake)	1 cupcake (1¼ oz.)		3.0			
Coconut (Tastykake)	1 cupcake (¾ oz.)		2.3			
Creme-filled, chocolate butter cream (Tastykake)	1 cupcake (1⅛ oz.)		6.7			
Devil's food cake (Hostess):						
2 to pkg.	1 cupcake (1¾ oz.)	194	4.6			
12 to pkg.	1 cupcake (1.3 oz.)	148	3.5			
Lemon, creme-filled (Tastykake)	1 cupcake (⅞ oz.)		5.7			
Orange (Hostess):						
2 to pkg.	1 cupcake (1.5 oz.)	124	4.3			
12 to pkg.	1 cupcake (1.3 oz.)	110	3.8			
Orange, creme-filled (Tastykake)	1 cupcake (⅞ oz.)		6.7			
Vanilla, creme-filled (Tastykake)	1 cupcake (⅞ oz.)		5.9			
Vanilla *Triplets* (Tastykake)	1 cupcake (.8 oz.)		3.6			

CUPCAKE MIX:

Food and Description	Measure or Quantity	Sodium (mg.)	Total	Satu- rated	Unsatu- rated	Choles- terol (mg.)
(USDA)[2]	4 oz.	676	15.4	3.	12.	
*Prepared with eggs, milk, without icing (USDA)[1]	2½" cupcake (.9 oz.)	113	3.0	<1.	2.	
*Prepared with eggs, milk, with chocolate icing (USDA)[3]	2½" cupcake (1.2 oz.)	121	4.5	2.	3.	
*(Flako)	1 large cupcake (1.3 oz., ¹/₁₂ of pkg.)	178	5.0			

CURRANT:

Food and Description	Measure or Quantity	Sodium (mg.)	Total	Satu- rated	Unsatu- rated	Choles- terol (mg.)
Fresh (USDA):						
Black European:						
Whole	1 lb. (weighed with stems)	13	.4			0
Stems removed	4 oz.	3	.1			0
Red & white:						
Whole	1 lb. (weighed with stems)	9	.9			0

(USDA): United States Department of Agriculture
*Prepared as Package Directs
[1]Principal sources of fat: vegetable shortening, egg & milk.
[2]Principal source of fat: vegetable shortening.
[3]Principal sources of fat: vegetable shortening, chocolate, egg & milk.

Food and Description	Measure or Quantity	Sodium (mg.)	— Fats in grams —			Choles- terol (mg.)
			Total	Satu- rated	Unsatu- rated	
Stems removed	1 cup (3.9 oz.)	2	.2			0
Dried, Zante (Del Monte)	½ cup (2.5 oz.)	20	.2			0
CURRY POWDER:						
(Crosse & Blackwell)	1 T. (6 grams)		.2			(0)
(Spice Islands)	1 tsp.	1				(0)
CUSK (USDA):						
Raw, drawn	1 lb. (weighed drawn, head & tail on)		.5			
Raw, meat only	4 oz.		.2			
Steamed	4 oz.	84	.8			
CUSTARD:						
Home recipe, baked (USDA)	½ cup (4.7 oz.)	104	7.3	4.	3.	139
Chilled (Sealtest)	4 oz.	53	3.8			
CUSTARD APPLE, bullock's heart, fresh (USDA):						
Whole	1 lb. (weighed with skin & seeds)		1.6			0
Flesh only	4 oz.		.7			0
CUSTARD, FROZEN (See **ICE CREAM**)						
CUSTARD PIE:						
Home recipe (USDA)[1]	⅙ of 9″ pie (5.4 oz.)	436	16.9			160
Frozen (Banquet)	5-oz. serving		9.1			
CUSTARD PUDDING MIX:						
Dry, with vegetable gum base (USDA)	1 oz.	84	<.1			
*Prepared with whole milk (USDA)	4 oz.	112	4.0	2.	2.	
*No egg yolk (Jell-O)	½ cup (5 oz.)	180	5.8			29
Real egg (Lynden Farms)	4-oz. pkg.	338	5.7	2.	3.	
*Regular (Royal)	½ cup (5.1 oz.)	125	4.9			14

(USDA): United States Department of Agriculture
*Prepared as Package Directs
[1]Principal source of fat: lard.

Food and Description	Measure or Quantity	Sodium (mg.)	—Fats in grams—			Cholesterol (mg.)
			Total	Saturated	Unsaturated	

D

DAIQUIRI COCKTAIL:
  (National Distillers)

Duet, 12½% alcohol	8 fl.-oz. can	Tr.	0.			(0)
Mix (Bar-tender's)	1 serving (⅝ oz.)	48	.2			(0)

DAMSON PLUM (See **PLUM**)

DANDELION GREENS, raw
  (USDA):

Trimmed	1 lb.	345	3.2			0
Boiled without salt, drained	½ cup (3.2 oz.)	40	.5			0

DANISH PASTRY (See
  COFFEE CAKE)

DANISH-STYLE VEGETABLES,

frozen (Birds Eye)	⅓ of 10-oz. pkg.	402	7.0			0

DANNY (See **YOGURT**)

DATE, dry:
  Domestic:

With pits (USDA)	1 lb. (weighed with pits)	4	2.0			0
Without pits (USDA)	4 oz.	1	.6			0
Without pits, chopped (USDA)	1 cup (6.1 oz.)	2	.9			0
Whole (Cal-Date)	1 date (.8 oz.)	<1	.1			
Diced (Cal-Date)	4 oz.	1	.7			
Chopped (Dromedary)	1 cup (5 oz.)	17	2.5			
Pitted (Dromedary)	1 cup (5 oz.)	17	1.3			
Imported, Iraq (Bordo):						
Whole	4 average dates (.9 oz.)	52	.6			
Diced	¼ cup (2 oz.)	114	1.4			

DELAWARE WINE:

(Gold Seal) 12% alcohol	3 fl. oz. (3.2 oz.)	3	0.			(0)
(Great Western) 12.5% alcohol	3 fl. oz.	2	0.			

DESSERT CUP (Del Monte):

Pudding 'n apricot	5-oz. container	241	3.3			

(USDA): United States Department of Agriculture
*Prepared as Package Directs

Food and Description	Measure or Quantity	Sodium (mg.)	— Fats in grams —			Choles- terol (mg.)
			Total	Satu- rated	Unsatu- rated	
Pudding 'n peach	5-oz. container	266	3.0			
Pudding 'n pineapple	5-oz. container	244	3.4			
DEVIL DOGS (Drakes)	1 piece (1.6 oz.)		7.6			
DEVIL'S FOOD CAKE:						
Home recipe (USDA)[1]:						
Without icing	3″ x 2″ x 1½″ (1.9 oz.)	162	9.5			
With chocolate icing, 2-layer	1/16 of 9″ cake (2.6 oz.)	176	12.3			32
With chocolate icing, 2-layer	1/16 of 10″ cake (4.2 oz.)	282	19.7			52
With uncooked white icing	1/16 of 10″ cake (4.2 oz.)	281	17.5			
Commercial, frozen:						
With chocolate icing (USDA)[2]	2 oz.	238	10.0	5.	5.	
With whipped-cream filling & chocolate icing (USDA)[3]	2 oz.	108	12.4	4.	8.	
(Pepperidge Farm)	1/6 of cake (3.1 oz.)	305	14.0			
Layer (Mrs. Smith's)	1/6 of 14-oz. cake	227	6.6			
DEVIL'S FOOD CAKE MIX:						
Dry (USDA)[4]	1 oz.	130	3.3	<1.	3.	
*With chocolate icing (USDA)[5]	1/16 of 9″ cake (2.4 oz.)	181	8.5	3.	5.	33
*(Duncan Hines)	1/12 of cake (2.7 oz.)	491	6.8			50
*(Swans Down)	1/12 of cake (2.4 oz.)	441	3.2			48
*Butter (Betty Crocker)	1/12 of cake	358	13.1			
*Layer (Betty Crocker)	1/12 of cake	320	5.9			
DEWBERRY, fresh (See **BLACKBERRY,** fresh)						
DEWBERRY PRESERVE (Bama)	1 T. (.7 oz.)	1	<.1			
DILL:						
Seed (Data from General Mills)	1 oz.	28				(0)
Seed (Spice Islands)	1 tsp.	<1				(0)
Weed (Spice Islands)	1 tsp.	<1				(0)

(USDA): United States Department of Agriculture
*Prepared as Package Directs
[1]Made with sodium aluminum sulfate-type baking powder.
[2]Principal sources of fat: butter, vegetable shortening, chocolate, egg & milk.
[3]Principal sources of fat: vegetable shortening, cream, chocolate, egg & milk.
[4]Principal source of fat: vegetable shortening.
[5]Principal sources of fat: vegetable shortening, chocolate, egg & milk.

Food and Description	Measure or Quantity	Sodium (mg.)	— Fats in grams —			Choles- terol (mg.)
			Total	Satu- rated	Unsatu- rated	

DING DONG (Hostess):
Dark chocolate, 2 to pkg.	1 piece (1.4 oz.)	96	9.1			
Dark chocolate, 12 to pkg.	1 piece (1.3 oz.)	93	8.8			
Milk chocolate, 2 to pkg.	1 piece (1.4 oz.)	101	8.7			
Milk chocolate, 12 to pkg.	1 piece (1.3 oz.)	98	8.4			

DINNER, frozen (See individual listings such as **BEEF DINNER, CHICKEN DINNER, CHINESE DINNER, ENCHILADA DINNER,** etc.)

DIP:
Bacon & horseradish:						
(Breakstone)	2 T. (1.1 oz.)	224	5.5			12
(Kraft) *Teez*	1 oz.	201	5.0			
(Kraft) *Ready Dip*, Neufchâtel cheese	1 oz.	227	6.7			
Bacon & smoke (Sealtest) *Dip'n Dressing*	1 oz.	157	3.9			
Blue cheese:						
(Breakstone)	2 T. (1.1 oz.)	228	5.8			15
(Kraft) *Ready Dip*, Neufchâtel cheese	1 oz.	272	6.0			
(Kraft) *Teez*	1 oz.	156	4.6			
(Sealtest) *Dip 'n Dressing*	1 oz.	205	4.2			
Casino (Sealtest) *Dip 'n Dressing*	1 oz.	164	3.5			
Chipped beef (Sealtest) *Dip 'n Dressing*	1 oz.	192	3.6			
Clam:						
(Kraft) *Ready Dip*, Neufchâtel cheese	1 oz.	170	5.8			
(Kraft) *Teez*	1 oz.	143	3.9			
Cucumber & onion (Breakstone)	2 T. (1.1 oz.)	184	4.7			12
Dill pickle (Kraft) *Ready Dip*, Neufchâtel cheese	1 oz.	215	5.8			
Garlic (Kraft) *Teez*	1 oz.	120	4.3			
Green goddess (Kraft) *Teez*	1 oz.	200	4.2			
Jalapeño bean (Fritos)	1 oz.	170	1.4	<1.	<1.	
Onion:						
(Borden) French	1 oz.		5.3			
(Breakstone)	2 T. (1.1 oz.)	174	5.1			13
(Kraft) French onion, *Teez*	1 oz.	145	3.9			

(USDA): United States Department of Agriculture
*Prepared as Package Directs

(151)

DIP (Continued)

Food and Description	Measure or Quantity	Sodium (mg.)	Fats in grams — Total	Satu- rated	Unsatu- rated	Choles- terol (mg.)
(Kraft) *Ready Dip*, Neufchâtel cheese	1 oz.	242	6.1			
(Sealtest) French onion, *Dip'n Dressing*	1 oz.	159	3.7			
& garlic (Sealtest) *Dip'n Dressing*	1 oz.	166	3.7			
Tasty Tartar (Borden)	1 oz.		4.1			
Western Bar B-Q (Borden)	1 oz.		4.1			
DIP MIX:						
Any flavor (Fritos)	1 pkg. (.6 oz.)					0
Green onion (Lawry's)	1 pkg. (.6 oz.)		.1			
Guacomole (Lawry's)	1 pkg. (.6 oz.)		3.5			
Toasted onion (Lawry's)	1 pkg. (.6 oz.)		.2			
DISTILLED LIQUOR, 80 proof to 100 proof (USDA)	1 fl. oz. (1 oz.)	<1				0
DOCK, including **SHEEP SORREL** (USDA):						
Raw, whole	1 lb. (weighed un-trimmed)	16	1.0			0
Raw, trimmed	4 oz.	6	.3			0
Boiled, without salt, drained	4 oz.	3	.2			0
DOGFISH, Spiny, raw, meat only (USDA)	4 oz.		10.2			
DOLLY VARDEN, raw, meat & skin (USDA)	4 oz.		7.4			
DOUGHNUT:						
Cake type:						
(USDA)[1]	1 piece (1.1 oz.)	160	6.0	1.	5.	
(Hostess) 10 to pkg.	1 piece (1¼ oz.)	178	6.7			
Powdered, frozen (Morton)	1 piece (.6 oz.)	38	4.8			
Sugar & spice, frozen (Morton)	1 piece (.6 oz.)	34	5.0			
Yeast-leavened (USDA)[1]	2 oz.	133	15.1	3.	12.	
DR. BROWN'S CEL-RAY TONIC, soft drink:						
Regular	6 fl. oz.	7	0.			0
Low calorie	6 fl. oz.	30	0.			0

(USDA): United States Department of Agriculture
*Prepared as Package Directs
[1]Principal sources of fat: vegetable shortening & egg.

152

Food and Description	Measure or Quantity	Sodium (mg.)	—Fats in grams—			Choles- terol (mg.)
			Total	Satu- rated	Unsatu- rated	
DR. PEPPER, soft drink:						
Regular, canned or bottled	6 fl. oz.	10+	0.			(0)
Sugar free	6 fl. oz.	18+	0.			(0)
DRUM, raw (USDA):						
Freshwater:						
Whole	1 lb. (weighed whole)	83	6.1			
Meat only	4 oz.	79	5.9			
Red:						
Whole	1 lb. (weighed whole)	102	.7			
Meat only	4 oz.	62	.5			
DUCK, raw (USDA):						
Domesticated:						
Ready-to-cook	1 lb. (weighed with bones)		106.4			
Meat & skin	4 oz.		32.4			
Meat only	4 oz.	84	9.3			
Wild:						
Dressed	1 lb. (weighed dressed)		41.6			
Meat, skin and giblets	4 oz.		17.9			
Meat only	4 oz.		5.9			

E

Food and Description	Measure or Quantity	Sodium (mg.)	Total	Satu- rated	Unsatu- rated	Choles- terol (mg.)
ECLAIR, home recipe, with custard filling & chocolate icing (USDA)[1]	4 oz.	93	15.4	5.	11.	
EEL (USDA):						
Raw, meat only	4 oz.		20.8	5.	16.	
Smoked, meat only	4 oz.		31.5	7.	25.	
EGG BEATERS (Fleischmann's)	¼ cup (2.1 oz.)		7.5			<1
EGG, BREAKFAST:						
Frozen, scrambled (Swanson):						
With coffee cake	6½-oz. breakfast	706	35.9			
With link sausage & coffee cake	5½-oz. breakfast	703	31.4			
Mix (Durkee):						
Scrambled, plain	1 pkg. (.8 oz.)	320	9.5	3.	7.	226

(USDA): United States Department of Agriculture
*Prepared as Package Directs
[1]Principal sources of fat: vegetable shortening, egg, milk, chocolate & butter.

Food and Description	Measure or Quantity	Sodium (mg.)	Fats in grams — Total	Satu- rated	Unsatu- rated	Choles- terol (mg.)
Scrambled, with bacon	1 pkg. (1¼ oz.)	476	13.4	4.	10.	357
Western omelet	1 pkg. (1¼ oz.)	489	10.5	3.	7.	259
EGG, CHICKEN (USDA):						
Raw:						
White only	1 large egg (1.2 oz.)	48	Tr.			0
White only	1 cup (9 oz.)	372	Tr.			0
Yolk only	1 large egg (.6 oz.)	9	5.2	2.	4.	250
Yolk only	1 cup (8.5 oz.)	125	73.4	24.	49.	3552
Whole, small	1 egg (1.3 oz.)	45	4.3	1.	3.	186
Whole, medium	1 egg (1.5 oz.)	53	5.0	2.	3.	220
Whole	1 cup (8.8 oz.)	306	28.9	10.	19.	1265
Whole, large	1 egg (1.8 oz.)	61	5.7	2.	4.	251
Whole, extra large	1 egg (2 oz.)	70	6.6	2.	4.	289
Whole, jumbo	1 egg (2.3 oz.)	79	7.4	3.	5.	325
Cooked:						
Boiled without salt	1 large egg (1.8 oz.)	61	5.7	2.	4.	251
Fried in butter[1]	1 large egg (1.6 oz.)	155	7.9	3.	5.	
Omelet, mixed with milk & cooked in fat[2]	1 large egg (2.2 oz.)	159	8.0	3.	5.	
Poached	1 large egg (1.7 oz.)	130	5.6	2.	4.	242
Scrambled, mixed with milk & cooked in fat[2]	1 cup (7.8 oz.)	565	28.4	11.	17.	904
Scrambled, mixed with milk & cooked in fat[2]	1 large egg (2.3 oz.)	164	8.3	3:	5.	263
Dried:						
White, flakes	1 oz.	293	<.1			
White, powder	1 oz.	313	<.1			
Yolk	1 cup (3.4 oz.)	96	54.3	17.	37.	2525
Whole	1 cup (3.8 oz.)	461	44.5	14.	30.	2052
Whole, glucose reduced	1 oz.	126	12.2	4.	8.	
Frozen, whole, raw	1 oz.	35	3.3	1.	2.	
EGG, DUCK, raw (USDA)	1 egg (2.8 oz.)	98	11.6			
EGG, GOOSE, raw (USDA)	1 egg (5.8 oz.)		21.8			
EGG, TURKEY, raw (USDA)	1 egg (3.1 oz.)		10.4			
EGG McMUFFIN, (McDonald's)	1 piece (4.5 oz.)	1130	11.3			

(USDA): United States Department of Agriculture
*Prepared as Package Directs
[1]Principal sources of fat: egg & butter.
[2]Principal sources of fat: egg, milk & vegetable fat.

Food and Description	Measure or Quantity	Sodium (mg.)	Total	Satu- rated	Unsatu- rated	Choles- terol (mg.)
			—Fats in grams—			

EGG NOG Dairy: | | | | | | |
(Borden) 4.69% fat[1]	½ cup (4.2 oz.)		5.9			
(Borden) 6% fat[1]	½ cup (4.3 oz.)		8.1			
(Borden) 8% fat[1]	½ cup (4.3 oz.)		10.5			
(Sealtest) 6% butterfat	½ cup (4.6 oz.)	80	8.7			
(Sealtest) 8% butterfat	½ cup (4.6 oz.)	73	11.3			

EGGPLANT: | | | | | | |
Raw, whole (USDA)	1 lb. (weighed un- trimmed)	7	.7			0
Boiled without salt, drained (USDA)	4 oz.	1	.2			0
Boiled without salt, drained, diced (USDA)	1 cup (7.1 oz.)	2	.4			0
Frozen, fried sticks (Mrs. Paul's)	7-oz. pkg.		27.6			
Frozen, parmesan (Mrs. Paul's)	11-oz. pkg.		35.3			
Frozen, parmigiana (Buitoni)	4 oz.		10.7			
Frozen, slices (Mrs. Paul's)	9-oz. pkg.		48.2			

EGG ROLL, shrimp, frozen (Hung's) | 1 roll | | 5.1 | | | |

***EGGSTRA** (Tillie Lewis) | 1 large egg (1.8 oz.) | 80 | 1.2 | | | 57 |

ELDERBERRY, fresh (USDA): | | | | | | |
| Whole | 1 lb. (weighed with stems) | | 2.1 | | | 0 |
| Stems removed | 4 oz. | | .6 | | | 0 |

ENCHILADA, beef, frozen: | | | | | | |
With cheese & chili gravy (Banquet)	8 enchiladas (2 lbs.)		60.4			
With rice (Swanson)	9⅝-oz. pkg.	1018	16.1			
With sauce (Banquet)	6-oz. bag		12.2			

ENCHILADA DINNER, frozen: | | | | | | |
(Banquet):						
Meat compartment	6¼ oz.		12.4			
Rice compartment	3 oz.		.1			
Beans compartment	3¼ oz.		4.0			
Complete dinner	12-oz. dinner		16.6			
(Patio) 5-compartment	13-oz. dinner		17.0			
(Patio) 3-compartment	13-oz. dinner		16.0			
(Swanson)	15-oz. dinner	2420	27.2			

(USDA): United States Department of Agriculture
*Prepared as Package Directs
[1]Principal sources of fat: milk and egg yolk.

Food and Description	Measure or Quantity	Sodium (mg.)	—Fats in grams— Total	Satu- rated	Unsatu- rated	Choles- terol (mg.)
Cheese:						
(Banquet):						
Meat compartment	6¼ oz.		14.0			
Rice compartment	3 oz.		.1			
Beans compartment	3¼ oz.		2.4			
Complete dinner	12½-oz. dinner		16.6			
(Patio) 5-compartment	12-oz. dinner		15.0			
(Patio) 3-compartment	12-oz. dinner		13.0			
ENCHILADA MIX (Lawry's)	1.6-oz. pkg.		1.8			
ENDIVE, BELGIAN or FRENCH (See **CHICORY, WITLOOF**)						
ENDIVE, CURLY, or ESCAROLE, raw (USDA):						
Untrimmed	1 lb. (weighed un-trimmed)	56	.4			0
Trimmed	½ lb.	32	.2			0
Cut up or shredded	1 cup (2.5 oz.)	10	<.1			0
ESCAROLE (See **ENDIVE**)						
EULACHON or SMELT, raw, meat only (USDA)	4 oz.		7.0			
FARINA (See also **CREAM OF WHEAT**):						
Regular:						
Dry:						
(USDA)	1 cup (6 oz.)	3	1.5			0
Cream, enriched (H-O)	1 cup (6.1 oz.)	3	1.4			0
Pearls of Wheat (Albers)	1 cup		1.4			(0)
Cooked:						
(USDA)	1 cup (8.4 oz.)	343	.2			0
(Quaker)	1 cup (1 oz. dry)	<1	.2			(0)
Quick-cooking (USDA):						
Dry	1 oz.	71	.3			0
Cooked	1 cup (8.6 oz.)	466	.2			0
Instant-cooking (USDA):						
Dry	1 oz.	2	.3			0
Cooked	4 oz.	213	.1			0

F

(USDA): United States Department of Agriculture
*Prepared as Package Directs

Food and Description	Measure or Quantity	Sodium (mg.)	Total	Fats in grams Satu- rated	Unsatu- rated	Choles- terol (mg.)
FAT, COOKING:						
Lard (USDA)	1 cup (7.2 oz.)	0.	205.	78.	127.	
Lard (USDA)	1 T. (.5 oz.)	0.	13.0	5.	8.	
Vegetable (USDA)	1 cup (7.1 oz.)	0.	200.0	50.	150.	0
Vegetable (USDA)	1 T. (.4 oz.)	0.	12.0	3.	9.	0
Crisco	1 T. (.4 oz.)	0.	11.7	3.	9.	0
Fluffo	1 T. (.4 oz.)	0.	11.7	4.	7.	Tr.
Light Spry	1 T. (.4 oz.)	0.	10.6	3.	8.	0.
Light Spry	¼ lb.	0.	113.4	32.	82.	0.
FENNEL LEAVES, raw, (USDA):						
Untrimmed	1 lb. (weighed un- trimmed)		1.7			0
Trimmed	4 oz.		.5			0
FENNEL SEED (Spice Islands)	1 tsp.	1				(0)
FIG:						
Fresh:						
(USDA)	1 lb.	9	1.4			0
Small (USDA)	1.3-oz. fig (1½″)	<1	.1			0
Candied (USDA)	1 oz.		<.1			0
Candied (Bama)	1 T. (.7 oz.)	<1	<.1			
Canned, regular pack, solids & liq.:						
Light syrup (USDA)	4 oz.	2	.2			0
Heavy syrup:						
(USDA)	½ cup (4.4 oz.)	3	.3			0
(USDA)	3 figs & 2 T. syrup	2	.2			0
(Del Monte)	½ cup (4.4 oz.)	1	.4			0
(Stokely-Van Camp)	½ cup (4.2 oz.)		.2			(0)
Extra heavy syrup (USDA)	4 oz.	2	.2			0
Canned, unsweetened or dietetic pack, solids & liq.:						
Water pack (USDA)	4 oz.	2	.2			0
Kadota (Diet Delight)	½ cup (4.4 oz.)	4	.1			(0)
(Tillie Lewis)	½ cup (4.5 oz.)	<10	.2			(0)
Whole:						
(S and W) *Nutradiet,* low calorie	6 whole figs (3.5 oz.)	2	.1			(0)
(S and W) *Nutradiet,* un- sweetened	6 whole figs (3.5 oz.)	2	1			(0)
Dried:						
Chopped (USDA)	1 cup (6 oz.)	58.	2.2			0

(USDA): United States Department of Agriculture
*Prepared as Package Directs

Food and Description	Measure or Quantity	Sodium (mg.)	Total	Fats in grams — Satu- rated	Unsatu- rated	Choles- terol (mg.)
(USDA)	.7-oz. fig (2″ x 1″)	7	.3			0
Calimyrna (Del Monte)	1 cup (5.4 oz.)	27	3.0			(0)
Mission (Del Monte)	1 cup (5.4 oz.)	30	3.5			(0)
FIG PRESERVE, sweetened (Bama)	1 T.	1	<.1			(0)
FILBERT or HAZELNUT (USDA):						
Whole	1 lb. (weighed in shell)	4	130.2	7.	123.	0
Shelled	1 oz.	<1	17.7	<1.	17.	0
FINE HERBES (Spice Islands)	1 tsp.	<1				(0)
FINNAN HADDIE (See **HADDOCK, SMOKED**)						
FISH (See individual listings)						
FISH CAKE:						
Home recipe,[1] fried (USDA)	2 oz.		4.5			
Frozen:						
Fried, reheated (USDA)	2 oz.		10.1			
(Mrs. Paul's)	4 oz.		4.6			
Thins (Mrs. Paul's)	10-oz. pkg.		17.3			
FISH & CHIPS, frozen:						
(Gorton)	½ of 1-lb. pkg.	55	17.0			
(Mrs. Paul's)	14-oz. pkg.		21.4			
(Swanson)	5-oz. pkg.	558	11.5			
FISH CHOWDER, New England						
(Snow)	8 oz.		6.0			
FISH DINNER, frozen:						
(Morton)	8¾-oz. dinner	556	13.5			
Filet of ocean fish (Swanson)	11½-oz. dinner	1111	14.9			
With French fries (Swanson)	9¾-oz. dinner	1368	17.7			
With green beans & peach (Weight Watchers)	18-oz. dinner		2.0			
With pineapple chunks (Weight Watchers)	9½-oz. luncheon		2.2			
FISH FILLET						
Sandwich (McDonald's)	1 sandwich (4.8 oz.)	759	21.7			

(USDA): United States Department of Agriculture
*Prepared as Package Directs
[1]Prepared with canned flaked fish, potato & egg.

Food and Description	Measure or Quantity	Sodium (mg.)	— Fats in grams —			Choles- terol (mg.)
			Total	Satu- rated	Unsatu- rated	
Frozen:						
Breaded, fried (Mrs. Paul's)	14-oz. pkg.		26.0			
Buttered (Mrs. Paul's)	10-oz. pkg.		19.0			
Crisps (Gorton)	½ of 8-oz. pkg.	36	14.0			
FISH FLAKES, canned (USDA)	4 oz.		.7			
FISH LOAF, home recipe (USDA)[1]	4 oz.		4.2			
FISH PUFFS, frozen (Gorton)	½ of 8-oz. pkg.	4.1	17.0			
FISH STICK, frozen:						
Cooked, commercial, 3¾″ x 1″ x ½″ sticks (USDA)	10 sticks (8-oz. pkg.)		20.2			
(Gorton)	½ of 8-oz. pkg.	80	10.0			
Breaded, fried (Mrs. Paul's)	14-oz. pkg.		28.7			
FLAN PUDDING, chilled (Breakstone)	5-oz. container	175	3.7			10
FLOUNDER:						
Raw (USDA):						
Whole	1 lb. (weighed whole)	117	1.2			75
Meat only	4 oz.	88	.9			57
Baked (USDA)	4 oz.	269	9.3			
Frozen:						
(Gorton)	1-lb. pkg.	351	3.6			
Dinner (Weight Watchers)	18-oz. dinner		3.0			
& broccoli (Weight Watchers)	9½-oz. luncheon		4.8			
FLOUR:						
Buckwheat, dark, sifted (USDA)	1 cup (3.5 oz.)		2.4			0
Buckwheat, light, sifted (USDA)	1 cup (3.5 oz.)		1.2			0
Carob or St. John's-bread (USDA)	1 oz.		.4			0
Chestnut (USDA)	1 oz.	3	1.0			0
Corn, sifted (USDA)	1 cup (3.9 oz.)	1	2.9	Tr.	3.	0
Cottonseed (USDA)	1 oz.		1.9	<1.	1.	0
Cottonseed (Data from General Mills)	1 oz.	8	1.2			(0)

(USDA): United States Department of Agriculture
*Prepared as Package Directs
[1]Prepared with canned flaked fish, bread crumbs, eggs, tomatoes, onion & fat.

Food and Description	Measure or Quantity	Sodium (mg.)	— Fats in grams —			Choles- terol (mg.)
			Total	Satu- rated	Unsatu- rated	
Fish, from whole fish (USDA)	1 oz.	48	<.1			
Fish, from fillets (USDA)	1 oz.	11	.1			
Fish, from fillet waste (USDA)	1 oz.	62	.6			
Lima bean (USDA)	1 oz.		.4			0
Peanut, defatted (USDA)	1 oz.	3	2.6	<1.	2.	0
Potato (USDA)	1 oz.	10	.2			0
Rice, stirred, spooned (USDA)	1 cup (5.6 oz.)	8	.8			0
Rye:						
Light (USDA):						
Unsifted, spooned	1 cup (3.6 oz.)	1	1.0			0
Sifted, spooned	1 cup (3.1 oz.)	<1	.9			0
Medium (USDA)	1 oz.	<1	.5			0
Dark (USDA):						
Unstirred	1 cup (4.5 oz.)	1	3.3			0
Stirred	1 cup (4.5 oz.)	1	3.3			0
Soybean (USDA):						
Defatted, stirred	1 cup (3.6 oz.)	1	.9			0
Low fat, stirred	1 cup (3.1 oz.)	<1	5.9	<1.	5.	0
Full fat, stirred	1 cup (2.5 oz.)	<1	14.6	2.	12.	0
High fat	1 oz.	<1	3.4	<1.	3.	0
Sunflower seed, partially defat- ted (USDA)	1 oz.	16	1.0	Tr.	1.	0
Tapioca, unsifted, spooned (USDA)	1 cup (3.8 oz.)	3	.2			0
Wheat:						
All-purpose:						
(USDA)	1 oz.	<1	.3			0
Unsifted, dipped (USDA)	1 cup (5 oz.)	3	1.4			0
Unsifted, spooned (USDA)	1 cup (4.4 oz.)	3	1.3			0
Sifted, spooned (USDA)	1 cup (4.1 oz.)	2	1.2			0
Bread:						
(USDA)	1 oz.	<1	.3			0
Unsifted, dipped (USDA)	1 cup (4.8 oz.)	3	1.5			0
Unsifted, spooned (USDA)	1 cup (4.3 oz.)	2	1.4			0
Sifted, spooned (USDA)	1 cup (4.1 oz.)	2	1.3			0
Cake or pastry:						
(USDA)	1 oz.	<1	.2			0
Unsifted, dipped (USDA)	1 cup (4.2 oz.)	2	1.0			0
Unsifted, spooned (USDA)	1 cup (3.9 oz.)	2	.9			0
Sifted, spooned (USDA)	1 cup (3.5 oz.)	2	.8			0
Gluten:						
(USDA)	1 oz.	<1	.5			0
Unsifted, dipped (USDA)	1 cup (5 oz.)	3	2.7			0
Unsifted, spooned (USDA)	1 cup (4.8 oz.)	3	2.6			0

(USDA): United States Department of Agriculture
*Prepared as Package Directs

Food and Description	Measure or Quantity	Sodium (mg.)	—Fats in grams—			Choles- terol (mg.)
			Total	Satu- rated	Unsatu- rated	
Sifted, spooned (USDA)	1 cup (4.8 oz.)	3	2.6			0
Self-rising:						
(USDA)	1 oz.	306	.3			0
Unsifted, dipped (USDA)	1 cup (4.6 oz.)	1403	1.3			0
Unsifted, spooned (USDA)	1 cup (4.5 oz.)	1370	1.3			0
Sifted, spooned (USDA)	1 cup (3.7 oz.)	1144	1.1			0
Whole wheat:						
(USDA)	1 oz.	<1	.6			0
Stirred, spooned (USDA)	1 cup (4.8 oz.)	4	2.7			0
Aunt Jemima, self-rising						
(Quaker Oats)	1 cup (4 oz.)	1400	1.2			(0)
Gold Medal (Betty Crocker):						
Regular	1 cup	3	1.4			(0)
Better-for-bread	1 cup	3	1.4			(0)
Self-rising	1 cup	1743	1.3			(0)
Wondra	1 cup	3	1.4			(0)
Presto, self-rising	1 cup (3.9 oz.)	1320	.9			0
(Quaker)	1 cup (4 oz.)	4	1.2			(0)
Softasilk for cakes (Betty Crocker)	1 cup	2	.9			(0)
FOURNIER NATURE, wine (Gold Seal) 12% alcohol	3 fl. oz.	3	0.			(0)
FRANKENBERRY, cereal (General Mills)	1 cup (1 oz.)	150	.8			(0)
FRANKFURTER or WIENER:						
Raw:						
All kinds (USDA)	1.6-oz. frankfurter	499	12.5			
All meat (USDA)	1.6-oz. frankfurter		11.6			29
With cereal (USDA)	1.6-oz. frankfurter		9.3			
With nonfat dry milk (USDA)	1.6-oz. frankfurter		11.6			
With nonfat dry milk & cereal (USDA)	1.6-oz. frankfurter		9.8			
(Armour Star) all meat	1.6-oz. frankfurter		15.1			
(Hormel) all beef:						
10 per 12-oz. pkg.	1.2-oz. frankfurter	275	9.3	4.	5.	23
10 per 1-lb. pkg.	1.6-oz. frankfurter	364	12.3	5.	6.	31
12 per 12-oz. pkg.	1-oz. frankfurter	231	7.8	3.	4.	20
(Hormel) all meat:						
10 per 12-oz. pkg.	1.2-oz. frankfurter	258	9.8	3.	5.	16
10 per 1-lb. pkg.	1.6-oz. frankfurter	342	13.0	4.	7.	21

(USDA): United States Department of Agriculture
*Prepared as Package Directs

Food and Description	Measure or Quantity	Sodium (mg.)	— Fats in grams —			Choles-terol (mg.)
			Total	Satu-rated	Unsatu-rated	
12 per 12-oz. pkg. (Oscar Mayer) all meat,	1-oz. frankfurter	217	8.2	2.	4.	13
imperial size, 5 per lb. (Oscar Mayer) all meat,	3.2-oz. frankfurter	884	26.3			
Little Wiener, 16 per 5½ oz. (Oscar Mayer) all meat, 1883,	.7-oz. frankfurter	244	7.5			
6 per lb. (Oscar Mayer) all meat wieners:	2.7-oz. frankfurter	678	18.9			
8 per lb.	2-oz. frankfurter	544	16.5	6.	10.	25
10 per lb. (Oscar Mayer) pure beef:	1.6-oz. frankfurter	430	13.0	5.	8.	20
8 per lb.	2-oz. frankfurter	508	16.5	7.	9.	27
10 per lb. (Oscar Mayer) pure beef,	1.6-oz. frankfurter	401	13.0	6.	7.	22
Machiaeh Brand	2-oz. frankfurter	531	15.9			
(Wilson) all beef	1.6-oz. frankfurter	499	12.2	6.	6.	28
(Wilson) skinless, all meat	1.6-oz. frankfurter	499	12.7	5.	7.	27
Cooked, all kinds, 10 per lb. raw (USDA)	1 frankfurter		12.2			28
Canned (USDA)	2 oz.		10.3			
Canned (Hormel)	12-oz. can		88.1			

FRANKS & BEANS (See **BEANS & FRANKS**)

FRANKS-N-BLANKETS, frozen

(Durkee)	1 piece (.4 oz.)		3.8			

FRENCH TOAST, frozen:

(Aunt Jemima)	1 slice (1.5 oz.)	220	2.2			
With link sausage (Swanson)	4½-oz. breakfast	664	15.0			

FRESCA, soft drink

	6 fl. oz.	31	Tr.			0

FROG LEGS, raw (USDA):

Bone in	1 lb. (weighed with bone)		.9			147
Meat only	4 oz.		.3			57

FROOT LOOPS, cereal (Kellogg's)

	1 cup (1 oz.)	65	.8			(0)

FROSTED RICE KRINKLES, cereal

(Post)	⅞ cup (1 oz.)	204	.1			0

(USDA): United States Department of Agriculture
*Prepared as Package Directs

Food and Description	Measure or Quantity	Sodium (mg.)	Total	Fats in grams — Satu- rated	Unsatu- rated	Choles- terol (mg.)
FROSTED SHAKE, canned, any flavor						
(Borden)	9¼-fl.-oz. can	300	12.5			
FROSTED TREAT (Weight Watchers)	1 serving (4¾ oz.)		1.4			
FROSTING (See **CAKE ICING**)						
FROSTY O's, cereal (General Mills)	1 cup (1 oz.)	155	1.2			(0)
FROZEN CUSTARD (See **ICE CREAM**)						
FROZEN DESSERT:						
Charlotte Freeze (Borden):						
Chocolate	⅓ pt. (3 oz.)		5.2			
Vanilla	⅓ pt. (3 oz.)		5.0			
Cherry (SugarLo):						
4% fat, ice milk	⅓ pt. (3.5 oz.)	69	4.0	2.	2.	13
10% fat, ice cream	⅓ pt. (3.5 oz.)	67	10.2	6.	5.	32
Chocolate:						
(Borden)	⅓ pt. (3.4 oz.)	68	8.0			
Chocolate chip, mint (Borden)	⅓ pt. (3.4 oz.)	68	9.1			
(SugarLo) 4.6% fat, ice milk	⅓ pt. (3.5 oz.)	69	4.6	2.	2.	13
(SugarLo) 10.8% fat, ice cream	⅓ pt. (3.5 oz.)	67	10.8	6.	5.	32
Coffee (SugarLo):						
4% fat, ice milk	⅓ pt. (3.5 oz.)	69	4.0	2.	2.	13
10% fat, ice cream	⅓ pt. (3.5 oz.)	67	10.2	6.	5.	32
Lemon chiffon (SugarLo):						
4% fat, ice milk	⅓ pt. (3.5 oz.)	69	4.0	2.	2.	13
10% fat, ice cream	⅓ pt. (3.5 oz.)	67	10.2	6.	5.	32
Maple (SugarLo):						
4% fat, ice milk	⅓ pt. (3.5 oz.)	69	4.0	2.	2.	13
10% fat, ice cream	⅓ pt. (3.5 oz.)	67	10.2	6.	5.	32
Orange-Pineapple (SugarLo):						
3.6% fat, ice milk	⅓ pt. (3.5 oz.)	64	3.6	2.	2.	11
9% fat, ice cream	⅓ pt. (3.5 oz.)	61	9.0	5.	4.	28
Raspberry, black (Borden)	⅓ pt. (3.4 oz.)	68	7.7			
Shake (SugarLo) chocolate	⅓ pt. (3.2 oz.)	112	.9	Tr.	Tr.	3
Shake (SugarLo) vanilla & strawberry	⅓ pt. (3.2 oz.)	108	.9	Tr.	Tr.	3
Strawberry (SugarLo):						
3.6%, ice milk	⅓ pt. (3.5 oz.)	64	3.6	2.	2.	11
9%, ice cream	⅓ pt. (3.5 oz.)	61	9.0	5.	4.	28

(USDA): United States Department of Agriculture
*Prepared as Package Directs

Food and Description	Measure or Quantity	Sodium (mg.)	— Fats in grams —			Choles- terol (mg.)
			Total	Satu- rated	Unsatu- rated	
Vanilla:						
(Borden)	⅓ pt. (3.4 oz.)	68	7.7			
(SugarLo) 4% fat, ice milk	⅓ pt. (3.5 oz.)	69	4.0	2.	2.	13
(SugarLo) 10% fat, ice cream	⅓ pt. (3.5 oz.)	67	10.2	6.	5.	32
Chocolate spin (Borden)	⅓ pt. (3.4 oz.)	68	8.4			
Fudge or raspberry swirl:						
(SugarLo) 3.9% fat, ice milk	⅓ pt. (3.5 oz.)	56	3.9	2.	2.	12
(SugarLo) 10% fat, ice cream	⅓ pt. (3.5 oz.)	59	10.0	6.	4.	31
Vanilla-coated:						
Ice cream bar (SugarLo)	2½-oz. bar	57	9.6	5.	4.	22
Ice milk bar (SugarLo)	2½-oz. bar	46	7.0	4.	3.	9
FRUIT CAKE (USDA):						
Dark, home recipe[1]	1-lb. loaf	717	69.4			206
Dark, home recipe[1]	1/30 of 8″ loaf (.5 oz.)	24	2.3			7
Dark, home recipe[1]	2″ x 2″ x ½″ slice (1.1 oz.)	47	4.6			14
Light, home recipe, made with butter[2]	1-lb. loaf	875	71.2	26.	45.	
Light, home recipe, made with butter[2]	2″ x 2″ x ½″ slice (1.1 oz.)	58	4.7	2.	3.	
Light, home recipe, made with butter[2]	1/30 of 8″ loaf (.5 oz.)	29	2.4	<1.	1.	
Light, home recipe, made with vegetable shortening[3]	1-lb. loaf	875	74.8	16.	59.	
Light, home recipe, made with vegetable shortening[3]	2″ x 2″ x ½″ slice (1.1 oz.)	58	5.0	1.	4.	
Light, home recipe, made with vegetable shortening[3]	1/30 of 8″ loaf (.5 oz.)	29	2.5	<1.	2.	
FRUIT COCKTAIL:						
Canned, regular pack, solids & liq.:						
Light syrup (USDA)	4 oz.	6	.1			0
Heavy syrup (USDA)	½ cup (4.5 oz.)	6	.1			0
Heavy syrup (Del Monte)	½ cup (4.3 oz.)	11	.1			0
Heavy syrup (Dole)	½ cup (including 3 T. syrup)	6	.1			0
Heavy syrup (Hunt's)	½ cup (4.5 oz.)	6	.1			(0)

(USDA): United States Department of Agriculture
*Prepared as Package Directs
[1]Made with sodium aluminum sulfate-type baking powder.
[2]Principal sources of fat: butter, almonds & cream.
[3]Principal sources of fat: vegetable shortening, almonds & cream.

| Food and Description | Measure or Quantity | Sodium (mg.) | — Fats in grams — | | Choles- terol (mg.) |
			Total	Satu- rated	Unsatu- rated	
Heavy syrup (Stokely-Van Camp)	½ cup (4 oz.)		.1		(0)
Extra heavy syrup (USDA)	4 oz.	6	.1		0
Canned, unsweetened or dietetic pack, solids & liq.:					
Water pack (USDA)	4 oz.	6	.1		0
(S and W) *Nutradiet,* low calorie	4 oz.	3	.1		(0)
(Diet Delight)	½ cup (4.4 oz.)	5	<.1		(0)
(S and W) *Nutradiet,* unsweetened	4 oz.	3	<.1		(0)
(Tillie Lewis)	½ cup (4.3 oz.)	<10	.1		0

FRUIT CUP, solids & liq.

(Del Monte):					
Fruit cocktail	5¼-oz. container	13	.1		(0)
Mixed fruits	5-oz. container	18	0.		(0)
Peaches, diced	5¼-oz. container	15	0.		(0)
Pineapple, in its own juice	4¼-oz. container	23	.2		(0)

FRUIT MIX, soft drink:

(Hoffman)	6 fl. oz.	14	0.		0
(Nedick's)	6 fl. oz.	14	0.		0

FRUIT, MIXED:

Dried (Del Monte)	1 cup (6.2 oz.)	56	.5		(0)
Frozen, quick thaw (Birds Eye)	½ cup (5 oz.)	Tr.	Tr.		0

FRUIT PUNCH:

(Del Monte) tropical	6 fl. oz.	2	Tr.		(0)
Soft drink, sweetened (Nehi)	6 fl. oz.	8+	0.		0

FRUIT SALAD:

Bottled, chilled (Kraft)	4 oz.	115	.1		(0)
Canned, regular pack, solids & liq.:					
Light syrup (USDA)	4 oz.	1	.1		0
Heavy syrup (USDA)	½ cup (4.3 oz.)	1	.1		0
Heavy syrup (Del Monte):					
Fruits for salad	½ cup (4.3 oz.)	30	.1		0
Tropical	½ cup (4.4 oz.)	30	.4		0
Heavy syrup (Stokely-Van Camp)	½ cup (4.2 oz.)		.1		(0)
Extra heavy syrup (USDA)	4 oz.	1	.1		0
Canned, unsweetened or dietetic pack:					
Water pack (USDA)	4 oz.	1	.1		0
(Diet Delight)	½ cup (4.4 oz.)	6	<.1		(0)
(S and W) *Nutradiet,* low calorie	4 oz.	5	.1		(0)
(S and W) *Nutradiet,* unsweetened	4 oz.	3	.1		(.0)

(USDA): United States Department of Agriculture
*Prepared as Package Directs

Food and Description	Measure or Quantity	Sodium (mg.)	Total	Satu- rated	Unsatu- rated	Choles- terol (mg.)
			—Fats in grams—			
FRUIT TREATS, apple (Mott's):						
& apricots	½ cup (4.4 oz.)		.1			(0)
& cherries	½ cup (4.6 oz.)		.1			(0)
& pineapple	½ cup (5.4 oz.)		.2			(0)
& raspberries	½ cup (4.7 oz.)		.1			(0)
& strawberries	½ cup (4.4 oz.)		.1			(0)
FRUITY PEBBLES, cereal (Post)	⅞ cup (1 oz.)	128	.1			0
FUDGE CAKE MIX:						
*Butter recipe (Duncan Hines)	¹/₁₂ of cake (3.3 oz.)	439	14.8			
*Cherry (Betty Crocker)	¹/₁₂ of cake	320	5.9			
*Dark chocolate (Betty Crocker)	¹/₁₂ of cake	293	5.9			
*Marble (Duncan Hines)	¹/₁₂ of cake (2.7 oz.)	381	6.1			50
*Sour cream, chocolate flavor layer (Betty Crocker)	¹/₁₂ of cake	243	5.9			
FUDGE ICE BAR:						
Fudgesicle (Popsicle Industries)	2½-fl. oz. bar (2.8 oz.)	52	.3			
(Sealtest)	2½-fl.-oz. bar (2.6 oz.)	55	.2			
FUDGE PUDDING, canned (Thank You)	½ cup (4.5 oz.)		5.5			

G

GARBANZO, dry (See **CHICK PEA,** dry)						
GARBANZO SOUP, canned (Hormel)	15-oz. can		23.8			
GARLIC, raw (USDA):						
Whole	2 oz. (weighed with skin)	10	.1			0
Peeled	1 oz.	5	<.1			0
GARLIC:						
Dried, chips (Spice Islands)	1 tsp.	1				(0)
Powdered (Spice Islands)	1 tsp.	1				(0)
GARLIC SPREAD (Lawry's)	1 T. (.5 oz.)		8.2			

(USDA): United States Department of Agriculture
*Prepared as Package Directs

Food and Description	Measure or Quantity	Sodium (mg.)	— Fats in grams —			Choles- terol (mg.)
			Total	Satu- rated	Unsatu- rated	

GAZPACHO SOUP, canned
| (Crosse & Blackwell) | ½ can (6½ oz.) | | 2.4 | | | |

GEFILTE FISH:
(Manischewitz):
2-lb. jar	1 piece (2.4 oz.)		2.7			
24-oz. jar	1 piece (2.6 oz.)		3.0			
1-lb. jar	1 piece (2.2 oz.)		2.5			
4-piece can	1 piece (3.7 oz.)		4.2			
2-piece can	1 piece (3.5 oz.)		4.0			
Fish balls	1 piece (1.5 oz.)		1.7			
Fishlets	1 piece (7 grams)		.3			

Whitefish & pike:
2-lb. jar	1 piece (1.7 oz.)		1.2			
1-lb. jar	1 piece (1.5 oz.)		1.1			
4-piece can	1 piece (3.8 oz.)		2.7			
2-piece can	1 piece (3.5 oz.)		2.5			
Fishlets	1 piece (7 grams)		.2			

GELATIN, unflavored, dry:
| (USDA) | 1 envelope (7 grams) | | Tr. | | | 0 |
| (Knox) | 1 envelope (7 grams) | 0 | 0. | | | (0) |

GELATIN DESSERT POWDER:
Regular:
Dry (USDA)	3-oz. pkg.	270	0.			0
Dry (USDA)	½ cup (3.3 oz.)	297	0.			0
*Prepared with water (USDA)	½ cup (4.2 oz.)	61	0.			0
*Prepared with fruit added (USDA)	½ cup (4.3 oz.)	41	.2			0
*All flavors (Jells Best)	½ cup	23	0.			0
*All flavors (Royal)	½ cup (4.2 oz.)	90				0
*Regular flavors (Jell-O)	½ cup (4.9 oz.)	46	Tr.			0
*Wild flavors (Jell-O)	½ cup (4.9 oz.)	64	Tr.			0
*Dietetic, all flavors (D-Zerta)	½ cup (4.3 oz.)	6	Tr.			0

GELATIN DRINK, plain or flavored
| (Knox) | 1 envelope (.7 oz.) | 0 | 0. | | | (0) |

GEL CUP (Del Monte):
Lemon-lime with pineapple	5-oz. container	120	0.			(0)
Orange with peaches	5-oz. container	122	0.			(0)
Strawberry with peaches	5-oz. container	126	0.			(0)

(USDA): United States Department of Agriculture
*Prepared as Package Directs

Food and Description	Measure or Quantity	Sodium (mg.)	—Fats in grams—			Cholesterol (mg.)
			Total	Satu- rated	Unsatu- rated	
GERMAN DINNER, frozen						
(Swanson)	11-oz. dinner	1372	14.7			
GINGER (Spice Islands):						
Whole	1 average piece	1				(0)
Ground	1 tsp.	1				(0)
GINGER ALE, soft drink:						
Sweetened:						
(Canada Dry) bottled	6 fl. oz.	0+	0.			0
(Canada Dry) cannned	6 fl. oz.	Tr.+	0.			0
(Clicquot Club)	6 fl. oz.	<1	0.			0
(Cott)	6 fl. oz.	<1	0.			0
(Dr. Brown's)	6 fl. oz.	8	0.			0
(Fanta)	6 fl. oz.	13	0.			0
(Hoffman) pale dry	6 fl. oz.	24	0.			0
(Key Food)	6 fl. oz.	8	0.			0
(Kirsch) pale dry	6 fl. oz.	<1	0.			0
(Mission)	6 fl. oz.	<1	0.			0
(Nedick's) pale dry	6 fl. oz.	8	0.			0
(Schweppes)	6 fl. oz.	Tr.	0.			0
(Shasta)	6 fl. oz.	10	0.			0
(Waldbaum)	6 fl. oz.	8	0.			0
(Yukon Club) golden	6 fl. oz.	3	0.			0
(Yukon Club) pale dry	6 fl. oz.	8	0.			0
Low calorie:						
(Canada Dry) bottle or can	6 fl. oz.	11+	0.			0
(Clicquot Club)	6 fl. oz.	32	0.			0
(Cott)	6 fl. oz.	32	0.			0
(Dr. Brown's) pale dry	6 fl. oz.	8	0.			0
(Hoffman) pale dry	6 fl. oz.	8	0.			0
(Key Food) pale dry	6 fl. oz.	8	0.			0
(Mission)	6 fl. oz.	32	0.			0
(No-Cal)	6 fl. oz.	11	0			0
(Shasta)	6 fl. oz.	37	0.			0
(Waldbaum) pale dry	6 fl. oz.	8	0.			0
(Yukon Club) pale dry	6 fl. oz.	25	0.			0
GINGER BEER, soft drink, regular (Schweppes)	6 fl. oz.	31	0.			0
GINGERBREAD, home recipe (USDA)[1]:						

(USDA): United States Department of Agriculture
*Prepared as Package Directs
[1]Made with sodium aluminum sulfate-type baking powder.

Food and Description	Measure or Quantity	Sodium (mg.)	Total	Fats in grams — Saturated	Unsaturated	Cholesterol (mg.)
Made with butter[1]	1.9-oz. piece (2″ x 2″ x 2″)	130	5.3	3.	3.	
Made with vegetable shortening[2]	1.9-oz. piece (2″ x 2″ x 2″)	130	5.9	2.	4.	
GINGERBREAD MIX:						
Dry (USDA)[3]	1 oz.	131	2.9	<1.	2.	
*Prepared with water (USDA)[4]	1/9 of 8″ sq. (2.2 oz.)	192	4.3	<1.	4.	<1
*(Betty Crocker)	1/9 of cake	320	4.8			
*(Dromedary)	1.2-oz. piece (2″ x 2″)	180	2.2			
GINGER ROOT, fresh (USDA):						
With skin	1 oz.	2	.3			0
Without skin	1 oz.	2	.3			0
GOLD 'n CRUST (Adolph's)	1 tsp. (4 grams)	18	Tr.	Tr.	Tr.	0
GOOD HUMOR:						
Toasted almond bar	1 piece (2.1 oz.)	73	13.6	6.	8.	16
Vanilla ice cream bar	1 piece (2.4 oz.)	27	13.8	8.	6.	16
GOOSE, domesticated (USDA):						
Raw, ready-to-cook	1 lb. (weighed with bones)		104.3			
Raw, total edible	1 lb.		142.9			
Raw, meat & skin	1 lb.		152.4			
Raw, meat only	1 lb.		32.2			
Roasted, total edible	4 oz.		40.8			
Roasted, meat & skin	4 oz.		43.2			
Roasted, meat only	4 oz.	141	11.1			
GOOSEBERRY (USDA):						
Fresh	1 lb.	5	.9			0
Fresh	1 cup (5.3 oz.)	2	.3			0
Canned, solids & liq.:						
Regular pack, heavy or extra heavy syrup	4 oz.	1	.1			0
Water pack	4 oz.	1	.1			0

(USDA): United States Department of Agriculture
*Prepared as Package Directs
[1]Principal sources of fat: butter, egg & milk.
[2]Principal sources of fat: vegetable shortening, egg & milk.
[3]Principal source of fat: vegetable shortening.
[4]Principal sources of fat: vegetable shortening & egg.

Food and Description	Measure or Quantity	Sodium (mg.)	— Fats in grams —			Cholesterol (mg.)
			Total	Saturated	Unsaturated	
GOOSE, GIBLET, raw (USDA)	4 oz.		7.9			
GOOSE GIZZARD, raw (USDA)	4 oz.		6.0			
GOULASH DINNER (Chef Boy-Ar-Dee)	7⅓-oz. pkg.	1216	10.2			
GRANOLA:						
Sun Country, regular or honey almond	½ cup (2 oz.)		9.0			(0)
Vita-Crunch:						
Regular	½ cup (2.4 oz.)	328	9.4			(0)
Date	½ cup	308	9.4			(0)
Raisin	½ cup	308	8.0			(0)
Toasted almonds	½ cup	295	9.3			(0)
GRAPE:						
Fresh:						
American type (slip skin), Concord, Delaware, Niagara, Catawba & Scuppernong, pulp only:						
(USDA)	½ lb. (weighed with stem, skin & seeds)	4	1.4			0
(USDA)	½ cup (2.7 oz.)	2	.8			0
(USDA)	3½" x 3" bunch (3.5 oz.)	2	.6			0
European type (adherent skin), Malaga, Muscat, Thompson seedless, Emperor & Flame Tokay, with skin:						
(USDA)	½ lb. (weighed with stems & seeds)	6	.6			0
Whole (USDA)	20 grapes (¾" dia.)	2	.2			0
Whole (USDA)	½ cup (3.1 oz.)	3	.3			0
Halves (USDA)	½ cup (3 oz.)	3	.3			0
Canned, solids & liq. (USDA):						
Thompson seedless, heavy syrup	4 oz.	5	.1			0
Thompson seedless, water pack	4 oz.	5	.1			0
GRAPEADE, chilled (Sealtest)	6 fl. oz. (6.5 oz.)	<1				(0)
GRAPE BERRY, juice drink (Ocean Spray)	½ cup (5 oz.)	4	.4			0

(USDA): United States Department of Agriculture
*Prepared as Package Directs

170

Food and Description	Measure or Quantity	Sodium (mg.)	— Fats in grams —			Cholesterol (mg.)
			Total	Saturated	Unsaturated	
GRAPE DRINK:						
Canned:						
(Del Monte)	6 fl. oz. (6.5 oz.)	16	Tr.			(0)
(Hi-C)	6 fl. oz. (4.2 oz.)	<1	Tr.			0
GRAPE JAM:						
(Bama)	1 T. (.7 oz.)	3	<.1			(0)
(Smucker's)	1 T. (.7 oz.)	5	<.1			(0)
GRAPE JELLY, low calorie:						
(Kraft)	1 oz.	37	<.1			(0)
(Diet Delight) Concord	1 T. (.6 oz.)	21	Tr.			(0)
(S and W) *Nutradiet*, Concord	1 T. (.5 oz.)		<.1			(0)
(Slenderella)	1 T. (.7 oz.)	34	<.1			(0)
(Smucker's)	1 T. (.7 oz.)	Tr.	Tr.			(0)
(Tillie Lewis)	1 T. (.5 oz.)	4	Tr.			0
GRAPE JUICE:						
Canned:						
(USDA)	½ cup (4.4 oz.)		Tr.			0
(Heinz)	5½-fl.-oz. can	2	.2			(0)
(S and W) *Nutradiet*	4 oz. (by wt.)	2	.1			(0)
Frozen, concentrate, sweetened:						
(USDA)	6-fl.-oz. can (7.6 oz.)	6	Tr.			0
Diluted with 3 parts water						
(USDA)	½ cup (4.4 oz.)	1	Tr.			0
*(Minute Maid)	½ cup (4.2 oz.)	1	Tr.			0
*(Snow Crop)	½ cup (4.2 oz.)	1	Tr.			0
GRAPE JUICE DRINK, canned (USDA) approximately 30% grape juice	1 cup (8.8 oz.)	2	Tr.			0
GRAPE-NUTS, cereal (Post)	¼ cup (1 oz.)	147	.1			0
GRAPE-NUTS FLAKES, cereal (Post)	⅔ cup (1 oz.)	150	.3			0
GRAPE PIE (Tastykake)	4-oz. pie		16.2			
GRAPE SOFT DRINK:						
Sweetened:						
(Canada Dry) bottle or can	6 fl. oz.	13+	0.			0
(Dr. Brown's)	6 fl. oz.	15	0.			0

(USDA): United States Department of Agriculture
*Prepared as Package Directs

Food and Description	Measure or Quantity	Sodium (mg.)	— Fats in grams —			Choles-terol (mg.)
			Total	Satu-rated	Unsatu-rated	
(Fanta)	6 fl. oz.	6	0.			0
(Hoffman)	6 fl. oz.	15	0.			0
(Key Food)	6 fl. oz.	15	0.			0
(Kirsch)	6 fl. oz.	<1	0.			0
(Nedick's)	6 fl. oz.	15	0.			0
(Nehi)	6 fl. oz. (6.6 oz.)	0+	0.			0
(Shasta)	6 fl. oz.	18	0.			0
(Waldbaum)	6 fl. oz.	15	0.			0
(Yukon Club)	6 fl. oz.	15	0.			0
Low calorie:						
(Dr. Brown's)	6 fl. oz.	71	0.			0
(Hoffman)	6 fl. oz.	71	0			0
(Key Food)	6 fl. oz.	71	0.			0
(No-Cal)	6 fl. oz.	12	0.			0
(Shasta)	6 fl. oz.	45	0.			0
(Waldbaum)	6 fl. oz.	71	0.			0
(Yukon Club)	6 fl. oz.	69	0			0
GRAPE SYRUP, low calorie						
(No-Cal)	1 tsp. (5 grams)	<1	0.			0
GRAPEFRUIT:						
Fresh, pulp only:						
Pink & red:						
Seeded type (USDA)	1 lb. (weighed with seeds and skin)	2	.2			0
Seeded type (USDA)	½ med. grapefruit (3¾", 8.5 oz.)	1	.1			0
Seedless type (USDA)	1 lb. (weighed with skin)	2	.2			0
Seedless type (USDA)	½ med. grapefruit (8.5 oz.)	1	.1			0
White:						
Seeded type (USDA)	1 lb. (weighed with seeds & skin)	2	.2			0
Seeded type (USDA)	½ med. grapefruit (3¾", 8.5 oz.)	1	.1			0
Seedless type (USDA)	1 lb. (weighed with skin)	2	.2			0
Seedless type, sections (USDA)	1 cup (7 oz.)	2	.2			0
Seedless type (USDA)	½ med. grapefruit (3¾", 8.5 oz.)	1	.1			0
(Sunkist)	½ grapefruit (8.5 oz.)	1	Tr.			0

(USDA): United States Department of Agriculture
*Prepared as Package Directs

Food and Description	Measure or Quantity	Sodium (mg.)	— Fats in grams —			Choles- terol (mg.)
			Total	Satu- rated	Unsatu- rated	
Bottled, chilled, sweetened sections (Kraft)	4 oz.	115	.1			(0)
Bottled, chilled, unsweetened sections (Kraft)	4 oz.	115	.1			(0)
Canned, sections, syrup pack, solids & liq.:						
(USDA)	½ cup (4.5 oz.)	1	.1			0
(Del Monte)	½ cup (4.5 oz.)	2				0
Light syrup (Stokely-Van Camp)	½ cup (4 oz.)		.1			(0)
Canned, sections, unsweetened or dietetic pack, solids & liq.:						
Water pack (USDA)	½ cup (4.2 oz.)	5	.1			0
Juice pack (Del Monte)	½ cup (4.5 oz.)	1	.5			0
(Diet Delight) unsweetened	½ cup (4.3 oz.)	5	<.1			(0)
(S and W) *Nutradiet*, low calorie	4 oz.	2	.1			(0)
(S and W) *Nutradiet*, unsweetened	4 oz.	2	.2			(0)
(Tillie Lewis)	½ cup (4.4 oz.)	<10	.1			(0)

GRAPEFRUIT JUICE:

Food and Description	Measure or Quantity	Sodium (mg.)	Total	Satu- rated	Unsatu- rated	Choles- terol (mg.)
Fresh, pink, red or white, all varieties (USDA)	½ cup (4.3 oz.)	1	.1			0
Bottled, chilled, sweetened (Kraft)	½ cup (4.3 oz.)	1	.1			(0)
Bottled, chilled, unsweetened (Kraft)	½ cup (4.3 oz.)	1	.1			(0)
Canned:						
Sweetened:						
(USDA)	½ cup (4.4 oz.)	1	.1			0
(Del Monte)	½ cup (4.3 oz.)	1	Tr.			0
(Heinz)	5½-fl.-oz. can	2	.5			(0)
(Stokely-Van Camp)	½ cup (4.4 oz.)		.1			(0)
Unsweetened:						
(USDA)	½ cup (4.4 oz.)	1	.1			0
(Del Monte)	½ cup (4.3 oz.)	5	.2			0
(Diet Delight)	½ cup (4 oz.)	7	Tr.			(0)
(Heinz)	5½-fl.-oz. can	2	.5			(0)
(Stokely-Van Camp)	½ cup (4.5 oz.)		.1			(0)
Frozen, concentrate:						
Sweetened:						
(USDA)	6-fl.-oz. can (7.4 oz.)	6	.6			0

(USDA): United States Department of Agriculture
*Prepared as Package Directs

Food and Description	Measure or Quantity	Sodium (mg.)	Fats in grams — Total	Satu- rated	Unsatu- rated	Choles- terol (mg.)
*Diluted with 3 parts water						
(USDA)	½ cup (4.4 oz.)	1	.1			0
*(Minute Maid)	½ cup (4.2 oz.)	<1	Tr.			0
*(Snow Crop)	½ cup (4.2 oz.)	<1	Tr.			0
Unsweetened:						
(USDA)	6-fl.-oz. can (7.3 oz.)	8	.8			0
*Diluted with 3 parts water						
(USDA)	½ cup (4.4 oz.)	1	.1			0
*(Florida Diet)	½ cup (4.3 oz.)		.6			(0)
*(Minute Maid)	½ cup (4.2 oz.)	<1	Tr.			0
*(Snow Crop)	½ cup (4.2 oz.)	<1	Tr.			0
Dehydrated, crystals:						
(USDA)	4-oz. can	11	1.1			0
*Reconstituted (USDA)	½ cup (4.4 oz.)	1	.1			0

GRAPEFRUIT-ORANGE JUICE
(See **ORANGE-GRAPEFRUIT JUICE**)

GRAPEFRUIT PEEL, CANDIED

(USDA)	1 oz.		<.1			0

GRAPEFRUIT SOFT DRINK:

Sweetened:						
(Clicquot Club)	6 fl. oz.	11	0.			0
(Cott)	6 fl. oz.	11	0.			0
(Fanta)	6 fl. oz.	12	Tr.			0
(Hoffman)	6 fl. oz.	14	0.			0
(Mission)	6 fl. oz.	11	0.			0
(Shasta)	6 fl. oz.	22	0.			0
Low calorie:						
(Canada Dry) golden or pink,						
bottle or can	6 fl. oz.	13+	0.			0
(Clicquot Club)	6 fl. oz.	43	0.			0
(Cott)	6 fl. oz.	43	0.			0
(Hoffman)	6 fl. oz.	64	0.			0
(Mission)	6 fl. oz.	43	0.			0
(No-Cal) pink	6 fl. oz.	14	0.			0
(Shasta)	6 fl. oz.	37	0.			0

GRAVY, canned:

Beef (Franco-American)	¼ cup	278	2.0			
Brown with onion (Franco-American)	¼ cup	324	1.2			

(USDA): United States Department of Agriculture
*Prepared as Package Directs

Food and Description	Measure or Quantity	Sodium (mg.)	— Fats in grams —			Choles- terol (mg.)
			Total	Satu- rated	Unsatu- rated	
Chicken (Franco-American)	¼ cup	284	3.6			
Chicken giblet (Franco-American)	¼ cup	348	1.4			
Mushroom:						
(B in B)	¼ cup		.8			
Dawn Fresh, brown	5¾-oz. can	668	1.6			
(Franco-American)	¼ cup	306	1.4			
Ready Gravy	¼ cup (2.2 oz.)	344	.8			
GRAVY MASTER	1 fl. oz. (1.3 oz.)	560	0.			0
GRAVY with MEAT or TURKEY, frozen:						
Giblet & sliced turkey (Banquet):						
Cooking bag	5-oz. bag		6.9			
Buffet	2-lb. pkg.		20.1			
Sliced beef, frozen (Banquet):						
Cooking bag	5-oz. bag		6.9			
Buffet	2-lb. pkg.		49.9			
Sliced beef (Morton House)	6¼ oz.	995	12.3	6.	6.	42
Sliced pork (Morton House)	6¼ oz.	975	11.9	5.	7.	43
Sliced turkey (Morton House)	6¼ oz.	1076	6.5	2.	4.	47
GRAVY MIX:						
*Au jus (Durkee)	1 cup (1-oz. pkg.)	2256	.2			
Au jus (French's)	¾-oz. pkg.	2400	.6			
*Au jus (French's)	¼ cup	300	.1			
Beef:						
(Swiss Products)	1¼-oz. pkg.		.8			
(Swiss Products)	⅞-oz. pkg.		.5			
*(Wyler's)	2-oz. serving		.4			
Brown:						
*(Durkee)	1 cup (.8-oz. pkg.)	1512	.8			
(French's)	¾-oz. pkg.	940	1.8			
*(French's)	¼ cup	235	.5			
(Kraft)	2-oz. serving	242	.5			
(Lawry's)	1¼-oz. pkg.		5.3			
(McCormick)	⅞-oz. pkg.	680	1.2			
*(McCormick)	2-oz. serving	168	.4			
Cheese:						
(McCormick)	1¼-oz. pkg.	685	12.6			
*(McCormick)	2-oz. serving	200	5.0			
Chicken:						
*(Durkee)	1 cup (1-oz. pkg.)	2168	3.2			

(USDA): United States Department of Agriculture
*Prepared as Package Directs

Food and Description	Measure or Quantity	Sodium (mg.)	Total	Fats in grams — Satu- rated	Unsatu- rated	Choles- terol (mg.)
(French's)	1¼-oz. pkg.	900	4.6			
*(French's)	¼ cup	225	1.2			
*(Kraft)	2-oz. serving	211	1.5			
(Lawry's)	1-oz. pkg.		4.2			
*(McCormick)	2-oz. serving	325	3.0			
(Swiss)	1¼-oz. pkg.		.6			
(Swiss)	⅞-oz. pkg.		.4			
*(Wyler's)	2-oz. serving		.7			
*Herb (McCormick)	2-oz. serving	160	.8			
Homestyle (French's)	⅞-oz. pkg.	1540	3.0			6
*Homestyle (French's)	¼ cup	380	1.0			2
Mushroom:						
*(Durkee)	1 cup (.8-oz. pkg.)	1696	.8			
(French's)	¾-oz. pkg.	1000	2.0			
*(French's)	¼ cup	250	.5			
(Lawry's)	1.3-oz. pkg.		6.6			
*(McCormick)	2-oz. serving	170	.3			
*(Wyler's)	2-oz. serving		.6			
Onion:						
*(Durkee)	1 cup (1-oz. pkg.)	1408	.8			
(French's)	1-oz. pkg.	1100	1.6			
*(French's)	¼ cup	275	.4			
*(Kraft)	2-oz. serving	175	.5			
*(McCormick)	2-oz. serving	364	1.6			
*(Wyler's)	2-oz. serving		.5			
Pork:						
(French's)	¾-oz. pkg.	1180	2.2			
*(French's)	¼ cup	295	.6			
Turkey:						
(French's)	⅞-oz. pkg.	1250	3.2			
*(French's)	¼ cup	313	.8			

GREAT HONEY CRUNCHERS:

Rice	1 cup (1 oz.)	50	1.0			(0)
Wheat	1 cup (1 oz.)	50	1.4			(0)

GREEN PEA (See **PEA**)

GRITS (See **HOMINY GRITS**)

GROUND-CHERRY, Poha or Cape
Gooseberry, fresh (USDA):

Whole	1 lb. (weighed with husks & stems)		2.9			0
Flesh only	4 oz.		.8			0

(USDA): United States Department of Agriculture
*Prepared as Package Directs

Food and Description	Measure or Quantity	Sodium (mg.)	Total	Fats in grams — Satu- rated	Unsatu- rated	Choles- terol (mg.)
GROUPER, raw (USDA):						
Whole	1 lb. (weighed whole)		1.0			
Meat only	4 oz.		.6			
GUAVA, COMMON, fresh (USDA):						
Whole	1 lb. (weighed untrimmed)	18	2.6			0
Whole	1 guava (2.8 oz.)	3	.5			0
Flesh only	4 oz.	5	.7			0
GUAVA, STRAWBERRY, fresh (USDA):						
Whole	1 lb. (weighed untrimmed)	18	2.7			0
Flesh only	4 oz.	5	.7			0
GUINEA HEN, raw (USDA):						
Ready-to-cook	1 lb.(weighed with bones)		24.4			
Meat & skin	4 oz.		7.3			
Giblets	2 oz.		4.0			
GUM (See **CHEWING GUM**)						

H

Food and Description	Measure or Quantity	Sodium (mg.)	Total	Satu- rated	Unsatu- rated	Choles- terol (mg.)
HADDOCK:						
Raw (USDA):						
Whole	1 lb. (weighed whole)	133	.2			131
Meat only	4 oz.	69	.1			68
Fried, dipped in egg, milk & bread crumbs (USDA)	4″ x 3″ x ½″ fillet (3.5 oz.)	177	6.4			
Frozen (Gorton)	⅓ of 1-lb. pkg.	93	.2			
Smoked, canned or not (USDA)	4 oz.		.5			
HADDOCK MEALS, frozen:						
(Banquet):						
Fish compartment	5 oz.		7.7			
Potato compartment	1.6 oz.		7.6			

(USDA): United States Department of Agriculture
*Prepared as Package Directs

Food and Description	Measure or Quantity	Sodium (mg.)	— Fats in grams — Total	Satu- rated	Unsatu- rated	Choles- terol (mg.)
Peas compartment	2.2 oz.		1.4			
Complete dinner	8.8-oz. dinner		16.7			
(Weight Watchers)	18-oz. dinner		1.5			
& spinach (Weight Watchers)	9½-oz. luncheon		4.7			
HAKE, raw (USDA):						
Whole	1 lb. (weighed whole)	144	.8			
Meat only	4 oz.	84	.5			
HALF & HALF, milk & cream (See **CREAM**)						
HALF & HALF SOFT DRINK:						
Sweetened:						
(Dr. Brown's)	6 fl. oz.	14	0.			0
(Hoffman)	6 fl. oz.	14	0.			0
(Key Food)	6 fl. oz.	14	0.			0
(Kirsch)	6 fl. oz.	<1	0.			0
(Waldbaum)	6 fl. oz.	14	0.			0
(Yukon Club)	6 fl. oz.	14	0.			0
Low calorie (Hoffman)	6 fl. oz.	63	0.			0
HALIBUT:						
Atlantic & Pacific:						
Raw (USDA):						
Whole	1 lb. (weighed whole)	145	3.2			134
Meat only, not dipped in brine	4 oz.	61	1.4			57
Meat only, dipped in brine (USDA)	4 oz.	408	1.4			57
Broiled with vegetable shortening (USDA)	6½" x 2½" x 8" or 4" x 3" x ½" steak (4.4 oz.)	168	8.8			75
Smoked (USDA)	4 oz.		17.0			
California, raw, meat only (USDA)	4 oz.		1.6			
HAM (See also **PORK**):						
Boiled:						
Luncheon meat (USDA)[1]	1 oz.		4.8	2.	3.	

(USDA): United States Department of Agriculture
*Prepared as Package Directs
[1]Principal source of fat: pork.

Food and Description	Measure or Quantity	Sodium (mg.)	—Fats in grams—			Choles-terol (mg.)
			Total	Satu-rated	Unsatu-rated	
Luncheon meat, chopped						
(USDA)[1]	1 cup (4.8 oz.)		23.1	8.	15.	
Luncheon meat, diced (USDA)[1]	1 cup (5 oz.)		24.0	8.	16.	
(Hormel)	1 oz.	319	1.4	Tr.	<1.	15
Chopped, sliced (Hormel)	1 oz.	289	5.8			
Minced (Oscar Mayer)	1 slice (10 per ½ lb.)	251	4.4			
Smoked (Oscar Mayer)	1 slice (8 per 6 oz.)	246	1.5			
Smoked, thin sliced (Oscar Mayer)	1 slice (10 per 3 oz.)	87	.6			
Canned:						
(USDA)	1 oz.	312	3.5	1.	2.	
(Armour Golden Star)	1 oz.		1.4			
(Armour Star)	1 oz.		3.5			
(Hormel)	1 oz. (8-lb. can)		4.2			
(Hormel)	1 oz. (6-lb. can)		2.6			
(Hormel)	1 oz. (4-lb. can)		3.1			
(Hormel)	1 oz. (1-lb. 8-oz. can)		3.1			
(Oscar Mayer) *Jubilee*, bone in	1 lb.	5307	54.4			
(Oscar Mayer) *Jubilee*, boneless	1 lb.	5307	54.4			
(Oscar Mayer) *Jubilee*, boneless	½-lb. slice	2654	15.9			
(Oscar Mayer) *Jubilee*, special trim, as purchased	1 oz.	276	2.3			
(Oscar Mayer) *Jubilee*, special trim, cooked	1 oz.	298	1.4	<1.	<1.	8
(Oscar Mayer) steak	1 slice (8 to lb.)	663	3.4			
(Swift)	1¾-oz. slice (5" x 2¼" x ¼")	422	7.8			
(Swift) *Hostess*	1 oz. (4-lb. can)	269	1.6			
(Wilson)	1 oz.	272	2.8	1.	2.	16
(Wilson) *Tender Made*	1 oz.	305	2.6	<1.	1.	18
Chopped or minced, canned:						
(USDA)[1]	1 oz.		4.8	2.	3.	
(Armour Star)	1 oz.		7.3			
(Hormel)	1 oz. (8-lb. can)	369	7.9	3.	5.	13
(Oscar Mayer)	1-oz. slice	310	4.8			
Chopped, spiced or unspiced, canned:						
(USDA)[1]	1 oz.	350	7.1	3.	5.	
Chopped (USDA)[1]	1 cup (4.8 oz.)	1678	33.9	12.	22.	
Diced (USDA)[1]	1 cup (5 oz.)	1740	35.1	13.	22.	
(Hormel)	1 oz. (5-lb. can)		6.6			

(USDA): United States Department of Agriculture
*Prepared as Package Directs
[1]Principal source of fat: pork.

Food and Description	Measure or Quantity	Sodium (mg.)	Total	Fats in grams Satu- rated	Unsatu- rated	Choles- terol (mg.)
Deviled, canned:						
(USDA)[1]	1 oz.		9.2	3.	6.	
(USDA)[1]	1 T. (.5 oz.)		4.2	2.	3.	
(Armour Star)	1 oz.		6.7			
(Hormel)	1 oz. (3-oz. can)		6.1			
(Underwood)	4½-oz. can	1156	40.8			
(Underwood)	1 T. (.5 oz.)	122	4.3			
Freeze-dry, diced, canned						
(Wilson) *Campsite:*						
Dry	¾-oz. can	490	6.2	2.	4.	41
*Reconstituted	1 oz.	326	4.1	2.	3.	28
HAM & CHEESE:						
Loaf (Oscar Mayer)	1-oz. slice	287	5.4			
Roll (Oscar Mayer)	1 oz.	287	4.8			
Spread (Oscar Mayer)	1 oz.	287	6.0			
HAM CROQUETTE, home recipe						
(USDA)[2]	4 oz.	388	17.1	7.	10.	
HAM DINNER:						
Frozen:						
(Banquet)						
Meat compartment	5.2 oz.		7.4			
Apple compartment	2.8 oz.		1.1			
Peas & carrots compartment	2.1 oz.		.2			
Complete dinner	10-oz. dinner		8.8			
(Morton)	10-oz. dinner	278	18.1			
(Swanson)	10¼-oz. dinner	1258	11.9	3.	9.	
*Mix, au gratin (Jeno's)						
Add 'n Heat	35-oz. pkg.		83.3			
HAM SPREAD, salad						
(Oscar Mayer)	1 oz.	254	4.0			
HAMBURGER (See also **BEEF,**						
Ground):						
Regular (McDonald's)	1 hamburger (3.4 oz.)	542	9.6			
Regular, cheese (McDonald's)	1 hamburger (3.9 oz.)	821	13.9			
¼ pound, (McDonald's)	1 hamburger (5.5 oz.)	690	19.3			
¼ pound, cheese (McDonald's)	1 hamburger (6.6 oz.)	1173	27.7			

(USDA): United States Department of Agriculture
*Prepared as Package Directs
[1]Principal source of fat: pork.
[2]Principal sources of fat: butter, ham & vegetable shortening.

Food and Description	Measure or Quantity	Sodium (mg.)	— Fats in grams —			Choles- terol (mg.)
			Total	Satu- rated	Unsatu- rated	
Freeze dry, canned (Wilson)						
Campsite:						
Dry	3¼-oz. can	1413	35.6	18.	18.	216
*Reconstituted	4 oz.	261	17.9	9.	9.	109
HAWAIIAN DINNER MIX						
(Hunt's) *Skillet*[1]	1-lb. pkg.	3041	6.2	3.	3.	
HAWAIIAN-STYLE VEGETABLES,						
frozen (Birds Eye)	⅓ of 10-oz. pkg.	416	5.0			0
HAWS, SCARLET, raw (USDA):						
Whole	1 lb. (weighed with core)		2.5			0
Flesh & skin	4 oz.		.8			0
HAZELNUT (See **FILBERT**)						
HEADCHEESE:						
(USDA)[2]	1 oz.		6.2	2.	4.	
(Oscar Mayer)	1 slice (8 per ½ lb.)	332	3.1			
HEART (USDA):						
Beef:						
Lean, raw	1 lb.	390	16.3			680
Lean, braised	4 oz.	118	6.5			311
Lean, braised, chopped or diced	1 cup (5.1 oz.)	151	8.3			397
Lean with visible fat, raw	1 lb.		93.9			
Lean with visible fat, braised	4 oz.		32.9			
Calf, raw	1 lb.	426	26.8			
Calf, braised	4 oz.	128	10.3			
Chicken, raw	1 lb.	358	27.2			771
Chicken, simmered	1 heart (5 grams)	3	.4			12
Chicken, simmered, chopped or diced	1 cup (5.1 oz.)	100	10.4			335
Hog, raw	1 lb.	245	20.0			
Hog, braised	4 oz.	74	7.8			
Lamb, raw	1 lb.		43.5			
Lamb, braised	4 oz.		16.3			
Turkey, raw	1 lb.	313	50.8			680
Turkey, simmered	4 oz.	69	15.0			270
Turkey, simmered, chopped or diced	1 cup (5.1 oz.)	88	19.1			345

(USDA): United States Department of Agriculture
*Prepared as Package Directs
[1]Principal source of fat: almonds.
[2]Principal source of fat: pork.

Food and Description	Measure or Quantity	Sodium (mg.)	— Fats in grams —			Choles-terol (mg.)
			Total	Satu-rated	Unsatu-rated	
HEARTLAND, cereal (Pet)	1 oz.		3.7			(0)
HERRING:						
Raw (USDA):						
Atlantic, whole	1 lb. (weighed whole)		26.1	5.	21.	197
Atlantic, meat only	4 oz.		12.8	2.	11.	96
Pacific, meat only	4 oz.	84	2.9	Tr.	3.	
Canned:						
Plain, solids & liq. (USDA)	4 oz.		15.4			110
Plain, solids & liq. (USDA)	15-oz. can		57.8			412
Bismark, drained (Vita)	5-oz. jar		15.6			
Cocktail, drained (Vita)	8-oz. jar		13.7			
In cream sauce (Vita)	8-oz. jar		25.3			
In tomato sauce, solids & liq. (USDA)	4 oz.		11.9			
In wine sauce, drained (Vita)	8-oz. jar		22.0			
Lunch, drained (Vita)	8-oz. jar		30.9			
Matjis, drained (Vita)	8 oz.		8.7			
Party snacks, drained (Vita)	8-oz. jar		22.0			
Tastee Bits, drained (Vita)	8-oz. jar		17.8			
Pickled, Bismarck type (USDA)	4 oz.		17.1			
Salted or brined (USDA)	4 oz.		17.2			
Smoked (USDA):						
Bloaters	4 oz.		14.1			
Hard	4 oz.	7066	17.9			
Kippered	4 oz.		14.6			
HICKORY NUT (USDA):						
Whole	1 lb. (weighed in shell)		109.1	9.	100.	0
Shelled	4 oz.		77.9	7.	71.	0
HO-HO (Hostess):						
2 to pkg.	1 piece (.9 oz.)	59	5.0			
10 to pkg.	1 piece (.9 oz.)	61	5.1			
HOMINY GRITS:						
Dry:						
Degermed (USDA)	½ cup (2.8 oz.)	<1	.6			0
Instant (Quaker)	.8-oz. packet	347	.1			
Cooked:						
Degermed (USDA)	1 cup (8.6 oz.)	502	.2			0

(USDA): United States Department of Agriculture
*Prepared as Package Directs

Food and Description	Measure or Quantity	Sodium (mg.)	—Fats in grams—			Choles- terol (mg.)
			Total	Satu- rated	Unsatu- rated	
(Albers)	1 cup		.2			
(Aunt Jemima/Quaker)	²/₃ cup	<1	.2			
HONEY, strained:						
(USDA)	½ cup (5.7 oz.)	8	0.			0
(USDA)	1 T. (.7 oz.)	1	0.			0
HONEYCOMB, cereal (Post)	1⅓ cups (1 oz.)	150	.1			0
HONEYDEW, fresh (USDA):						
Whole	1 lb. (weighed whole)	34	.9			0
Wedge	2″ x 7″ wedge (5.3 oz.)	11	.3			0
Flesh only	4 oz.	14	.3			0
Flesh only, diced	1 cup (5.9 oz.)	20	.5			0
HORSERADISH:						
Raw (USDA):						
Whole	1 lb. (weighed un- pared)	26	1.0			0
Pared	1 oz.	2	<.1			0
Dehydrated (Heinz)	1 T.	95	.2			(0)
Dry (Spice Islands)	1 tsp.	<1				(0)
Prepared:						
(USDA)	1 oz.	27	<.1			0
(Kraft)	1 oz.	312	<.1			(0)
Cream style (Kraft)	1 oz.	312	.5			(0)
Oil style (Kraft)	1 oz.	312	1.9			(0)
HOTCHAS (General Mills)	15 pieces (.5 oz.)	148	3.1			(0)
*****HOT DOG BEAN SOUP,** canned						
(Campbell)	1 cup	904	3.9	2.	2.	
HYACINTH BEAN (USDA):						
Young pod, raw:						
Whole	1 lb. (weighed un- trimmed)	8	1.2			0
Trimmed	4 oz.	2	.3			0
Dry seeds	4 oz.		1.7			0

(USDA): United States Department of Agriculture
*Prepared as Package Directs

Food and Description	Measure or Quantity	Sodium (mg.)	— Fats in grams —			Choles- terol (mg.)
			Total	Satu- rated	Unsatu- rated	
ICE CREAM and FROZEN CUS- TARD (See also listing by flavor or brand name, e.g., **CHOCOLATE ICE CREAM** or *DREAMSICLE* and **FROZEN DESSERT**) with salt added (USDA):						
10% fat, regular ice cream	3-fl.-oz. container (1.8 oz.)	32	5.3	3.	2.	20
10% fat, regular ice cream	1 cup (4.7 oz.)	84	14.1	8.	6.	53
10% fat, frozen custard or French ice cream	1 cup (4.7 oz.)	84	14.1			97
10% fat, frozen custard or French ice cream	3-fl.-oz. container (1.8 oz.)	32	5.3			36
12% fat, ice cream	2½-oz. slice (⅛ of qt. brick)	28	8.9	5.	4.	
12% fat, ice cream	3½-fl.-oz. container (2.2 oz.)	25	7.8	4.	3.	
12% fat, ice cream	1 cup (5 oz.)	57	17.8	10.	8.	
16% fat, rich ice cream	1 cup (5.2 oz.)	49	23.8	13.	11.	84
ICE CREAM BAR, chocolate-coated (Sealtest)	2½-fl.-oz. bar (1.7 oz.)	24	10.5			
ICE CREAM CONE, cone only:						
(USDA)[1]	1 piece (5 grams)	12	.1	Tr.	Tr.	
(Comet)	1 piece (4 grams)	4	.2			
Assorted colors (Comet)	1 piece (4 grams)	5	.2			
Rolled sugar (Comet)	1 piece (.4 oz.)	54	.5			
ICE CREAM CUP, cup only:						
(Comet)	1 piece (5 grams)	5	.2			
Assorted colors (Comet)	1 piece (5 grams)	6	.2			
Pilot (Comet)	1 piece (4 grams)	4	.2			
ICE CREAM SANDWICH (Sealtest)	3 fl. oz. (2.2 oz.)	92	6.2			
ICE MILK:						
Hardened, with salt added (USDA)	1 cup (4.6 oz.)	89	6.7	4.	3.	26

(USDA): United States Department of Agriculture
*Prepared as Package Directs
[1]Principal source of fat: vegetable shortening.

Food and Description	Measure or Quantity	Sodium (mg.)	—Fats in grams—			Choles- terol (mg.)
			Total	Satu- rated	Unsatu- rated	
Soft-serve, with salt added						
(USDA)	1 cup (6.2 oz.)	119	8.9	5.	4.	35
Any flavor (Borden) 2.5% fat	¼ pt. (2.3 oz.)	39	1.6			
Any flavor (Borden) 3.25% fat	¼ pt. (2.4 oz.)		2.2			
Any flavor (Borden) *Lite-line*	¼ pt.		1.9			
Light n' Lively (Sealtest):						
Banana	¼ pt. (2.4 oz.)	48	1.9			
Banana strawberry twirl	¼ pt. (2.5 oz.)	45	1.7			
Buttered almond	¼ pt. (2.4 oz.)	130	4.0			
Caramel nut	¼ pt. (2.4 oz.)	102	3.9			
Cherry pineapple	¼ pt. (2.4 oz.)	46	1.8			
Chocolate	¼ pt. (2.4 oz.)	47	2.2			
Coffee	¼ pt. (2.4 oz.)	56	2.1			
Lemon	¼ pt. (2.4 oz.)	56	2.2			
Lemon chiffon	¼ pt. (2.4 oz.)	46	1.8			
Orange pineapple	¼ pt. (2.4 oz.)	51	2.0			
Peach	¼ pt. (2.4 oz.)	47	1.8			
Raspberry	¼ pt. (2.4 oz.)	53	2.0			
Strawberry	¼ pt. (2.4 oz.)	47	1.8			
Strawberry royale	¼ pt. (2.5 oz.)	50	1.9			
Toffee	¼ pt. (2.4 oz.)	78	2.8			
Toffee crunch	¼ pt. (2.4 oz.)	85	3.3			
Vanilla	¼ pt. (2.3 oz.)	56	2.2			
Vanilla fudge royale	¼ pt. (2.5 oz.)	65	2.4			
ICE MILK BAR, chocolate-coated						
(Sealtest)	2½-fl.-oz. bar					
	(1.8 oz.)	31	7.7			
ICE STICK, Twin Pops (Sealtest)	3 fl. oz. (3.1 oz.)		Tr.			
ICES (See **LIME ICE**)						
ICING (See **CAKE ICING**)						
INCONNU or SHEEFISH, raw:						
Whole (USDA)	1 lb. (weighed whole)		19.4			
Meat only (USDA)	4 oz.		7.7			
INDIAN PUDDING, New England						
(B & M)	½ cup (4 oz.)	189	.3			

(USDA): United States Department of Agriculture
*Prepared as Package Directs

(185)

Food and Description	Measure or Quantity	Sodium (mg.)	Total	Fats in grams — Satu- rated	Unsatu- rated	Choles- terol (mg.)
INSTANT BREAKFAST (See individual brand name or company listings)						
IRISH WHISKEY (See **DISTILLED LIQUOR**)						
ITALIAN DINNER, frozen:						
(Banquet):						
Meat compartment	2¾ oz.		10.3			
Mostaccioli compartment	7¾ oz.		5.6			
Bread compartment	5 oz.		1.8			
Complete dinner	11-oz. dinner		17.7			
(Swanson)	13½-oz. dinner	1244	18.1			
ITALIAN HERBS (Spice Islands)	1 tsp.	<1				(0)
ITALIAN-STYLE VEGETABLES, frozen (Birds Eye)	⅓ of 10-oz. pkg.	468	6.8			0

J

Food and Description	Measure or Quantity	Sodium (mg.)	Total	Satu- rated	Unsatu- rated	Choles- terol (mg.)
JACKFRUIT, fresh (USDA):						
Whole	1 lb. (weighed with seeds & skin)	3	.4			0
Flesh only	4 oz.	2	.3			0
JACK MACKEREL, raw, meat only (USDA)	4 oz.		6.4			
JACK ROSE MIX (Bar-Tender's)	1 serving (⅝ oz.)	46	.2			(0)
JAM, sweetened (See also individual listings by flavor):						
(USDA)	1 oz.	3	<.1			0
(USDA)	1 T. (.7 oz.)	2	Tr.			0
JAPANESE-STYLE VEGETABLES, frozen (Birds Eye)	⅓ of 10-oz. pkg.	107	8.2			0
JELLY, sweetened (See also individual listings by flavor):						
(USDA)	1 oz.	5	<.1			0
(USDA)	1 T. (.6 oz.)	3	Tr.			0

(USDA): United States Department of Agriculture
*Prepared as Package Directs

Food and Description	Measure or Quantity	Sodium (mg.)	—Fats in grams—			Choles- terol (mg.)
			Total	Satu- rated	Unsatu- rated	
All flavors (Bama)	1 T. (.7 oz.)	2	<.1			(0)
All flavors (Kraft)	1 oz.	<1	<.1			
All flavors (Smucker's)	1 T. (.7 oz.)	<1	0.			(0)

JERUSALEM ARTICHOKE
(USDA):

| Unpared | 1 lb. (weighed with skin) | | .3 | | | 0 |
| Pared | 4 oz. | | .1 | | | 0 |

JORDAN ALMOND (See **CANDY**)

JUICE (See individual flavors)

JUJUBE or CHINESE DATE
(USDA):

Fresh, whole	1 lb. (weighed with seeds)	13	.8			0
Fresh, flesh only	4 oz.	3	.2			0
Dried, whole	1 lb. (weighed with seeds)		4.4			0
Dried, flesh only	1 oz.		.3			0

JUNIOR FOOD (See **BABY FOOD**)

JUNIORS (Tastykake):

Chocolate	2¾-oz. pkg.		9.9			
Chocolate devil food	2¾-oz. pkg.		9.7			
Coconut	2¾-oz. pkg.		6.5			
Coconut devil food	2¾-oz. pkg.		6.8			
Jelly square	3¼-oz. pkg.		4.4			
Koffee Kake	2½-oz. pkg.		13.5			
Lemon	2¾-oz. pkg.		6.9			

| **JUNIPER BERRY** (Spice Islands) | 1 berry | Tr. | | | | (0) |

JUNKET **RENNET MIX** (See individual flavors)

K

| *KABOOM,* cereal (General Mills) | 1 cup (1 oz.) | 196 | .7 | | | (0) |

(USDA): United States Department of Agriculture
*Prepared as Package Directs

Food and Description	Measure or Quantity	Sodium (mg.)	—Fats in grams— Total	Satu- rated	Unsatu- rated	Choles- terol (mg.)
KALE:						
Raw, leaves only (USDA)	1 lb. (weighed un- trimmed)	218	2.3			0
Raw, leaves including stems (USDA)	1 lb. (weighed trimmed)	252	2.7			0
Boiled without salt, leaves only (USDA)	4 oz.	49	.8			0
Boiled without salt, including stems (USDA)	½ cup (1.9 oz.)	24	.4			0
Frozen:						
Not thawed (USDA)	4 oz.	29	.6			0
Boiled, drained (USDA)	½ cup (3.2 oz.)	19	.5			0
Chopped (Birds Eye)	½ cup (3.3 oz.)	24	.5			0
KARO, syrup:						
Dark corn	1 cup (11.7 oz.)	447	0.			0
Dark corn	1 T. (.7 oz.)	28	0.			0
Imitation maple	1 cup (11.6 oz.)	328	0.			0
Imitation maple	1 T. (.7 oz.)	20	0.			0
Light corn	1 cup (11.7 oz.)	381	0.			0
Light corn	1 T. (.7 oz.)	23	0.			0
Pancake & waffle	1 cup (11.5 oz.)	331	0.			0
Pancake & waffle	1 T. (.7 oz.)	20	0.			0
KASHA (See **BUCKWHEAT,** Groats)						
KETCHUP (See **CATSUP**)						
KIDNEY (USDA):						
Beef, raw	4 oz.	200	7.6			425
Beef, braised	4 oz.	287	13.6			912
Beef braised, ¼" slices	1 cup (4.9 oz.)	354	16.8			1126
Calf, raw	4 oz.		5.2			425
Hog, raw	4 oz.	130	4.1			425
Lamb, raw	4 oz.	257	3.7			425
KIELBASA (Oscar Mayer)	6-oz. link	1658	4.4			
KINGFISH, raw (USDA):						
Whole	1 lb. (weighed whole)	166	6.0			
Meat only	4 oz.	94	3.4			

(USDA): United States Department of Agriculture
*Prepared as Package Directs

Food and Description	Measure or Quantity	Sodium (mg.)	—Fats in grams—			Choles- terol (mg.)
			Total	Satu- rated	Unsatu- rated	
KING VITAMAN (Quaker)	¾ cup (1 oz.)	210	2.0			(0)
KIPPERS (See **HERRING**)						
KIX, cereal (General Mills)	1½ cups (1 oz.)	330	.6			(0)
KNOCKWURST:						
(USDA)	1 oz.		6.6			
Chubbies, all meat (Oscar Mayer)	1 link (2.4 oz.)	663	19.0			
KOHLRABI (USDA):						
Raw, whole	1 lb. (weighed with skin, without leaves)	26	.3			0
Raw, diced	1 cup (4.9 oz.)	11	.1			0
Boiled without salt, drained	1 cup (5.5 oz.)	9	.2			0
*KOOL-AID (General Foods):						
Regular	1 cup (9.3 oz.)	13	Tr.			0
Sugar sweetened	1 cup (9.3 oz.)	13	Tr.			0
KOOL-POPS (General Foods)	1 bar (1.5 oz.)	14	Tr.			0
KOTTBULLAR, canned (Hormel)	1 oz. (1-lb. can)		3.5			
KRIMPETS (Tastykake):						
Apple spice	.9-oz. cake		2.9			
Butterscotch	.9-oz. cake		2.7			
Chocolate	.9-oz. cake		2.8			
Jelly	.9-oz. cake		1.4			
Lemon	.9-oz. cake		2.4			
Orange	.9-oz. cake		2.4			
KRUMBLES, cereal (Kellogg's)	¾ cup (1 oz.)	167	.4			(0)
KUMQUAT, fresh (USDA):						
Whole	1 lb. (weighed with seeds)	30	.4			0
Flesh & skin	4 oz.	8	.1			0

L

Food and Description	Measure or Quantity	Sodium (mg.)	Total	Satu- rated	Unsatu- rated	Choles- terol (mg.)
LAKE COUNTRY WINE (Taylor):						
Red dinner, 12½% alcohol	3 fl. oz.					0
White dinner, 12½% alcohol	3 fl. oz.					0

(USDA): United States Department of Agriculture
*Prepared as Package Directs

Food and Description	Measure or Quantity	Sodium (mg.)	—Fats in grams— Total	Satu- rated	Unsatu- rated	Choles- terol (mg.)
LAKE HERRING, raw (USDA):						
Whole	1 lb. (weighed whole)	111	5.4			
Meat only	4 oz.	53	2.6			
LAKE TROUT, raw (USDA):						
Drawn	1 lb. (weighed with head, fins & bone)		16.8			
Meat only	4 oz.		11.3			
LAKE TROUT or SISCOWET, raw (USDA):						
Less than 6.5 lb. whole	1 lb. (weighed whole)		33.4			
Less than 6.5 lb. whole	4 oz. (meat only)		22.6			
More than 6.5 lb. whole	1 lb. (weighed whole)		88.8			
More than 6.5 lb. whole	4 oz. (meat only)		61.7			
LAMB, choice grade (USDA):						
Chop, broiled:						
Loin. One 5-oz. chop (weighed with bone before cooking) will give you:						
Lean & fat	2.8 oz.	55	22.9	13.	10.	76
Lean only	2.3 oz.	46	4.9	3	2	65
Rib. One 5-oz. chop (weighed with bone before cooking) will give you:						
Lean & fat	2.9 oz.	57	29.2	16.	13.	80
Lean only	2 oz.	39	5.9	3.	3.	56
Fat, separable, cooked	1 oz.		21.4	12.	10.	
Leg:						
Raw, lean & fat	1 lb. (weighed with bone)	280	61.7	35.	27.	271
Roasted, lean & fat	4 oz.	79	21.4	12.	9.	111
Roasted, lean only	4 oz.	79	7.9	4.	4.	113
Shoulder:						
Raw, lean & fat	1 lb. (weighed with bone)	280	92.0	52.	40.	274
Roasted, lean & fat	4 oz.	79	30.8	17.	14.	111
Roasted, lean only	4 oz.	79	11.3	6.	5.	113

(USDA): United States Department of Agriculture
*Prepared as Package Directs

Food and Description	Measure or Quantity	Sodium (mg.)	Total	Satu-rated	Unsatu-rated	Choles-terol (mg.)
				— Fats in grams —		

LAMB'S QUARTERS (USDA):

Food and Description	Measure or Quantity	Sodium (mg.)	Total	Satu-rated	Unsatu-rated	Choles-terol (mg.)
Raw, trimmed	1 lb.		3.6			0
Boiled, drained	4 oz.		.8			0
LAMB STEW, canned (B & M)	1 cup (8.1 oz.)	911	8.9			
LARD:						
(USDA)	1 lb.	0	454.0	172.	282.	431
(USDA)	1 cup (7.2 oz.)	0	205.0	78.	127.	195
(USDA)	1 T. (.5 oz.)	0	13.0	5.	8.	12
LASAGNE:						
Canned (Chef Boy-Ar-Dee)	⅕ of 40-oz. can	1459	12.9			
Canned (Nalley's)	8 oz.		7.3			
Frozen (Buitoni)	⅐ of 56-oz. pkg.		2.6			
Frozen (Buitoni)	½ of 15-oz. pkg.		10.8			
Frozen (Celeste)	¼ of 2-lb. pkg.	1090	26.7			
*Mix, dinner (Chef Boy-Ar-Dee)	8¾-oz. pkg.	1819	7.9			
*Mix, dinner (Jeno's)						
Add 'n Heat	30-oz. pkg.		89.3			
Mix (Hunt's) *Skillet*	1-lb. 2-oz. pkg.	1933	21.7	7.	15.	
Seasoning mix (Lawry's)	1.1-oz. pkg.		.2			
LEEKS, raw (USDA):						
Whole	1 lb. (weighed un-trimmed)	12	.7			0
Trimmed	4 oz.	6	.3			0
LEMON, fresh (USDA):						
Fruit, including peel	1 lb. (weighed whole)	13	1.3			0
Fruit, including peel	2⅛" lemon (3.8 oz., seeds removed)	3	.3			0
Peeled fruit	1 med. lemon (2⅛")	1	.2			0
LEMONADE:						
Chilled (Sealtest)	½ cup (4.4 oz.)	<1	.1			(0)
Frozen, concentrate, sweetened:						
(USDA)	6-fl.-oz. can (7.7 oz.)	4	.2			0
*Diluted with 4⅓ parts water (USDA)	½ cup (4.4 oz.)	Tr.	Tr.			0
*(Minute Maid)	½ cup (4.2 oz.)	<1	Tr.			(0)

(USDA): United States Department of Agriculture
*Prepared as Package Directs

Food and Description	Measure or Quantity	Sodium (mg.)	Total	—Fats in grams— Satu- rated	Unsatu- rated	Choles- terol (mg.)
(ReaLemon)	6-oz. can	4	.2			(0)
*(Snow Crop)	½ cup (4.2 oz.)	<1	Tr.			(0)
*Low calorie (Weight Watchers)	6 fl. oz. (5.8 oz.)		<.1			(0)
LEMON CAKE MIX:						
*(Duncan Hines)	$^1/_{12}$ of cake (2.7 oz.)	385	6.1			50
*Chiffon (Betty Crocker)	$^1/_{16}$ of cake	148	3.5			
*Layer (Betty Crocker)	$^1/_{12}$ of cake	271	5.6			
*Pudding cake (Betty Crocker)	$^1/_6$ of cake	277	4.7			
LEMON DRINK, chilled (Sealtest)	6 fl. oz. (6.5 oz.)	Tr.	<.1			
LEMON JUICE:						
Fresh:						
(USDA)	½ cup (4.3 oz.)	1	.2			0
(USDA)	1 T. (.5 oz.)	<1	<.1			0
(Sunkist)	1 lemon (3.9 oz.)	1	Tr.			0
(Sunkist)	1 T. (.5 oz.)	Tr.	Tr.			0
Canned, unsweetened:						
(USDA)	½ cup (4.3 oz.)	1	.1			0
(USDA)	1 T. (.5 oz.)	<1	<.1			0
Plastic container:						
(USDA)	½ cup (4 oz.)	1	.1			0
(ReaLemon)	1 T. (.5 oz.)	4	Tr.			(0)
Frozen, unsweetened:						
Concentrate (USDA)	½ cup (5.1 oz.)	7	1.3			0
Single strength (USDA)	½ cup (4.3 oz.)	1	.2			0
Full strength, already reconstituted (Minute Maid)	½ cup (4.2 oz.)	1	<.1			0
Full strength, already reconstituted (Snow Crop)	½ cup (4.2 oz.)	1	<.1			0
LEMON-LIMEADE, sweetened, concentrate, frozen:						
*(Minute Maid)	½ cup	Tr.	Tr.			0
*(Snow Crop)	½ cup	Tr.	Tr.			0
LEMON-LIME SOFT DRINK:						
Sweetened:						
(Dr. Brown's)	6 fl. oz.	32	0.			0
(Hoffman)	6 fl. oz.	32	0.			0

(USDA): United States Department of Agriculture
*Prepared as Package Directs

Food and Description	Measure or Quantity	Sodium (mg.)	— Fats in grams —			Choles- terol (mg.)
			Total	Satu- rated	Unsatu- rated	
Key Food)	6 fl. oz.	32	0.			0
(Nedick's)	6 fl. oz.	32	0.			0
(Shasta)	6 fl. oz.	16	0.			0
(Waldbaum)	6 fl. oz.	32	0.			0
(Yukon Club)	6 fl. oz.	32	0.			0
Low calorie:						
Diet Rite	6 fl. oz.	45+	0.			0
(Hoffman)	6 fl. oz.	72	0.			0
(Shasta)	6 fl. oz.	37	0.			0
(Yukon Club)	6 fl. oz.	72	0.			0
LEMON PEEL:						
Raw (USDA)	1 oz.	2	<.1			0
Dried (Spice Islands)	1 tsp.	<1				(0)
Candied (USDA)	1 oz.		<.1			0
LEMON PIE:						
(Hostess)	4½-oz. pie	605	13.7			
(Tastykake)	4-oz. pie		15.5			
Chiffon, home recipe, made with lard (USDA)[1]	1/6 of 9″ pie (3.8 oz.)	282	13.6	5.	9.	183
Chiffon, home recipe, made with vegetable shortening (USDA)[2]	1/6 of 9″ pie (3.8 oz.)	282	13.6	4.	10.	
Cream, frozen:						
(Banquet)	2½-oz. serving		7.9			
(Morton)	¼ of 14.4-oz. pie	190	13.9			
(Mrs. Smith's)	1/6 of 8″ pie (2.3 oz.)	75	12.3			
Meringue, home recipe, 1-crust made with lard (USDA)[1]	1/6 of 9″ pie (4.9 oz.)	395	14.3	5.	9.	130
Meringue, home recipe, 1-crust made with vegetable shortening (USDA)[2]	1/6 of 9″ pie (4.9 oz.)	395	14.3	4.	10.	
Meringue, frozen (Mrs. Smith's)	1/6 of 8″ pie (3.7 oz.)	263	12.8			
Tart, frozen (Pepperidge Farm)	1 pie tart (3 oz.)	218	18.2			
LEMON PIE FILLING, canned:						
(Comstock)	½ cup (5.4 oz.)	<1	.3			
(Lucky Leaf)	8 oz.	472	4.8			
(Wilderness)	22-oz. can	395				

LEMON PIE FILLING MIX
(See **LEMON PUDDING MIX**)

(USDA): United States Department of Agriculture
*Prepared as Package Directs
[1]Principal sources of fat: lard & butter.
[2]Principal sources of fat: vegetable shortening & butter.

Food and Description	Measure or Quantity	Sodium (mg.)	—Fats in grams—			Choles- terol (mg.)
			Total	Satu- rated	Unsatu- rated	
LEMON PUDDING, canned:						
(Betty Crocker)	½ cup	129	3.8			
(Hunt's)[1]	5-oz. can	90	4.2	<.1	3.	
(Thank You)	½ cup (4.5 oz.)		4.1			
LEMON PUDDING or PIE MIX:						
Regular:						
*(Jell-O)	½ cup (5.1 oz.)	114	2.0			128
*(Royal)	⅛ of 9″ pie (includ- ing crust, 4.2 oz.)	260	12.2			14
Instant:						
*(Jell-O)	½ cup (5.3 oz.)	406	4.7			65
*(Royal)	½ cup (5.1 oz.)	280	4.5			14
LEMON RENNET MIX:						
Powder:						
Dry (Junket)	1 oz.	9	<.1			
*(Junket)	4 oz.	54	3.9			
Tablet:						
Dry (Junket)	1 tablet (9 grams)	197	Tr.			
*& sugar (Junket)	4 oz.	98	3.9			
LEMON SOFT DRINK:						
Sweetened:						
(Canada Dry) bottle or can	6 fl. oz.	13+	0.			0
(Canada Dry) *Hi-Spot,* bottle or can	6 fl. oz.	19+	0.			0
(Clicquot Club)	6 fl. oz.	22	0.			0
(Cott)	6 fl. oz.	22	0.			0
(Mission)	6 fl. oz.	22	0.			0
(Royal Crown)	6 fl. oz.	10+	0.			0
Low calorie:						
(Canada Dry) bottle or can	6 fl. oz.	7+	0.			
(Clicquot Club)	6 fl. oz.	34	0.			0
(Cott)	6 fl. oz.	34	0.			0
(Mission)	6 fl. oz.	34	0.			0
(No-Cal)	6 fl. oz.	11	0.			0
LEMON TURNOVER, frozen						
(Pepperidge Farm)	1 turnover (3.3 oz.)	284	21.9			

(USDA): United States Department of Agriculture
*Prepared as Package Directs
[1]Principal source of fat: soybean oil.

Food and Description	Measure or Quantity	Sodium (mg.)	—Fats in grams— Total	Satu- rated	Unsatu- rated	Choles- terol (mg.)
LENTIL:						
Whole:						
Dry:						
(USDA)	½ lb.	68	2.5			0
(USDA)	1 oz.	9	.3			0
(USDA)	1 cup (6.7 oz.)	57	2.1			0
Cooked, drained (USDA)	½ cup (3.6 oz.)		Tr.			0
Split, dry, without seed coat						
(USDA)	½ lb.		2.0			0
LENTIL SOUP, canned:						
*(Manischewitz)	1 cup		2.4			
With ham (Crosse & Blackwell)	6½ oz. (½ can)	460	5.3			
LETTUCE (USDA):						
Bibb, untrimmed	1 lb. (weighed un- trimmed)	30	.7			0
Bibb, untrimmed	7.8-oz. head (4″ dia.)	15	.3			0
Boston, untrimmed	1 lb. (weighed un- trimmed)	30	.7			0
Boston, untrimmed	7.8-oz. head (4″ dia.)	15	.3			0
Butterhead varieties (See Bibb)						
Cos (See Romaine)						
Dark green (See Romaine)						
Grand Rapids	1 lb. (weighed un- trimmed)	26	.9			0
Grand Rapids	2 large leaves (1.8 oz.)	4	.2			0
Great Lakes, untrimmed	1 lb. (weighed un- trimmed)	39	.4			0
Great Lakes, trimmed	1-lb. head (4¾″ dia., weighed trimmed)	41	.5			0
Iceberg:						
Untrimmed	1 lb. (weighed un- trimmed)	39	.4			0
Trimmed	1-lb. head (4¾″ dia., weighed trimmed)	41	.5			0
Leaves	1 cup (2.3 oz.)	6	<.1			0
Chopped	1 cup (2 oz.)	5	<.1			0
Chunks	1 cup (2.6 oz.)	7	<.1			

(USDA): United States Department of Agriculture
*Prepared as Package Directs

| Food and Description | Measure or Quantity | Sodium (mg.) | — Fats in grams — | | | Cholesterol (mg.) |
			Total	Satu-rated	Unsatu-rated	
Loose leaf varieties (See Salad Bowl):						
New York	1 lb. (weighed un-trimmed)	39	.4			0
New York	1-lb. head (4¾" dia., weighed trimmed)	41	.5			0
Romaine:						
Untrimmed	1 lb. (weighed un-trimmed)	26	.9			0
Shredded & broken into pieces	½ cup (.8 oz.)	2	<.1			0
Salad Bowl:						
Untrimmed	1 lb. (weighed un-trimmed)	26	.9			0
Trimmed	2 large leaves (1.8 oz.)	4	.2			0
Simpson:						
Untrimmed	1 lb. (weighed un-trimmed)	26	.9			0
Trimmed	2 large leaves (1.8 oz.)	4	.2			0
White Paris (See Romaine)						
LIFE, cereal (Quaker)	⅔ cup (1 oz.)	176	.6			(0)
LIKE, soft drink	6 fl. oz.	Tr.+	0.			0
LIMA BEAN (See **BEAN, LIMA**)						
LIME, fresh, whole:						
(USDA)	1 lb. (weighed with skin & seeds)	8	.8			0
(USDA)	1 med. (2" dia., 2.4 oz.)	1	.1			0
LIMEADE, concentrate, sweetened, frozen:						
(USDA)	6-fl.-oz. can (7.7 oz.)	Tr.	.2			0
*Diluted with 4⅓ parts water						
(USDA)	½ cup (4.4 oz.)	Tr.	Tr.			0
(ReaLemon)	6-oz. can	4	.2			
*(Minute Maid)	½ cup (4.2 oz.)	Tr.	Tr.			0
*(Snow Crop)	½ cup (4.2 oz.)	Tr.	Tr.			0

(USDA): United States Department of Agriculture
*Prepared as Package Directs

Food and Description	Measure or Quantity	Sodium (mg.)	Total	Satu- rated	Unsatu- rated	Choles- terol (mg.)
			—Fats in grams—			
LIME ICE, home recipe (USDA)	8 oz. (by wt.)	Tr.	Tr.			0
LIME JUICE:						
Fresh (USDA)	1 cup (8.7 oz.)	2	.2			0
Canned or bottled, unsweetened:						
(USDA)	1 T. (.5 oz.)	Tr.	Tr.			0
(USDA)	1 cup (8.7 oz.)	2	.2			0
ReaLime	1 T.	5	Tr.			(0)
LIME PIE, Key lime, cream, frozen (Banquet)	2½-oz. serving		9.7			
*__LIME PIE FILLING MIX,__ Key lime (Royal)	⅛ of 9″ pie (in- cluding crust, 4.2 oz.)	250	12.2			65
LIME SOFT DRINK (Yukon Club)	6 fl. oz.	14	0.	0	0	0
LINGCOD, raw (USDA):						
Whole	1 lb. (weighed whole)	91	1.2			
Meat only	4 oz.	67	.9			
LITCHI NUT (USDA):						
Fresh:						
Whole	4 oz. (weighed in shell with seeds)	2	.2			0
Flesh only	4 oz.	3	.3			0
Dried:						
Whole	4 oz. (weighed in shell with seeds)	2	.6			0
Flesh only	2 oz.	2	.7			0
LIVER:						
Beef, raw (USDA)	1 lb.	617	17.2			1361
Beef, fried (USDA)	4 oz.	209	12.0			497
Beef, fried (USDA)	6½″ x 2⅜″ x ⅜″ slice (3 oz.)	150	9.0			372
Calf, raw (USDA)	1 lb.	331	21.3			1361
Calf, fried (USDA)	4 oz.	134	15.0			497
Calf, fried (USDA)	6½″ x 2⅜″ x ⅜″ slice (3 oz.)	100	11.2			372

(USDA): United States Department of Agriculture
*Prepared as Package Directs

Food and Description	Measure or Quantity	Sodium (mg.)	Total	Fats in grams — Satu-rated	Unsatu-rated	Choles-terol (mg.)
Chicken, raw (USDA)	1 lb.	318	16.8			2517
Chicken, raw, frozen (Swanson)	8-oz. pkg.	138	6.1			
Chicken, simmered (USDA)	4 oz.	69	5.0			846
Chicken, simmered (USDA)	2″ x 2″ x ⅝″ liver (.9 oz.)	15	1.1			186
Goose, raw (USDA)	1 lb.	635	45.4			
Hog, raw (USDA)	1 lb.	331	16.8	6.	11.	1301
Hog, fried (USDA)	4 oz.	126	13.0	3.	10.	497
Hog, fried (USDA)	6½″ x 2⅜″ x ⅜″ slice (3 oz.)	94	9.8	3.	7.	372
Lamb, raw (USDA)	1 lb.	236	17.7			1361
Lamb, broiled (USDA)	4 oz.	96	14.1			497
Lamb, broiled (USDA)	6½″ x 2⅜″ x ⅜″ slice (3 oz.)	72	10.5			372
Turkey, raw (USDA)	1 lb.	286	18.1			1973
Turkey, simmered (USDA)	4 oz.	62	5.4			679
Turkey, simmered, chopped (USDA)	1 cup (4.9 oz.)	77	6.7			839

LIVER PATE (See PATE)

LIVER SAUSAGE or LIVER-WURST:

Fresh (USDA)	1 oz.		7.3			
Sliced (Oscar Mayer)	.9-oz. slice (10 slices to 9 oz.)	264	8.8			
Ring (Oscar Mayer)	1 oz.	287	7.9			
Smoked (USDA)	1 oz.		7.8			

LIVERWURST SPREAD

(Underwood)	1 T. (.5 oz.)	105	3.8			

LOBSTER (USDA):

Raw:

Whole	1 lb. (weighed whole)		2.2			
Meat only	4 oz.	238	2.2			96
Cooked, meat only	4 oz.		1.7			
Cooked, meat only	1 cup (½″ cubes, 5.1 oz.)	304	2.2			123
Canned, meat only	4 oz.	238	1.7			

(USDA): United States Department of Agriculture
*Prepared as Package Directs

Food and Description	Measure or Quantity	Sodium (mg.)	— Fats in grams — Total	Satu- rated	Unsatu- rated	Choles- terol (mg.)
LOBSTER NEWBURG:						
Home recipe (USDA)[1]	4 oz.	260	12.0			206
Home recipe (USDA)[1]	1 cup (8.8 oz.)	572	26.5			455
Frozen (Stouffer's)	11½-oz. pkg.		55.0			
LOBSTER PASTE, canned (USDA)	1 oz.		2.7			
LOBSTER SALAD, home recipe						
(USDA)	4 oz.	141	7.3			
LOBSTER SOUP, canned,						
cream of (Crosse & Blackwell)	½ can (6½ oz.)		4.8			
LOGANBERRY (USDA):						
Fresh:						
Untrimmed	1 lb. (weighed with					
	caps)	4	2.6			0
Trimmed	1 cup (5.1 oz.)	1	.9			0
Canned, solids & liq.:						
Water pack	4 oz.	1	.5			0
Juice pack	4 oz.	1	.6			0
Light syrup	4 oz.	1	.5			0
Heavy syrup	4 oz.	1	.5			0
Extra heavy syrup	4 oz.	1	.5			0
LOG CABIN, syrup:						
Regular	1 T. (.7 oz.)	2	Tr.			0
Buttered	1 T. (.7 oz.)	10	.3			0
Country Kitchen, pancake &						
waffle	1 T.	2	Tr.			0
Maple-honey	1 T.	2	Tr.			0
LONGAN (USDA):						
Fresh:						
Whole	1 lb. (weighed with					
	shell & seeds)		.2			0
Flesh only	4 oz.		.1			0
Dried:						
Whole	1 lb. (weighed with					
	shell & seeds)		.7			0
Flesh	4 oz.		.5			0

(USDA): United States Department of Agriculture
*Prepared as Package Directs
[1]Prepared with butter, egg yolks, sherry & cream.

Food and Description	Measure or Quantity	Sodium (mg.)	— Fats in grams —			Choles- terol (mg.)
			Total	Satu- rated	Unsatu- rated	
LOQUAT, fresh (USDA):						
Whole	1 lb. (weighed with seeds)		.7			0
Flesh only	4 oz.		.2			0
LUCKY CHARMS, cereal (General Mills)	1 cup (1 oz.)	188	1.0			(0)
LUNCHEON MEAT (See also individual listings, e.g., **BOLOGNA**):						
All meat (Oscar Mayer)	1-oz. slice	332	8.8			
Bar-B-Q Loaf (Oscar Mayer)	1-oz. slice	321	2.3			
Cocktail loaf (Oscar Mayer)	1-oz. slice	365	3.7			
Ham & cheese (See **HAM & CHEESE**)						
Honey loaf (Oscar Mayer)	1-oz. slice	365	1.7			
Jellied:						
Beef loaf (Oscar Mayer)	1-oz. slice	321	1.1			
Corned beef loaf (Oscar Mayer)	1-oz. slice	299	1.1			
Luncheon roll, sausage, all meat (Oscar Mayer)	.8-oz. slice	260	1.2			
Luxury Loaf (Oscar Mayer)	1-oz. slice (8 per ½ lb.)	299	1.7			
Meat loaf (USDA)	1 oz.		3.7			
Minced roll sausage, all meat (Oscar Mayer)	.8-oz. slice	242	4.1			
Old fashioned loaf (Oscar Mayer)	1-oz. slice	321	4.0			
Olive loaf (Oscar Mayer)	1-oz. slice	365	4.3			
Peppered loaf (Oscar Mayer)	1-oz. slice	321	2.3			
Pickle & pimento:						
(Hormel)	1 oz. (6-lb. can)		7.2			
(Oscar Mayer)	1-oz. slice	365	3.7			
(Sugardale)	1-oz. slice					
Picnic loaf (Oscar Mayer)	1-oz. slice	310	4.8			
Plain loaf (Oscar Mayer)	1-oz. slice	221	6.0			
Pure beef (Oscar Mayer)	1-oz. slice	343	6.0			
Spiced (Hormel)	1 oz.	312	6.3			
LUNG, raw (USDA):						
Beef	1 lb.		10.4			
Calf	1 lb.		17.2			
Lamb	1 lb.		10.4			

(USDA): United States Department of Agriculture
*Prepared as Package Directs

Food and Description	Measure or Quantity	Sodium (mg.)	Fats in grams — Total	Satu- rated	Unsatu- rated	Choles- terol (mg.)

M

MACADAMIA NUT (USDA):

Whole	1 lb. (weighed in shell)		100.7			0
Shelled	4 oz.		81.2			0

MACARONI. Plain macaroni products are essentially the same in caloric value and carbohydrate content on the same weight basis. The longer they are cooked, the more water is absorbed and this affects the nutritive values.[1]

Dry:						
(USDA)	1 oz.	<1	.3			0
Elbow type	1 cup (4.8 oz.)	3	1.6			0
1-inch pieces	1 cup (3.8 oz.)	2	1.3			0
2-inch pieces	1 cup (3 oz.)	2	1.0			0
Cooked (USDA):						
8-10 minutes, firm	4 oz.	1	.6			0
8-10 minutes, firm	1 cup (4.6 oz.)	1	.6			0
14-20 minutes, tender	1 cup (4.9 oz.)	1	.6			0
14-20 minutes, tender	(4 oz.)	1	.5			· 0
20% Protein, dry (Buitoni)	1 oz.		.6			(0)

MACARONI & BEEF:

Canned, tiny meatballs & sauce (Buitoni)	4 oz.		4.5			
Canned, in tomato sauce (Franco-American)	1 cup	1386	8.4			
Frozen:						
(Banquet) buffet	2-lb. pkg.		52.9			
In tomato sauce (Kraft)	11½-oz. pkg.	936	19.9			
With tomatoes (Stouffer's)	11½-oz. pkg.	1133	19.0			
(Swanson)	11¼-oz. dinner	1499	10.9			

MACARONI & CHEESE:

Home recipe, baked (USDA)[2]	1 cup (7.1 oz.)	1086	22.2	10.	12.	42
Canned:						
(USDA)[3]	1 cup (8.5 oz.)	730	9.6	5.	5.	

(USDA): United States Department of Agriculture
*Prepared as Package Directs
[1]Cholesterol applies to this plain macaroni which is made without milk.
[2]Principal sources of fat: cheese, margarine, milk & butter.
[3]Principal sources of fat: cheese, corn oil & milk.

Food and Description	Measure or Quantity	Sodium (mg.)	— Fats in grams —			Cholesterol (mg.)
			Total	Satu-rated	Unsatu-rated	
(Franco-American)	1 cup	1000	9.5			
(Heinz)	8¼-oz. can	1253	9.7			
Frozen:						
(Banquet) cooking bag	8-oz. bag		10.2			
(Banquet) entrée	8-oz. entrée		9.9			
(Banquet) entrée	20-oz. pkg.		31.9			
(Kraft)	12½-oz. pkg.	1593	32.2			
(Morton) casserole	8-oz. pkg.	934	13.7			
(Stouffer's)	12-oz. pkg.		20.3			

*MACARONI & CHEESE MIX

Food and Description	Measure or Quantity	Sodium (mg.)	Total	Satu-rated	Unsatu-rated	Cholesterol (mg.)
cheddar sauce (Betty Crocker)	1 cup	1292	8.9			

MACARONI DINNER:

Food and Description	Measure or Quantity	Sodium (mg.)	Total	Satu-rated	Unsatu-rated	Cholesterol (mg.)
& beef, frozen (Morton)	11-oz. dinner	989	12.1			
& cheese:						
*(Chef Boy-Ar-Dee)	4½-oz. pkg.	199	3.3			
*(Kraft)	4 oz.	287	8.1			
*(Kraft) deluxe	4 oz.	469	6.1			
Frozen (Banquet):						
Macaroni compartment	8 oz.		7.9			
Peas compartment	1.9 oz.		1.1			
Carrots compartment	2.1 oz.		1.2			
Complete dinner	12-oz. dinner		10.2			
Frozen (Morton)	12¾-oz. dinner	1182	13.9			
Frozen (Swanson)	12¾-oz. dinner	1451	13.7			
Creole, with mushrooms (Heinz)	8¾-oz. can	1577	3.9			
*Italian-style (Kraft)	4 oz.	345	2 4			
*Mexican-style (Kraft)	4 oz.	296	2.2			
*_Monte Bello_ (Betty Crocker)	1 cup	1205	12.9			

MACARONI ENTREE, shells in

Food and Description	Measure or Quantity	Sodium (mg.)	Total	Satu-rated	Unsatu-rated	Cholesterol (mg.)
meat sauce (Buitoni)	4 oz.		3.5			

MACARONI SALAD,

Food and Description	Measure or Quantity	Sodium (mg.)	Total	Satu-rated	Unsatu-rated	Cholesterol (mg.)
canned (Nalley's)	4 oz.		14.8			

Food and Description	Measure or Quantity	Sodium (mg.)	Total	Satu-rated	Unsatu-rated	Cholesterol (mg.)
MACE, ground (Spice Islands)	1 tsp.	2				(0)

(USDA): United States Department of Agriculture
*Prepared as Package Directs

Food and Description	Measure or Quantity	Sodium (mg.)	—Fats in grams—			Cholesterol (mg.)
			Total	Saturated	Unsaturated	

MACKEREL (USDA):
  Atlantic:
    Raw:

Food and Description	Measure or Quantity	Sodium (mg.)	Total	Saturated	Unsaturated	Cholesterol (mg.)
Whole	1 lb. (weighed whole)		29.9			233
Meat only	4 oz.		13.8			108
Broiled with butter or margarine	8½″ x 2½″ x ½″ fillet (3.7 oz.)		16.6			
Broiled with vegetable shortening	8½″ x 2½″ x ½″ fillet (3.7 oz.)		16.6			106
Canned, solids & liq.	4 oz.		12.6			107
Canned, solids & liq.	15-oz. can		47.2			400
Pacific:						
Raw:						
Dressed	1 lb. (weighed with bones & skin)		23.8			
Meat only	4 oz.		8.3			
Canned, solids & liq.	4 oz.		11.3			
Salted	4 oz.		28.5			
Smoked	4 oz.		14.7			

MACKEREL, JACK (See **JACK MACKEREL**)

MAI TAI COCKTAIL:
  Canned (National Distillers)

Food and Description	Measure or Quantity	Sodium (mg.)	Total	Saturated	Unsaturated	Cholesterol (mg.)
Duet, 12½% alcohol	8-fl.-oz. can	<1	0.			0
Dry mix (Bar-Tender's)	1 serving (⅝ oz.)	106	.3			

Food and Description	Measure or Quantity	Sodium (mg.)	Total	Saturated	Unsaturated	Cholesterol (mg.)
MALT, dry (USDA)	1 oz.		.5			(0)

MALTED MILK MIX:

Food and Description	Measure or Quantity	Sodium (mg.)	Total	Saturated	Unsaturated	Cholesterol (mg.)
Dry powder, "unfortified" (USDA)	1 oz. (3 heaping tsps.)	125	2.4			
*Prepared with whole milk (USDA)	1 cup (8.3 oz.)	214	10.3			
Chocolate, instant (Borden)	2 heaping tsps. (.7 oz.)	64	.9			
Chocolate (Carnation)[1]	3 heaping tsps. (.7 oz.)	50	.8	Tr.	Tr.	

(USDA): United States Department of Agriculture
*Prepared as Package Directs
[1]Principal sources of fat: wort solids, milk, cocoa & lecithin.

Food and Description	Measure or Quantity	Sodium (mg.)	—Fats in grams— Total	Saturated	Unsaturated	Cholesterol (mg.)
Chocolate, dry (Kraft)	2 heaping tsps. (.4 oz.)	55	.7			
*Chocolate (Kraft)	1 cup (8.7 oz.)	213	9.2			
Natural, instant (Borden)	2 heaping tsps. (.7 oz.)	67	1.7			
Natural (Carnation)[1]	3 heaping tsps. (.7 oz.)	98	1.7	<1.	<1.	
Natural, dry (Kraft)	2 heaping tsps. (.4 oz.)	50	1.0			
*Natural (Kraft)	1 cup (8.6 oz.)	201	9.6			
MALTEX, cereal	1 oz.	<1	.4			(0)
MALT EXTRACT, dried (USDA)	1 oz.	23	Tr.			(0)
***MALT-O-MEAL**, cereal	¾ cup (1 oz. dry)	<1	.2			0
MAMEY or MAMMEE APPLE, fresh (USDA):						
Whole	1 lb. (weighed with skin & seeds)	42	1.4			0
Flesh only	4 oz.	17	.6			0
MANDARIN ORANGE, CANNED:						
Light syrup (Del Monte)	½ cup (4.5 oz.)	8	Tr.			0
Low calorie, solids & liq. (Diet Delight)	½ cup (4.3 oz.)	5	Tr.			(0)
Unsweetened, solids & liq. (S and W) *Nutradiet*	4 oz.	2	Tr.			(0)
MANDARIN ORANGE, FRESH (See **TANGERINE**)						
MANGO, fresh (USDA):						
Whole	1 lb. (weighed with seeds & skin)	21	1.2			0
Whole	1 med. (7.1 oz.)	9	.5			0
Flesh only, diced or sliced	½ cup (2.9 oz.)	6	.3			0
MANHATTAN COCKTAIL:						
Canned: (National Distillers) *Duet,* 20% alcohol	8-fl.-oz. can	Tr.	0.			0

(USDA): United States Department of Agriculture
*Prepared as Package Directs
[1]Principal sources of fat: wort solids, milk & lecithin.

Food and Description	Measure or Quantity	Sodium (mg.)	Total	Satu- rated	Unsatu- rated	Choles- terol (mg.)
			— Fats in grams —			
Brandy (National Distillers)						
Duet, 20% alcohol	8-fl.-oz. can	Tr.	0.			0
Dry mix (Bar-Tender's)	1 serving (¹/₃ oz.)	Tr.	<.1			(0)
MANICOTTI, frozen:						
Dinner (Celeste)	2 manicotti (with sauce)	1510	19.6			
Without sauce (Buitoni)	4 oz.		10.0			
With sauce (Buitoni)	4 oz.		5.8			
MAPLE RENNET MIX:						
Powder:						
Dry (Junket)	1 oz.	14	<.1			
*(Junket)	4 oz.	57	3.9			
Tablet:						
Dry (Junket)	1 tablet (<1 gram)	197	Tr.			
*& sugar (Junket)	4 oz.	98	3.9			
MAPLE SYRUP (See also individual brand names):						
(USDA)	1 T. (.7 oz.)	2				0
(Cary's)	1 T. (.8 oz.)	2	0.			0
Dietetic (Tillie Lewis)	1 T. (.5 oz.)	4	Tr.			(0)
MARBLE CAKE MIX:						
Dry (USDA)[1]	1 oz.	108	3.8	<1.	3.	
*Prepared with eggs, boiled white icing (USDA)[2]	4 oz.	294	9.9	3.	6.	
*Layer (Betty Crocker)	¹/₁₂ of cake	279	5.7			
MARGARINE:						
Salted:						
Made with hydrogenated fat, regular or soft:						
(USDA)	1 lb.	4477	367.4	82.	286.	0
(USDA)	4 oz. (1 stick)	1119	91.9	20.	70.	0
(USDA)	1 cup or 1 tub (8 oz.)	2239	183.7	41.	143.	0
(USDA)	1 T. (⅛ of stick, .5 oz.)	138	11.3	3.	9.	0
(USDA)	1 pat (1″ x ⅓″ x 1″, 5 grams)	49	4.0	<1.	3.	0

(USDA): United States Department of Agriculture
*Prepared as Package Directs
[1]Principal source of fat: vegetable shortening.
[2]Principal sources of fat: vegetable shortening, chocolate, egg & milk.

Food and Description	Measure or Quantity	Sodium (mg.)	— Fats in grams —			Choles- terol (mg.)
			Total	Satu- rated	Unsatu- rated	
Made with liquid oil, regular or soft:						
(USDA)	1 lb.	4477	367.4	86.	281.	0
(USDA)	4 oz. (1 stick)	1119	91.9	22.	70.	0
(USDA)	1 cup or 1 tub (8 oz.)	2239	183.7	43.	141.	0
(USDA)	1 T. (.5 oz.)	138	11.3	3.	9.	0
(USDA)	1 pat (1″ x ⅓″ x 1″, 5 grams)	49	4.0	<1.	3.	0
Made with two-thirds animal fat & one-third vegetable fat:						
(USDA)	1 lb.	4477	367.4			227
(USDA)	4 oz. (1 stick)	1119	91.9			57
(USDA)	1 cup or 1 tub (8 oz.)	2239	183.7			113
(USDA)	1 T. (⅛ of stick, .5 oz.)	138	11.3			7
(USDA)	1 pat (1″ x ⅓″ x 1″, 5 grams)	49	4.0			2
(Blue Bonnet) regular	1 T. (.5 oz.)	110	11.2	2.	9.	0
(Blue Bonnet) soft	1 T. (.5 oz.)	110	11.2	2.	9.	0
(Borden) Danish flavor	1 T. (.5 oz.)	112	11.2	8.	3.	
(Fleischmann's) regular	1 T. (.5 oz.)	110	11.2	2.	9.	0
(Fleischmann's) soft	1 T. (.5 oz.)	110	10.2	2.	8.	0
(Golden Glow)	1 T. (.4 oz.)	86	10.0			
(Good Luck) soft	1 T. (.4 oz.)	86	10.0	2.	8.	0
(Holiday)	1 T. (.5 oz.)	172	11.2			0
(Imperial) stick	1 T. (.5 oz.)	112	11.2	2.	9.	0
(Imperial) *Sof-Spread*	1 T. (.5 oz.)	96	11.2	2.	9.	0
(Mazola)[1]	1 T. (.5 oz.)	119	11.2	2.	9.	0
(Miracle) corn oil	1 T. (9 grams)	74	7.4			
(Nucoa)[1]	1 T. (.5 oz.)	125	11.2	3.	9.	0
(Nucoa) soft[1]	1 T. (.4 oz.)	109	10.0	2.	8.	0
(Parkay) regular	1 T. (.5 oz.)	112	11.2			
(Parkay) soft cup	1 T. (.5 oz.)	106	10.6			
(Parkay) corn oil deluxe	1 T. (.5 oz.)	112	11.2			
(Parkay) corn oil soft	1 T. (.5 oz.)	106	10.6			
(Parkay) safflower oil, soft	1 T. (.5 oz.)	106	10.6			
(Parkay) squeeze	1 T. (.5 oz.)	112	11.2			
(Phenix)	1 T. (.5 oz.)	112	11.2			
(Promise) soft	1 T. (.5 oz.)	96	11.2	2.	10.	0
(Promise) stick	1 T. (.5 oz.)	112	11.2	2.	10.	0
(Saffola) cube	1 T. (.5 oz.)	120	11.2	2.	9.	0
(Saffola) soft	1 T. (.5 oz.)	114	11.2	2.	10.	0

(USDA): United States Department of Agriculture
*Prepared as Package Directs
[1]Principal source of fat: vegetable oil.

Food and Description	Measure or Quantity	Sodium (mg.)	—Fats in grams—			Choles- terol (mg.)
			Total	Satu- rated	Unsatu- rated	
Unsalted:						
Made with hydrogenated fat, regular or soft:						
(USDA)	1 lb.	45	367.4	82.	286.	0
(USDA)	4 oz. (1 stick)	11	91.9	20.	71.	0
(USDA)	1 cup or 1 tub (8 oz.)	23	183.7	41.	143.	0
(USDA)	1 T. (⅛ of stick, .5 oz.)	1	11.3	3.	9.	0
(USDA)	1 pat (1″ x ⅓″ x 1″, 5 grams)	<1	4.0	<1.	3.	0
Made with liquid oil, regular or soft:						
(USDA)	1 lb.	45	367.4	86.	281	0
(USDA)	4 oz. (1 stick)	11	91.9	22.	70.	0
(USDA)	1 cup or 1 tub (8 oz.)	23	183.7	43.	141.	0
(USDA)	1 T. (⅛ stick, .5 oz.)	1	11.3	3.	9.	0
(USDA)	1 pat (1″ x ⅓″ x 1″, 5 grams)	<1	4.0	<1.	3	0
Made with two-thirds animal fat & one-third vegetable fat:						
(USDA)	1 lb.	45	367.4			227
(USDA)	4 oz. (1 stick)	11	91.9			57
(USDA)	1 cup or 1 tub (8 oz.)	23	183.7			113
(USDA)	1 T. (⅛ stick, .5 oz.)	1	11.3			7
(USDA)	1 pat (1″ x ⅓″ x 1″, 5 grams)	<1	4.0			2
(Fleischmann's) regular	1 T. (.5 oz.)	1	11.2	2.	9.	0
(Mazola)[1]	1 T. (.5 oz.)	Tr.	11.3	2.	9.	0
MARGARINE, IMITATION, diet, salted:						
(Fleischmann's)	1 T. (.5 oz.)	110	5.6	1.	5.	0
(Imperial)	1 T. (.5 oz.)	136	5.5	Tr.	5.	(0)
(Mazola)	1 cup (8.3 oz.)	2338	92.6	17.	76.	0
(Mazola)	1 T. (.5 oz.)	144	5.7	1.	5.	0
(Parkay)	1 T. (.5 oz.)	120	5.9			
MARGARINE, WHIPPED:						
Salted:						
(USDA)	1 stick or ½ cup (2.7 oz.)	750	61.6	14.	48.	0
(Blue Bonnet)	1 T. (9 grams)	70	7.4	2.	6.	0

(USDA): United States Department of Agriculture
*Prepared as Package Directs
[1]Principal source of fat: vegetable oil.

Food and Description	Measure or Quantity	Sodium (mg.)	Fats in grams — Total	Satu- rated	Unsatu- rated	Choles- terol (mg.)
(Imperial)	1 T. (9 grams)	62	7.2	1.	6.	(0)
(Miracle) cottonseed-soybean	1 T. (9 grams)	74	7.4			
(Parkay) cup	1 T. (9 grams)	74	7.4			
Unsalted (USDA)	1 stick or ½ cup (2.7 oz.)	8	61.6	14.	47.	0
MARGARITA COCKTAIL, dry						
mix (Bar-Tender's)	1 serving (⅝ oz.)	42	.2			(0)
MARINADE MIX:						
(Adolph's) instant, meat	.8-oz. pkg.	4068	.5	Tr.	Tr.	
*(Durkee) meat	6 T. (.9-oz. pkg.)	4022	.3			
(Lawry's) beef	1.6-oz. pkg.		.2			
(Lawry's) lemon pepper	2.7-oz. pkg.		3.3			
MARJORAM (Spice Islands)	1 tsp.	<1				(0)
MARMALADE:						
Sweetened:						
(USDA)	1 T. (.7 oz.)	3	<.1			0
(Bama)	1 T. (.7 oz.)	2	<.1			(0)
(Kraft)	1 oz.	<1	<.1			(0)
(Smucker's) bitter	1 T. (.7 oz.)	8	Tr.			(0)
(Smucker's) sweet	1 T. (.7 oz.)	4	Tr.			(0)
Low calorie:						
(Kraft)	1 oz.	15	<.1			(0)
(S and W) *Nutradiet*	1 T. (.5 oz.)		<.1			(0)
(Slenderella)	1 T. (.6 oz.)	<1	0.			(0)
MARMALADE PLUM (See **SAPOTE**)						
MARTINI COCKTAIL, canned (National Distillers) *Duet,* 20%						
alcohol	8-fl.-oz. can	Tr.	0.			0
*MASA HARINA** (Quaker)	2 tortillas (6″ dia.)	5	1.5			
*MASA TRIGO** (Quaker)	2 tortillas (6″ dia.)	300	4.0			
MATZO:						
Regular (Manischewitz)	1 matzo (1.1 oz.)	<1	.2			
American (Manischewitz)	1 matzo (1 oz.)		2.0			

(USDA): United States Department of Agriculture
*Prepared as Package Directs

Food and Description	Measure or Quantity	Sodium (mg.)	—Fats in grams—			Choles- terol (mg.)
			Total	Satu- rated	Unsatu- rated	
Diet-10's (Goodman's)	1 small square ($^1/_9$ of matzo, 3 grams)	Tr.	Tr.			
Diet-10's (Goodman's)	1 matzo (1 oz.)	<1	.6			
Diet-thins (Manischewitz)	1 matzo (1 oz.)	<1	.3			
Egg (Manischewitz)	1 matzo (1.2 oz.)	5	1.1			
Egg 'n Onion (Manischewitz)	1 matzo (1 oz.)		.6			
Onion Tams (Manischewitz)	1 piece (2 grams)		.5			
Tam Tams (Manischewitz)	1 piece (3 grams)		.6			
Tasteas (Manischewitz)	1 matzo (1 oz.)	<1	1.3			
Tea (Goodman's)	1 matzo (.6 oz.)	<1	.4			
Tea (Goodman's) Midgetea	1 matzo (.4 oz.)	<1	.2			
Thin tea (Manischewitz)	1 matzo (1 oz.)	<1	.3			
Unsalted (Goodman's)	1 matzo (1 oz.)	<1	.6			
Whole wheat (Manischewitz)	1 matzo (1.2 oz.)	<1	.7			
MATZO MEAL (Manischewitz)	1 cup (4.1 oz.)	2	.9			
MAYONNAISE:						
(USDA)[1]	1 cup (7.8 oz.)	1319	176.6	31.	146.	155
(USDA)[1]	1 T. (.5 oz.)	84	11.2	2.	9.	10
(Bama)	1 T. (.5 oz.)	16	13.2			
(Bennett's)	1 T. (.5 oz.)	73	12.4			13
(Best Foods) *Real*[2]	1 T. (.5 oz.)	74	11.2	2.	10.	5
(Hellmann's) *Real*[2]	1 T. (.5 oz.)	74	11.2	2.	10.	5
(Kraft)	1 T. (.5 oz.)	75	11.4			
(Kraft) *Salad Bowl*	1 T. (.5 oz.)	73	11.4			
(Nalley's)	1 oz.		22.4			
Saffola	1 cup (7.2 oz.)	1224	166.5	15.	152.	112
Saffola	1 T. (.5 oz.)	78	10.6	<1.	10.	7
Salt free (Healthlife)	1 oz.	3	23.0	2.	21.	15
MAYPO, cereal, dry, any flavor:						
Instant	1 oz.	<1	1.3			
1-minute	1 oz.	<1	1.4			

MEAL (See **CORNMEAL** or
CRACKER MEAL or
MATZO MEAL)

MEATBALL:

In sauce, canned (Prince)	1 can (3.7 oz.)		11.4			

(USDA): United States Department of Agriculture
*Prepared as Package Directs
[1]Principal sources of fat: soybean oil, cottonseed oil, corn oil & egg.
[2]Principal sources of fat: vegetable oil & egg.

Food and Description	Measure or Quantity	Sodium (mg.)	Fats in grams — Total	Satu-rated	Unsatu-rated	Choles-terol (mg.)
Stew, canned (Chef Boy-Ar-Dee)	¼ of 30-oz. can	829	10.4			
With gravy, canned (Chef Boy-Ar-Dee)	¼ of 15¼-oz. can	404	7.8			
With gravy & whipped potato, frozen (Swanson)	9¼-oz. pkg.	524	16.5			
MEAT LOAF ENTREE, frozen:						
(Banquet)	5-oz. serving		19.9			
With tomato sauce (Swanson)	9-oz. pkg.	1175	17.2			
MEAT LOAF DINNER, frozen:						
With tomato sauce, mashed potato & peas (USDA)	12 oz.	1336	22.8			
Banquet:						
Meat compartment	6½ oz.		22.0			
Potato compartment	2¾ oz.		.4			
Peas compartment	2 oz.		1.1			
Complete dinner	11-oz. dinner		23.5			
(Kraft)	5 oz.	771	21.3			
(Morton)	11-oz. dinner	896	21.5			
(Morton) 3-course	1-lb. 1-oz. dinner	1564	28.1			
(Swanson)	10¾-oz. dinner	1006	19.3	7.	13.	
(Swanson) 3-course	16½-oz. dinner	1915	24.2			
In brown gravy (Morton House)	4¹/₆-oz. serving	663	12.1	6.	6.	41
In tomato sauce (Morton House)	4¹/₆-oz. serving	779	12.2	6.	6.	41
MEAT LOAF SEASONING MIX:						
(Contadina)	3¾-oz. pkg.	4190	2.6			
(Lawry's)	3½-oz. pkg.		2.4			
MEAT, POTTED:						
(Armour Star)	3-oz. can		14.6			
(Hormel)	3-oz. can		12.3			
(Van Camp)	½ cup (3.9 oz.)		21.1			
MEAT TENDERIZER:						
Unseasoned (Adolph's)	1 tsp. (5 grams)	1700	Tr.	Tr.	Tr.	0
Unseasoned (French's)	1 tsp.	1760	Tr.			
Seasoned (Adolph's)	1 tsp. (5 grams)	1600	<.1	Tr.	Tr.	0
Seasoned (French's)	1 tsp.	1520	Tr.			
MELBA TOAST:						
Garlic (Keebler)	1 piece (2 grams)	30	.2			

(USDA): United States Department of Agriculture
*Prepared as Package Directs

Food and Description	Measure or Quantity	Sodium (mg.)	Total	Fats in grams — Satu- rated	Unsatu- rated	Choles- terol (mg.)
Garlic, round (Old London)	1 piece (2 grams)	42	.2			
Onion (Keebler)	1 piece (2 grams)	32	.2			
Onion, round (Old London)	1 piece (2 grams)	33	.3			
Plain (Keebler)	1 piece (2 grams)	32	.2			
Pumpernickel (Old London)	1 piece (4 grams)	72	<.1			
Rye (Old London)	1 piece (4 grams)	44	<.1			
Rye, unsalted (Old London)	1 piece (4 grams)	<1	<.1			
Sesame (Keebler)	1 piece (2 grams)	24	.5			
Sesame, rounds (Old London)	1 piece (2 grams)	23	.4			
Wheat (Old London)	1 piece (4 grams)	35	<.1			
Wheat, unsalted (Old London)	1 piece (4 grams)	<1	<.1			
White:						
(Keebler)	1 piece (4 grams)	39	.1			
(Old London)	1 piece (4 grams)	37	<.1			
Unsalted (Old London)	1 piece (4 grams)	<1	<.1			
MELLORINE (Sealtest)	¼ pt. (2.3 oz.)	48	6.6			
MELON (See individual listings, e.g., **CANTALOUPE, WATER-MELON,** etc.)						
MELON BALL, in syrup, frozen (USDA)	½ cup (4.1 oz.)	10	.1			0
MENHADEN, Atlantic, canned, solids & liq. (USDA)	4 oz.		11.6			
MEXICAN DINNER:						
Mix (Hunt's) *Skillet*[1]	1-lb. 2-oz. pkg.	3801	14.6	4.	11.	
Combination dinner, frozen (Patio)	11-oz. dinner		29.0			
Mexican style, frozen:						
(Banquet):						
Meat compartment	9 oz.		17.3			
Rice compartment	3½ oz.		.1			
Beans compartment	3¾ oz.		3.6			
Complete dinner	16¼-oz. dinner		21.0			
(Patio) 5-compartment	12-oz. dinner		22.0			
(Patio) 3-compartment	12-oz. dinner		18.0			
(Swanson)	16¼-oz. dinner	1866	31.3	11.	20.	
(Swanson) 3-course	18-oz. dinner	1935	26.7			

(USDA): United States Department of Agriculture
*Prepared as Package Directs
[1]Principal sources of fat: cottonseed oil, egg & cheese.

Food and Description	Measure or Quantity	Sodium (mg.)	Total	Fats in grams — Satu- rated	Unsatu- rated	Choles- terol (mg.)
MEXICAN-STYLE VEGETA-						
BLES, frozen (Birds Eye)	⅓ of 10-oz. pkg.	80	7.9			0
MILK, CONDENSED, sweetened, canned:						
(USDA)	1 cup (10.8 oz.)	343	26.6	13.	13.	105
Dime Brand	1 T. (.7 oz.)	21	1.7			
Eagle Brand	1 T. (.7 oz.)	21	1.7			
Magnolia Brand	1 T. (.7 oz.)	21	1.7			
MILK, DRY:						
Whole:						
(USDA) packed	1 cup (5.1 oz.)	587	39.9	22.	18.	158
(USDA) spooned	1 cup (4.3 oz.)	490	33.3	18.	15.	132
Nonfat, instant:						
⅞ cup makes 1 qt. (USDA)	⅞ cup (3.2 oz.)	484	.7			20
1⅓ cups make 1 qt. (USDA)	1⅓ cups (3.2 oz.)	479	.7			20
(Carnation)	1 cup (2.4 oz.)	347	.5	Tr.	Tr.	2
*(Carnation)	1 cup (8.6 oz.)	115	.2	Tr.	Tr.	<1
Chocolate (Carnation)	1 cup (2.4 oz.)	371	5.2	3.	2.	
*Chocolate (Carnation)	1 cup (8.6 oz.)	184	2.6	1.	1.	
(Weight Watchers)	1 packet (3 grams)		<.1			
MILK, EVAPORATED, canned:						
Regular:						
Unsweetened (USDA)	1 cup (8.9 oz.)	297	19.9	10.	10.	79
(Borden)	14.5-oz. can	485	32.5			
(Carnation)	1 cup (8.9 oz.)	297	19.9	12.	8.	62
Skimmed:						
(Carnation)	1 cup (9 oz.)	329	.5	Tr.	Tr.	2
Sunshine (Defiance Milk)	1 cup (8.9 oz.)		.5			
MILK, FRESH:						
Whole:						
3.5% fat (USDA)	1 cup (8.6 oz.)	122	8.5	5.	4.	34
3.7% fat (USDA)	1 cup (8.5 oz.)	121	8.9	5.	4.	
3.25% fat, homogenized (Borden)	1 cup (8.6 oz.)		7.8			
3.25% fat (Sealtest)	1 cup (8.6 oz.)	110	7.9			
3.5% fat (Sealtest)	1 cup (8.6 oz.)	113	8.5			
3.7% fat (Sealtest)	1 cup (8.6 oz.)	115	9.0			
Multivitamin (Sealtest)	1 cup (8.6 oz.)	113	8.5			
Skim:						
(USDA)	1 cup (8.6 oz.)	127	.2			5

(USDA): United States Department of Agriculture
*Prepared as Package Directs

Food and Description	Measure or Quantity	Sodium (mg.)	Fats in grams — Total	Satu- rated	Unsatu- rated	Choles- terol (mg.)
1% fat with 1-2% nonfat milk solids added (USDA)	1 cup (8.7 oz.)		2.5			15
2% fat with 1-2% nonfat milk solids added (USDA)	1 cup (8.7 oz.)	150	4.9	2.	2.	22
(Borden)	1 cup (8.6 oz.)		.7			
(Sealtest)	1 cup (8.6 oz.)	116	.2			
Diet (Sealtest)	1 cup (8.6 oz.)	143	1.0			
Light n' Lively, low-fat (Sealtest)	1 cup (8.6 oz.)	140	2.4			
Lite-line, fortified, low-fat (Borden)	1 cup (8.6 oz.)		2.4			
Pro-Line, 2% fat (Borden)	1 cup (8.6 oz.)		4.8			
Skim-line, fortified (Borden)	1 cup (8.6 oz.)		.7			
Vita Lure, 2% fat (Sealtest)	1 cup (8.6 oz.)	140	4.9			
Buttermilk, cultured, fresh:						
(USDA)	1 cup (8.6 oz.)	318	.2			5
0.1% fat (Borden)	1 cup (8.6 oz.)	317	.2			
1.0% fat (Borden)	1 cup (8.6 oz.)	317	2.4			
3.5% fat (Borden)	1 cup (8.6 oz.)	122	8.5			
Golden Nugget (Sealtest)	1 cup (8.6 oz.)	288	2.0			
Light n' Lively (Sealtest)	1 cup (8.6 oz.)	244	2.0			
Low-fat (Sealtest)	1 cup (8.6 oz.)	281	4.9			
Skimmilk (Sealtest)	1 cup (8.6 oz.)	281	.2			
Buttermilk, cultured, dried (USDA)	1 cup (4.2 oz.)	608	6.4	4.	3.	
Chocolate milk drink, fresh:						
With whole milk:						
(USDA)	1 cup (8.8 oz.)	118	8.5	4.	5.	32
3.4% fat (Sealtest)	1 cup (8.6 oz.)	172	8.2			
With skim milk & 2% added butterfat (USDA)	1 cup (8.8 oz.)	115	5.8	2.	3.	20
With skim milk:						
0.5% fat (Sealtest)	1 cup (8.6 oz.)	174	1.2			
1% fat (Sealtest)	1 cup (8.6 oz.)	174	2.5			
2% fat (Sealtest)	1 cup (8.6 oz.)	173	4.8			
MILK, GOAT, whole (USDA)	1 cup (8.6 oz.)	83	9.8	5.	5.	
MILK, HUMAN (USDA)	1 oz. (by wt.)	5	1.1	<1.	<1.	
MILK, REINDEER (USDA)	1 oz. (by wt.)	45	5.6			
MILK SHAKE (McDonald's):						
Chocolate	1 serving (9.5 oz.)	296	7.3			

(USDA): United States Department of Agriculture
*Prepared as Package Directs

213

Food and Description	Measure or Quantity	Sodium (mg.)	—Fats in grams— Total	Satu- rated	Unsatu- rated	Choles- terol (mg.)
Strawberry	1 serving (9.4 oz.)	267	8.0			
Vanilla	1 serving (9.7 oz.)	274	6.6			
MILLET, whole-grain (USDA)	1 lb.		13.2	4.	9.	0
MINCEMEAT:						
(Crosse & Blackwell)	½ cup (5.2 oz.)	328	1.6			
Condensed (None Such)	9-oz. pkg.	1405	3.8			
Ready-to-use (None Such)	½ cup (5.1 oz.)	473	2.0			
With brandy & rum (None Such)	½ cup (5.3 oz.)	353	3.0			
MINCE PIE:						
Home recipe, 2-crust (USDA)	¹/₆ of 9″ pie (5.6 oz.)	708	18.2			
(Tastykake)	4-oz. pie		17.2			
Frozen:						
(Banquet)	5-oz. serving		14.6			
(Morton)	¹/₆ of 20-oz. pie	318	10.8			
(Morton)	⅛ of 46-oz. pie	460	31.4			
(Mrs. Smith's)	¹/₆ of 8″ pie (4.2 oz.)	395	16.1			
(Mrs. Smith's)	⅛ of 10″ pie (5.6 oz.)	538	20.9			
MINESTRONE SOUP:						
Canned:						
Condensed(USDA)	8 oz. (by wt.)	1844	6.4			
*Prepared with equal volume water (USDA)	1 cup (8.6 oz.)	995	3.4			
*(Campbell)	1 cup	905	2.7	Tr.	2.	
(Crosse & Blackwell)	6½ oz. (½ can)	380	.8			
MINT MIST PIE, frozen (Kraft)	¼ of 13-oz. pie	72	18.4			
MISO, cereal & soybeans (USDA)	4 oz.	3345	5.2	1.	4.	0
***MOCHA NUT PUDDING MIX,** instant (Royal)	½ cup (5.1 oz.)	345	5.7			14
MOCHA PIE, frozen (Kraft)	3-oz. serving	88	21.0			
MOLASSES:						
Blackstrap (USDA)	½ cup (5.4 oz.)	148				0
Blackstrap (USDA)	1 T. (.7 oz.)	18				0

(USDA): United States Department of Agriculture
*Prepared as Package Directs

Food and Description	Measure or Quantity	Sodium (mg.)	—Fats in grams— Total	Satu- rated	Unsatu- rated	Choles- terol (mg.)
Light (USDA)	½ cup (5.4 oz.)	23				0
Light (USDA)	1 T. (.7 oz.)	3				0
Medium (USDA)	½ cup (5.4 oz.)	57				0
Medium (USDA)	1 T. (.7 oz.)	7				0
Unsulphured (Grandma's)	1 T. (.7 oz.)	21				(0)
MOR (Wilson) canned luncheon meat	3 oz.	1010	23.1	10.	14.	51
MORTADELLA, sausage (USDA)	1 oz.		7.1			
MR. PiBB, soft drink	6 fl. oz.	8	0.			0
MRS. BUTTERWORTH'S SYRUP	1 T. (.7 oz.)	16	.3	Tr.	Tr.	Tr.

MUFFIN:

Food and Description	Measure or Quantity	Sodium (mg.)	Total	Satu- rated	Unsatu- rated	Choles- terol (mg.)
Blueberry, home recipe (USDA)[1]	3″ muffin (1.4 oz.)	253	3.7	1.	3.	
Blueberry, frozen (Morton)	1.6-oz. muffin	151	2.7			
Bran:						
Home recipe (USDA)[2]	3″ muffin (1.4 oz.)	179	3.9	2.	2.	
(Thomas') with raisins	1.9-oz. muffin	405	5.5			
Corn:						
Home recipe, prepared with whole ground cornmeal (USDA)[3]	2⅜″ muffin (1.4 oz.)	198	4.1	2.	3.	
Home recipe, prepared with degermed cornmeal (USDA)[3]	2⅜″ muffin (1.4 oz.)	192	4.0	2.	2.	
(Morton) frozen	1 muffin (1.7 oz.)	273	4.3			
(Thomas')	1 muffin (2 oz.)	335	7.7			
English:						
(Arnold)	1 muffin (2.2 oz.)		1.4			
(Newly Weds)	1 muffin (2.5 oz.)	475	.6			
(Thomas')	1 muffin (2.1 oz.)	225	.7			0
(Wonder)	1 muffin (2 oz.)	265	1.0			
Golden Egg Toasting (Arnold)	2.2-oz. muffin		3.3			
Plain, home recipe (USDA)[4]	3″ muffin (1.4 oz.)	176	4.0	<1.	3.	21
Scone:						
(Wonder)	1 piece (2 oz.)	197	1.3			
Raisin Round (Wonder)	1 piece (2.2 oz.)	214	1.4			

(USDA): United States Department of Agriculture
*Prepared as Package Directs
[1]Principal sources of fat: vegetable shortening, egg & milk.
[2]Principal sources of fat: butter, egg & milk.
[3]Principal sources of fat: lard, milk & egg.
[4]Principal sources of fat: vegetable shortening & egg.

Food and Description	Measure or Quantity	Sodium (mg.)	— Fats in grams — Total	Saturated	Unsaturated	Cholesterol (mg.)
MUFFIN MIX:						
*Apple cinnamon (Betty Crocker)	2¾" muffin	189	5.4			
*Banana nut (Betty Crocker)	2¾" muffin	165	6.8			
*Blueberry (Betty Crocker)	2¾" muffin	124	3.8			
*Blueberry (Duncan Hines)	1 muffin (1.2 oz.)	183	2.7			
*Butter pecan (Betty Crocker)	2¾" muffin	150	7.7			
Corn:						
With enriched flour (USDA)[1]	1 oz.	187	3.3	<1.	2.	
Prepared with egg & milk (USDA)[2]	2⅜" muffin (1.4 oz.)	136	3.1	<1.	2.	
With cake flour & nonfat dry milk (USDA)[1]	1 oz.	230	3.0	<1.	2.	
Prepared with egg & water (USDA)[3]	2⅜" muffin (1.4 oz.)	138	3.1	<1.	2.	
(Albers)	1 oz.	292	3.9			
*(Betty Crocker)	2¾" muffin	322	4.8			
*(Dromedary)	2½" muffin (1.4 oz.)	237	5.8			
*(Flako)	1.5-oz. muffin (¹⁄₁₂ of pkg.)	295	4.3			
*Date nut (Betty Crocker)	2¾" muffin	164	6.4			
*Honey bran (Betty Crocker)	2¾" muffin	165	4.9			
*Lemon (Betty Crocker)	1 muffin	150	5.4			
*Oatmeal (Betty Crocker)	2¾" muffin	205	6.6			
*Orange (Betty Crocker)	1 muffin	146	4.8			
*Spice (Betty Crocker)	1 muffin	150	5.6			
MULLET, raw (USDA):						
Whole	1 lb. (weighed whole)	195	16.6			
Meat only	4 oz.	92	7.8			
MUNG BEANSPROUT (See **BEAN SPROUT**)						
MUSCATEL WINE (Gold Seal)						
19% alcohol	3 fl. oz. (3.3 oz.)	3	0.			(0)
MUSHROOM:						
Raw (USDA):						
Whole	½ lb. (weighed un-trimmed)	33	.6			0

(USDA): United States Department of Agriculture
*Prepared as Package Directs
[1]Principal source of fat: vegetable shortening.
[2]Principal sources of fat: vegetable shortening, milk & egg.
[3]Principal sources of fat: vegetable shortening & egg.

Food and Description	Measure or Quantity	Sodium (mg.)	—Fats in grams—			Choles-terol (mg.)
			Total	Satu-rated	Unsatu-rated	
Trimmed, slices	½ cup (1.2 oz.)	5	.1			0
Powdered (Spice Islands)	1 tsp.	<1				(0)
Canned, solids & liq.:						
(USDA)	½ cup (4.3 oz.)	488	.1			0
Sliced, chopped or whole, broiled in butter (B in B)	6-oz. can	710	2.5			
Whole or sliced (Green Giant)	4-oz. can	533	.2			(0)
Frozen, whole, in butter sauce (Green Giant)	⅓ of 6-oz. pkg.	120	2.0			
MUSHROOM SOUP, canned:						
*Barley (Manischewitz)	8 fl. oz.		1.9			
Bisque (Crosse & Blackwell)	6½ oz. (½ can)		6.1			
Cream of:						
Condensed (USDA)[1]	8 oz. (by wt.)	1803	18.1	2.	16.	
Prepared with equal volume water (USDA)[1]	1 cup (8.5 oz.)	955	9.6	2.	7.	
Prepared with equal volume milk (USDA)[2]	1 cup (8.6 oz.)	1039	14.2	5.	9.	
*(Campbell)	1 cup	1047	9.9	3.	7.	1
*(Heinz)	1 cup (8.5 oz.)	1030	8.0			
(Heinz) *Great American*	1 cup (8¾ oz.)	1170	7.1			
*Dietetic (Claybourne)	8 oz.	57	3.2			
*Dietetic (Slim-ette)	8 oz. (by wt.)	69	1.4			
Low sodium (Campbell)	7¼-oz. can	20	9.2			
*Golden (Campbell)	1 cup	966	4.0	1.	3.	
MUSHROOM SOUP MIX:						
(Wyler's)	1 oz.		2.2			
*Beef flavor (Lipton)	1 cup	929	.8	Tr.	<1.	
Cream of (Lipton) *Cup-a-Soup*	1 pkg. (.7 oz.)	527	4.5	4.	Tr.	
MUSKELLUNGE, raw (USDA):						
Whole	1 lb. (weighed whole)		5.6			
Meat only	4 oz.		2.8			
MUSKMELON (See **CANTA-LOUPE, CASABA** or **HONEY-DEW**)						
MUSKRAT, roasted (USDA)	4 oz.		4.6			

(USDA): United States Department of Agriculture
*Prepared as Package Directs
[1]Principal sources of fat: corn oil & cream.
[2]Principal sources of fat: corn oil, milk & cream.

Food and Description	Measure or Quantity	Sodium (mg.)	— Fats in grams —			Choles- terol (mg.)
			Total	Satu- rated	Unsatu- rated	
MUSSEL (USDA):						
Raw, Atlantic & Pacific, with liq., in shell	1 lb. (weighed in shell)		3.2			
Raw, Atlantic & Pacific, meat only	4 oz.	328	2.5			
Canned, Pacific, drained	4 oz.		3.7			
MUSTARD, dry, hot or mayonnaise (Spice Islands)	1 tsp.	<1				
MUSTARD, prepared:						
Brown:						
(USDA)	1 tsp. (9 grams)	118	.6			0
(French's) spicy	1 tsp.	53	.3			(0)
(Heinz)	1 tsp.	58	.4			(0)
German style (Kraft)	1 oz.	408	1.8			(0)
Grey Poupon	1 tsp. (6 grams)	70	.2			(0)
Horseradish (Best Foods)[1]	1 tsp. (5 grams)	91	.2			0
Horseradish (French's)	1 tsp.	93	.3			(0)
Horseradish (Kraft)	1 oz.	397	1.8			(0)
Medford (French's)	1 tsp.	80	.3			(0)
Onion (French's)	1 tsp.	53	.2			(0)
Ring Star (French's)	1 tsp.	63	.2			(0)
Salad:						
(French's)	1 tsp.	63	.2			(0)
Glass or squeeze pack (Kraft)	1 oz.	374	1.1			(0)
Plastic squeeze bottle (Kraft)	1 oz.	373	1.1			(0)
Yellow:						
(USDA)	1 tsp. (9 grams)	113	.4			(0)
(Heinz)	1 tsp.	71	.2			(0)
MUSTARD GREENS:						
Raw, whole (USDA)	1 lb. (weighed untrimmed)	102	1.6			0
Boiled without salt, drained (USDA)	1 cup (7.8 oz.)	40	.9			0
Frozen:						
Not thawed (USDA)	4 oz.	14	.5			0
Boiled, drained (USDA)	½ cup (3.8 oz.)	11	.4			0
Chopped (Birds Eye)	½ cup (3.3 oz.)	24	.4			0

(USDA): United States Department of Agriculture
*Prepared as Package Directs
[1]Principal source of fat: mustard seed.

Food and Description	Measure or Quantity	Sodium (mg.)	—Fats in grams—			Choles-terol (mg.)
			Total	Satu-rated	Unsatu-rated	
MUSTARD SEED						
(Spice Islands)	1 tsp.	<1				(0)
MUSTARD SPINACH (USDA):						
Raw	1 lb.		1.4			0
Boiled without salt, drained	4 oz.		.2			0

N

Food and Description	Measure or Quantity	Sodium (mg.)	Total	Satu-rated	Unsatu-rated	Choles-terol (mg.)
NATTO, fermented soybean						
(USDA)	4 oz.		8.4	1.	7.	
NATURAL CEREAL:						
100% (Quaker)	¼ cup (1 oz.)	19	6.5			
100%, with fruit (Quaker)	¼ cup (1 oz.)	16	6.0			
NEAR BEER (See **BEER, NEAR**)						
NEAPOLITAN CREAM PIE, frozen:						
(Banquet)	2½-oz. serving		7.9			
(Morton)	¼ of 14.4-oz. pie	193	15.0			
(Mrs. Smith's)	⅙ of 8″ pie (2.3 oz.)	95	12.8			
NECTARINE, fresh (USDA):						
Whole	1 lb. (weighed with pits)	25	Tr.			0
Flesh only	4 oz.	7	Tr.			0
NEW ZEALAND SPINACH (USDA):						
Raw	1 lb.	721	1.4			
Boiled without salt, drained	4 oz.	104	.2			
NIAGARA WINE, white						
(Pleasant Valley)	12.5% alcohol	24	0.			0
NOODLE (USDA):						
Dry, 1½″ strips[1]	1 cup (2.6 oz.)	4	3.4	<1.	3.	67
Dry, 1½″ strips[1]	1 oz.	1	1.3	Tr.	1.	27
Cooked[1]	1 cup (5.6 oz.)	3	2.4	<1.	2.	50
Cooked[1]	1 oz.	<1	.4	Tr.	Tr.	9

(USDA): United States Department of Agriculture
*Prepared as Package Directs
[1]Principal source of fat: egg.

Food and Description	Measure or Quantity	Sodium (mg.)	—Fats in grams— Total	Satu- rated	Unsatu- rated	Choles- terol (mg.)
NOODLE & BEEF:						
Canned (Heinz)	8½-oz. can	1219	6.4			
Canned (Nalley's)	8 oz.		7.3			
Frozen (Banquet) buffet	2-lb. pkg.		14.1			
NOODLE, CHOW MEIN, canned:						
(USDA)	1 cup (1.6 oz.)		10.6			5
(Hung's)	1 oz.		8.2			
NOODLE DINNER:						
Cantong dinner mix (Betty Crocker)	1 cup	224	21.2			
*Romanoff, mix (Kraft)	8 oz.	499	25.6			
*Stroganoff dinner mix (Betty Crocker)	1 cup	204	25.4			
*With cheese, mix (Kraft)	8 oz.	524	20.4			
*With chicken, mix (Kraft)	8 oz.	1021	5.9			
With chicken, frozen (Swanson)	11-oz. dinner	1410	15.0			
NOODLE MIX:						
*Almondine (Betty Crocker)	½ cup	621	9.4			
Egg Noodles Plus (Pennsylvania Dutch Brand):						
Beef sauce	½ cup	757	2.4	1.	1.	45
Butter sauce	½ cup	706	4.6	2.	3.	48
Cheese sauce	½ cup	440	3.4	1.	2.	45
Chicken sauce	½ cup	591	2.7	<1.	2.	53
Mushroom sauce	½ cup	802	2.2	<1.	2.	43
Onion sauce	½ cup	498	2.2	<1.	2.	43
*Italiano (Betty Crocker)	½ cup	833	8.0			
*Romanoff (Betty Crocker)	½ cup	593	12.5			
NOODLE SOUP:						
Beef (See **BEEF SOUP**)						
Chicken (See **CHICKEN SOUP**)						
*Canned, with ground beef (Campbell)	1 cup	804	4.2	2.	2.	
NUT, mixed (See also individual kinds):						
Dry roasted, salted:						
(Flavor House)	1 oz.	81	14.9			(0)
(Planters)	1 oz.	340	14.2	2.	12.	0

(USDA): United States Department of Agriculture
*Prepared as Package Directs

Food and Description	Measure or Quantity	Sodium (mg.)	Total	Satu-rated	Unsatu-rated	Choles-terol (mg.)
				—Fats in grams—		
(Skippy)	1 oz.	147	15.1	2.	13.	0
Oil roasted:						
With peanuts (Planters)	1 oz.	220	14.3			0
Without peanuts (Planters)	1 oz.	220	15.0			0

NUT LOAF (See **BREAD, CANNED**)

NUTMEG:

Whole (Spice Islands)	1 nut	<1				(0)
Ground (Spice Islands)	1 tsp.	<1				(0)

NUTRIMATO (Mott's)	4 oz.		.3			

O

OAT FLAKES, cereal (Post)	⅔ cup (1 oz.)	198	1.3			0

OATMEAL:
Instant:

(H-O)	1 cup (2.4 oz.)	1	3.9	1.	3.	0
(H-O)	1 T. (4 grams)	Tr.	.2	Tr.	Tr.	0
*(Quaker)	1-oz. packet (¾ cup cooked)	255	1.7			
(Ralston)	4 T. (1 oz.)	1	.6			
Sweet and Mellow (H-O)	1 packet (1.4 oz.)	272	1.7	Tr.	1.	0
With apple & cinnamon (Quaker)	1⅛-oz. packet (¾ cup cooked)	202	1.3			
With chocolate flavor (Quaker)	1¾-oz. packet (¾ cup cooked	276	2.0			
With dates & brown sugar (Quaker)	1⅜-oz. packet (¾ cup cooked)	228	1.2			
With dates & caramel (H-O)	1 packet (1.4 oz.)	232	1.4	Tr.	1.	0
With maple & brown sugar (Quaker)	1⅝-oz. packet (¾ cup cooked)	275	1.8			
With raisins & spice (H-O)	1 packet (1.6 oz.)	252	1.6	Tr.	1.	0
With raisins & spice (Quaker)	1½-oz. packet (¾ cup cooked)	253	1.3			

(USDA): United States Department of Agriculture
*Prepared as Package Directs

Food and Description	Measure or Quantity	Sodium (mg.)	—Fats in grams—			Choles- terol (mg.)
			Total	Satu- rated	Unsatu- rated	
Quick:						
Dry:						
(H-O)	1 cup (2.5 oz.)	2	4.1	1.	3.	0
(H-O)	1 T. (4 grams)	Tr.	.3	Tr.	Tr.	0
(Ralston Oats)	5 T. (1 oz.)	1	1.8			
Cooked:						
*(Albers)	1 cup		2.8			
*(Quaker)	⅔ cup (1 oz. dry)	1	1.7			
Regular:						
Dry:						
(USDA)	1 cup (2.5 oz.)	1	5.3	1.	4.	0
(USDA)	1 T. (4 grams)	<1	.3	Tr.	Tr.	0
(H-O) old fashioned	1 cup (2.6 oz.)	1	4.1	1.	3.	0
(H-O) old fashioned	1 T. (5 grams)	Tr.	.3	Tr.	Tr.	0
(Ralston)	3⅓ T. (1 oz.)	1	.6			
Cooked:						
*(USDA)	1 cup (8.5 oz.)	523	2.4			0
*(Albers) old fashioned	1 cup		2.8			(0)
*(Quaker) old fashioned	⅔ cup (1 oz. dry)	1	1.7			(0)

OCEAN PERCH:
Atlantic:						
Raw, whole (USDA)	1 lb. (weighed whole)	111	1.7			
Fried, dipped in egg, milk & bread crumbs (USDA)	4 oz.	174	15.1			
Frozen, breaded, fried, reheated (USDA)	4 oz.		21.4			
Pacific, raw:						
Whole (USDA)	1 lb. (weighed whole)	77	1.8			
Meat only (USDA)	4 oz.	71	1.7			
Frozen (Gorton)	⅓ of 1-lb. pkg.	120	1.8			

OCEAN PERCH MEALS, frozen:
(Banquet):						
Fish compartment	5 oz.		8.7			
Potato compartment	1.6 oz.		7.6			
Corn compartment	2.2 oz.		.9			
Complete dinner	8.8-oz. dinner		17.1			
(Weight Watchers)	18-oz. dinner		6.1			
& broccoli (Weight Watchers)	9½-oz. luncheon		7.9			

(USDA): United States Department of Agriculture
*Prepared as Package Directs

Food and Description	Measure or Quantity	Sodium (mg.)	— Fats in grams —			Cholesterol (mg.)
			Total	Saturated	Unsaturated	
OCTOPUS, raw, meat only						
(USDA)	4 oz.		.9			
OIL, salad or cooking:						
Corn:						
(USDA)	½ cup (3.9 oz.)	0	110.0	11.	99.	0
(USDA)	1 T. (.5 oz.)	0	14.0	1.	13.	0
(Kraft)	1 oz.	0	28.4	4.	24.	0
(Mazola)	1 cup (7.7 oz.)	0	221.0	29.	192.	0
(Mazola)	1 T. (.5 oz.)	0	14.0	2.	12.	0
Cottonseed (USDA)	½ cup (3.9 oz.)	0	110.0	28.	82.	0
Cottonseed (USDA)	1 T. (.5 oz.)	0	14.0	4.	10.	0
Crisco	1 T.	0	14.0	2.	12.	
Olive (USDA)	½ cup (3.9 oz.)	0	110.0	12.	98.	0
Olive (USDA)	1 T. (.5 oz.)	0	14.0	2.	12.	0
Peanut:						
(USDA)	½ cup (3.9 oz.)	0	110.0	20.	90.	0
(USDA)	1 T. (.5 oz.)	0	14.0	3.	11.	0
(Planters)	1 T. (.5 oz.)	Tr.	14.0	3.	11.	0
Safflower:						
(USDA)	½ cup (3.9 oz.)	0	110.0	9.	101.	0
(USDA)	1 T. (.5 oz.)	0	14.0	1.	13.	0
(Kraft)	1 oz.		28.4	3.	26.	0
(Saff-o-life)	1 T.	Tr.	14.0			(0)
Saffola	½ cup (3.9 oz.)	0	110.0	9.	101.	0
Saffola	1 T. (.5 oz.)	0	14.0	1.	13.	0
Sesame (USDA)	½ cup (3.9 oz.)	0	110.0	15.	95.	0
Sesame (USDA)	1 T. (.5 oz.)	0	14.0	2.	12.	0
Soybean (USDA)	½ cup (3.9 oz.)	0	110.0	16.	94.	0
Soybean (USDA)	1 T. (.5 oz.)	0	14.0	2.	12.	0
Vegetable (Kraft)	1 oz.	0	28.4	5.	24.	0
Wesson	1 T. (.5 oz.)	0	14.0	3.	11.	0
OKRA:						
Raw, whole (USDA)	1 lb. (weighed untrimmed)	12	1.2			0
Boiled without salt, drained (USDA):						
Whole	½ cup (3.1 oz.)	2	.3			0
Pods	8 pods (3″ x ⅝″, 3 oz.)	2	.3			0
Slices	½ cup (2.8 oz.)	2	.2			0
Frozen:						
Cut & pods, not thawed (USDA)	4 oz.	2	.1			0

(USDA): United States Department of Agriculture
*Prepared as Package Directs

223

OKRA (Continued)

Food and Description	Measure or Quantity	Sodium (mg.)	Total	Satu-rated	Unsatu-rated	Choles-terol (mg.)
Cut, boiled, drained (USDA)	½ cup (3.2 oz.)	2	<.1			0
Whole, boiled, drained (USDA)	½ cup (2.4 oz.)	1	<.1			0
Cut (Birds Eye)	½ cup (3.3 oz.)	2	Tr.			0
Whole (Birds Eye)	½ cup (2.5 oz.)	1	Tr.			0

OLD FASHIONED:

| Mix (Bar-Tender's) | 1 serving (⅙ oz.) | Tr. | <.1 | | | (0) |

OLEOMARGARINE (See **MARGARINE**)

OLIVE:
Greek style, salt-cured, oil-coated:

| Pitted (USDA) | 1 oz. | 932 | 10.1 | 1. | 9. | 0 |
| With pits, drained (USDA) | 4 oz. | 2992 | 32.6 | 4. | 29. | 0 |

Green, pitted & drained:

(USDA)	1 oz.	680	3.6			0
(USDA)	4 med. or 3 extra large or 2 giant (.6 oz.)	384	2.0			0
(USDA)	1 olive ($^{13}/_{16}$″ x $1^1/_{16}$″, 6 grams)	132	.7			0

Ripe, by variety, pitted & drained:

Ascalano, any size (USDA)	1 oz.	230	3.9	<1.	3.	0
Manzanilla, any size (USDA)	1 oz.	230	3.9	<1.	3.	0
Mission, any size (USDA)	1 oz.	213	5.7	<1.	5.	0
Mission (USDA)	3 small or 2 large (.4 oz.)	75	2.0	Tr.	2.	0
Mission, slices (USDA)	½ cup (2.2 oz.)	465	12.5	1.	11.	0
Sevillano, any size (USDA)	1 oz.	235	2.7	Tr.	2.	0

| *1 ● 2 ● 3,* dessert (Jello-O) | ⅔ cup (5.3 oz.) | 44 | 3.3 | | | 0 |

ONION:
Raw (USDA):

Whole	1 lb. (weighed untrimmed)	41	.4			0
Whole	2½″ onion (3.9 oz.)	10	.1			0
Chopped	½ cup (3 oz.)	9	<.1			0

(USDA): United States Department of Agriculture
*Prepared as Package Directs

Food and Description	Measure or Quantity	Sodium (mg.)	—Fats in grams—			Cholesterol (mg.)
			Total	Saturated	Unsaturated	
Chopped	1 T. (.4 oz.)	1	<.1			0
Grated	1 T. (.5 oz.)	1	<.1			0
Ground	1 T. (.5 oz.)	2	<.1			0
Slices	½ cup (2 oz.)	6	<.1			0
Boiled without salt (USDA):						
Whole	½ cup (3.7 oz.)	7	.1			0
Halves or pieces	½ cup (3.2 oz.)	6	<.1			0
Pearl onions	½ cup (3.2 oz.)	6	<.1			0
Canned, boiled, solids & liq. (Comstock-Greenwood)	4 oz.	176	.1			(0)
Dehydrated:						
Flakes (USDA)	1 cup (2.3 oz.)	56	.8			0
Flakes (USDA)	1 tsp. (1 gram)	1	Tr.			0
Instant, minced or powder (Spice Islands)	1 tsp.	2				(0)
Frozen:						
Chopped (Birds Eye)	¼ cup (1 oz.)	3	Tr.			0
Whole, small (Birds Eye)	½ cup (4 oz.)	6	.2			0
Small with cream sauce (Birds Eye)	⅓ of 9-oz. pkg.	307	.8			Tr.
Small with cream sauce (Green Giant)	⅓ of 10-oz. pkg.	326	.9			
French-fried rings:						
Canned (Durkee) O&C	3½-oz. can	451	48.6			
Frozen (Birds Eye)	2 oz.	415	10.0			Tr.
Frozen (Mrs. Paul's)	9-oz. pkg.		34.8			
Frozen (Mrs. Paul's)	5-oz. pkg.		25.0			
Pickled, cocktail (Crosse & Blackwell)	1 T. (.5 oz.)	227	0.			

ONION BOUILLON:

(Herb-Ox)	1 cube (4 grams)	510	.1			
(Herb-Ox) instant	1 packet (5 grams)	750	.2			
(Steero)	1 cube (4 grams)		<.1			
(Wyler's)	1 cube (4 grams)		.2			

ONION, GREEN (USDA):

Raw:

Whole	1 lb. (weighed untrimmed)	22	.9			0
Bulb & entire top	1 oz.	1	<.1			0
Bulb & white portion of top	3 small onions (.9 oz.)	1	<.1			0

(USDA): United States Department of Agriculture
*Prepared as Package Directs

Food and Description	Measure or Quantity	Sodium (mg.)	Total	Fats in grams — Satu-rated	Unsatu-rated	Choles-terol (mg.)
Slices, bulb & white portion of top	½ cup (1.8 oz.)	2	.1			0
Tops only	1 oz.	1	.1			0
Dry, shredded (Spice Islands)	1 tsp.	11				(0)
ONION SOUP:						
Condensed (USDA)	8 oz. (by wt.)	1984	4.8			
*Prepared with equal volume water (USDA)	1 cup (8.5 oz.)	1051	2.4			
*(Campbell)	1 cup	932	1.6	<1.	1.	
(Crosse & Blackwell)	6½ oz. (½ can)		1.7			
(Hormel)	15-oz. can		7.2			
ONION SOUP MIX:						
Dry (USDA)[1]	1½-oz. pkg.	2871	4.6			
*Prepared (USDA)	1 cup (8.1 oz.)	660	1.2			
*(Lipton)	1 cup	872	.8	Tr.	<1.	
(Lipton) *Cup-a-Soup*	1 pkg. (.4 oz.)	920	.5	Tr.	Tr.	
(Wyler's)	1 cup		.7			
ONION, WELSH, raw (USDA):						
Whole	1 lb. (weighed untrimmed)		1.2			0
Trimmed	4 oz.		.5			0
OPOSSUM, roasted, meat only (USDA)	4 oz.		11.6			
ORANGE, fresh:						
All varieties:						
Whole (USDA)	1 lb. (weighed with rind & seeds)	3	.7			0
Whole (USDA)	small orange (2½″ dia., 5.3 oz.)	1	.2			0
Whole (USDA)	med. orange (3″ dia., 5.5 oz.)	2	.3			0
Whole (USDA)	large orange (3⅜″ dia., 8.4 oz.)	2	.5			0
Diced or sliced, drained (USDA)	1 cup (7.7 oz.)	2	.4			0
Sections (USDA)	1 cup (8.5 oz.)	2	.5			0
Sections, sweetened, chilled, bottled (Kraft)	4 oz.	115	.2			(0)

(USDA): United States Department of Agriculture
*Prepared as Package Directs
[1]Principal source of fat: vegetable shortening.

Food and Description	Measure or Quantity	Sodium (mg.)	— Fats in grams —			Choles-terol (mg.)
			Total	Satu-rated	Unsatu-rated	
Sections, unsweetened, chilled, bottled (Kraft)	4 oz.	115	.1			(0)
California Navel:						
Whole (USDA)	1 lb. (weighed with rind & seeds)	3	.3			0
Whole (USDA)	2⁴/₅″ orange (6.3 oz.)	1	.1			0
Sections (USDA)	1 cup (8.5 oz.)	2	.2			0
California Navel or Valencia:						
Unpeeled, wedge or slice (Sunkist)	¹/₆ orange (1.1 oz.)	1	Tr.			0
Peeled, cut bite-size (Sunkist)	½ cup (6.3-oz. orange)	1	Tr.			0
California Valencia (USDA):						
Whole	1 lb. (weighed with rind & seeds)	3	1.0			0
Fruit, including peel	2⅝″ orange (6.3 oz.)	4	.5			0
Sections	1 cup (8.5 oz.)	2	.7			0
Florida, all varieties (USDA):						
Whole	1 lb. (weighed with rind & seeds)	3	.7			0
Whole	3″ orange (5.5 oz.)	2	.3			0
Sections	1 cup (8.5 oz.)	2	.5			0
ORANGEADE:						
Chilled (Sealtest)	½ cup (4.4 oz.)	<1	.1			(0)
Frozen:						
*(Minute Maid)	½ cup (4.2 oz.)	47	Tr.			0
*(Snow Crop)	½ cup (4.2 oz.)	47	Tr.			0
ORANGE-APRICOT JUICE DRINK, canned:						
(USDA) 40% fruit juices	1 cup (8.8 oz.)	Tr.	.2			0
(Del Monte)	1 cup (8.6 oz.)	Tr.	Tr.			(0)
ORANGE CAKE MIX:						
*Chiffon (Betty Crocker)	¹/₁₆ of cake	116	2.8			
*(Betty Crocker) layer	¹/₁₂ of cake	269	5.6			
*(Duncan Hines)	¹/₁₂ of cake (2.7 oz.)	374	6.1			50
ORANGE CREAM BAR (Sealtest)	2½-fl.-oz. bar	27	3.1			

(USDA): United States Department of Agriculture
*Prepared as Package Directs

Food and Description	Measure or Quantity	Sodium (mg.)	— Fats in grams —			Choles- terol (mg.)
			Total	Satu- rated	Unsatu- rated	
ORANGE DRINK:						
Chilled (Sealtest)	6 fl. oz. (6.5 oz.)	Tr.	<.1			(0)
Canned (Del Monte)	6 fl. oz. (6.5 oz.)	15	Tr.			(0)
Canned (Hi-C)	6 fl. oz. (6.3 oz.)	<1	Tr.			0
ORANGE-GRAPEFRUIT JUICE:						
Bottled, chilled (Kraft)	½ cup (4.5 oz.)	1	.1			(0)
Canned:						
Sweetened (USDA)	½ cup (4.4 oz.)	1	.1			0
Sweetened (Del Monte)	½ cup (4.3 oz.)	10	.1			0
Sweetened (Stokely-Van Camp)	½ cup (4.4 oz.)		.1			(0)
Unsweetened (USDA)	½ cup (4.3 oz.)	1	.2			0
Unsweetened (Stokely-Van Camp)	½ cup (4.4 oz.)		.2			(0)
Frozen, concentrate, unsweetened:						
(USDA)	6-fl.-oz. can (7.4 oz.)	4	.1			0
*Diluted with 3 parts water						
(USDA)	½ cup (4.4 oz.)	Tr.	.1			0
*(Minute Maid)	½ cup (4.2 oz.)	<1	<.1			0
*(Snow Crop)	½ cup (4.2 oz.)	<1	<.1			0
ORANGE ICE (Sealtest)	¼ pt. (3.2 oz.)	Tr.	Tr.			
ORANGE JUICE:						
Fresh:						
All varieties (USDA)	½ cup (4.4 oz.)	1	.2			0
California Navel (USDA)	½ cup (4.4 oz.)	1	.1			0
California Valencia (USDA)	½ cup (4.4 oz.)	1	.4			0
California Navel or Valencia (Sunkist)	½ cup (4.4 oz.)	<1	Tr.			0
Florida, early or mid-season (USDA)	½ cup (4.4 oz.)	1	.2			0
Florida Temple (USDA)	½ cup (4.4 oz.)	1	.2			0
Florida Valencia (USDA)	½ cup (4.4 oz.)	1	.2			0
Chilled, fresh (Kraft)	½ cup (4.4 oz.)	1	.2			(0)
Chilled, fresh (Sealtest)	½ cup (4.3 oz.)	1	.2			(0)
Canned or bottled, sweetened:						
(USDA)	½ cup (4.4 oz.)	1	.3			0
(Del Monte)	½ cup (4.3 oz.)	1	.1			0
(Heinz)	5½-fl.-oz. can	3	.3			(0)
(Stokely-Van Camp)	½ cup (4.4 oz.)		.2			(0)
Canned or bottled, unsweetened:						
(USDA)	½ cup (4.4 oz.)	1	.2			0

(USDA): United States Department of Agriculture
*Prepared as Package Directs

Food and Description	Measure or Quantity	Sodium (mg.)	—Fats in grams—			Choles- terol (mg.)
			Total	Satu- rated	Unsatu- rated	
(Del Monte)	½ cup (4.3 oz.)	1	.1			0
(Heinz)	5½-fl.-oz. can	3	.4			(0)
*Reconstituted (Kraft)	½ cup (4.4 oz.)	1	.1			(0)
(Stokely-Van Camp)	½ cup (4.4 oz.)		.2			(0)
Canned, concentrate, unsweetened:						
(USDA)	4 oz.	1	.3			0
*Diluted with 5 parts water						
(USDA)	½ cup (4.4 oz.)	1	.4			0
Dehydrated, crystals:						
(USDA)	4-oz. can	9	1.9			0
*Reconstituted (USDA)	½ cup (4.4 oz.)	1	.2			0
Frozen, concentrate:						
(USDA)	6-fl.-oz. can (7.5 oz.)	4	.4			0
*Diluted with 3 parts water						
(USDA)	½ cup (4.4 oz.)	1	.1			0
*(Birds Eye)	½ cup (4 oz.)	1	.1			0
*(Lake Hamilton)	½ cup (4.4 oz.)		.2			(0)
*(Minute Maid)	½ cup (4.2 oz.)	<1	<.1			0
*(Nature's Best)	½ cup (4.4 oz.)		.2			(0)
*(Snow Crop)	½ cup (4.2 oz.)	<1	<.1			0

ORANGE, MANDARIN (See **TANGERINE**)

ORANGE PEEL:

Raw (USDA)	1 oz.	1	.1			0
Dry (Spice Islands)	1 tsp.	<1				(0)
Candied (USDA)	1 oz.		.1			0

ORANGE-PINEAPPLE DRINK,

canned (Hi-C)	6 fl. oz. (6.3 oz.)	<1	Tr.			0

ORANGE-PINEAPPLE JUICE,

bottled, chilled (Kraft)	½ cup (4.4 oz.)	1	.1			(0)

ORANGE-PINEAPPLE PIE

(Tastykake)	4-oz. pie		14.8			

*__ORANGE PLUS__ (Birds Eye)	½ cup (4.4 oz.)	10	.1			0

ORANGE RENNET MIX:
Powder:

Dry (Junket)	1 oz.	14	Tr.			
*(Junket)	4 oz.	57	3.8			

(USDA): United States Department of Agriculture
*Prepared as Package Directs

Food and Description	Measure or Quantity	Sodium (mg.)	Total	— Fats in grams — Satu- rated	Unsatu- rated	Choles- terol (mg.)
Tablet:						
Dry (Junket)	1 tablet (<1 gram)	197	Tr.			
*& sugar (Junket)	4 oz.	98	3.9			

ORANGE SHERBET (See **SHERBET**)

ORANGE SOFT DRINK:
Sweetened:
(Canada Dry) *Sunripe,*

bottle or can	6 fl. oz.	15+	0.			0
(Clicquot Club)	6 fl. oz.	12	0.			0
(Cott)	6 fl. oz.	12	0.			0
(Dr. Brown's)	6 fl. oz.	28	0.			0
(Fanta)	6 fl. oz.	6	Tr.			0
(Hoffman)	6 fl. oz.	28	0.			0
(Key Food)	6 fl. oz.	28	0.			0
(Kirsch)	6 fl. oz.	<1	0.			0
(Mission)	6 fl. oz.	12	0.			0
(Nedick's)	6 fl. oz.	28	0.			0
(Nehi)	6 fl. oz. (6.6 oz.)	0+	0.			0
(Shasta) draft	6 fl. oz.	22	0.			0
(Waldbaum)	6 fl. oz.	28	0.			0
(Yukon Club)	6 fl. oz.	28	0.			0
Low calorie:						
(Canada Dry) bottle	6 fl. oz.	14+	0.			0
(Canada Dry) can	6 fl. oz.	15+	0.			0
(Clicquot Club)	6 fl. oz.	45	0.			0
(Cott)	6 fl. oz.	45	0.			0
Diet Rite	6 fl. oz.	41+	0.			0.
(Dr. Brown's)	6 fl. oz.	46	0.			0
(Hoffman)	6 fl. oz.	46	0.			0
(Key Food)	6 fl. oz.	46	0.			0
(Mission)	6 fl. oz.	45	0.			0
(No-Cal)	6 fl. oz.	13	0.			0
(Shasta)	6 fl. oz.	37	0			0
(Waldbaum)	6 fl. oz.	46	0.			0
(Yukon Club)	6 fl. oz.	69	0.			0

OREGANO (Spice Islands)	1 tsp.	<1				(0)

ORIENTAL DINNER (Hunt's)

Skillet	1-lb. 1-oz. pkg.	3006	.8			

(USDA): United States Department of Agriculture
*Prepared as Package Directs

Food and Description	Measure or Quantity	Sodium (mg.)	Total	Satu- rated	Unsatu- rated	Choles- terol (mg.)
			—Fats in grams—			

OVALTINE, dry:

Food and Description	Measure or Quantity	Sodium (mg.)	Total	Satu- rated	Unsatu- rated	Choles- terol (mg.)
Chocolate	1 oz.	96	1.0			
Malt	1 oz.	68	1.1			
OXTAIL SOUP (Crosse & Blackwell)	6½ oz. (½ can)	530	1.1			
OYSTER:						
Raw:						
Eastern, meat only:						
(USDA)	1 lb. (weighed with shell & liq.)	33	.8			23
(USDA)	12 oysters (weighed in shell, 4 lb.)	132	3.2			92
(USDA)	4 oz.	83	2.0			57
(USDA)	19-31 small or 13-19 med. oysters (1 cup, 8.5 oz.)	175	4.3			120
Pacific & Western, meat only:						
(USDA)	4 oz.		2.5			57
(USDA)	6-9 small or 4-6 med. oysters (1 cup, 8.5 oz.)		5.3			120
Canned, solids & liq. (USDA)	4 oz.		2.5			51
Fried, dipped in egg, milk & bread crumbs (USDA)	4 oz.	234	15.8			
Frozen, solids & liq. (USDA)	4 oz.	431	6.9			

OYSTER CRACKER (See **CRACKER**)

OYSTER STEW:

Food and Description	Measure or Quantity	Sodium (mg.)	Total	Satu- rated	Unsatu- rated	Choles- terol (mg.)
Home recipe (USDA):						
1 part oysters to 1 part milk by volume	1 cup (6-8 oysters, 8.5 oz.)		13.2			
1 part oysters to 2 parts milk by volume[1]	1 cup (8.5 oz.)	814	15.4			62
1 part oysters to 3 parts milk by volume[1]	1 cup (8.5 oz.)	487	12.7			58
*Canned (Campbell)	1 cup	824	8.0	2.	6.	
Frozen (USDA):						
Condensed	8 oz. (by wt.)	1542	14.3			

(USDA): United States Department of Agriculture
*Prepared as Package Directs
[1]Prepared with added salt & butter.

231

Food and Description	Measure or Quantity	Sodium (mg.)	—Fats in grams—			Choles- terol (mg.)
			Total	Satu- rated	Unsatu- rated	
*Prepared with equal volume water	1 cup (8.5 oz.)	816	7.7			
*Prepared with equal volume milk	1 cup (8.5 oz.)	878	11.8			

P

PANCAKE:

Home recipe, wheat (USDA)	4″ pancake (1 oz.)	115	1.9	<1.	1.	
Frozen, breakfast, with link sausage (Swanson)	6-oz. breakfast	972	24.0			

PANCAKE & WAFFLE MIX
(See also **PANCAKE & WAFFLE MIX, DIETETIC**):

Buckwheat:						
(USDA)	1 oz.	378	.5			
(USDA)	1 cup (4.8 oz.)	1801	2.6			
*Prepared with egg & milk (USDA)	4″ pancake (1 oz.)	125	2.5			
*(Aunt Jemima)	4″ pancake (1.2 oz.)	148	2.4			
Buttermilk:						
(USDA)	1 oz.	406	.5			
(USDA)	1 cup (4.8 oz.)	1935	2.4			
*Prepared with milk (USDA)[1]	4″ pancake (1 oz.)	122	1.5	<1.	1.	
*Prepared with milk & egg (USDA)[2]	4″ pancake (1 oz.)	182	2.0	<1.	1.	
*(Aunt Jemima)	4″ pancake (1 oz.)	245	3.0			
*(Duncan Hines)	4″ pancake (2 oz.)	354	1.7			
Flapjack (Albers)	1 cup		1.4			
Plain:						
(USDA)	1 oz.	406	.5			
(USDA)	1 cup (4.8 oz.)	1935	2.4			
*Prepared with milk (USDA)[1]	4″ pancake (1 oz.)	122	1.5	<1.	1.	
*Prepared with milk & egg (USDA)[2]	4″ pancake (1 oz.)	152	2.0	<1.	1.	20
*Prepared with milk & egg (USDA)[2]	6″ x ½″ pancake (7 T. batter)	412	5.3	2.	3.	54
*(Aunt Jemima) Complete	4″ pancake (1.2 oz.)	193	.8			

(USDA): United States Department of Agriculture
*Prepared as Package Directs
[1]Principal sources of fat: vegetable shortening & milk.
[2]Principal sources of fat: vegetable shortening, milk & egg.

Food and Description	Measure or Quantity	Sodium (mg.)	— Fats in grams —			Choles-terol (mg.)
			Total	Satu-rated	Unsatu-rated	
*(Aunt Jemima) Easy Pour	4" pancake (1.2 oz.)	215	2.7			
*(Aunt Jemima) Original	4" pancake (1 oz.)	150	2.3			
Dry (Golden Mix)	1 cup (4.3 oz.)		3.0			
*PANCAKE & WAFFLE MIX, DIETETIC, buttermilk or plain (Tillie Lewis)	4" pancake (1.2 oz.)		.2			2
PANCAKE & WAFFLE SYRUP (See also individual brand names): Sweetened:						
Cane & maple (USDA)	1 T. (.7 oz.)	<1	0.			0
Chiefly corn, light & dark (USDA)	1 T. (.7 oz.)	14	0.			0
(Bama)	1 T. (.7 oz.)	6	0.			
(Golden Griddle)	1 T. (.7 oz.)	1	0.			0
Dietetic or low calorie:						
(Diet Delight)	1 T. (.6 oz.)	10	Tr.			
Boysenberry (Cary's)	1 T. (.6 oz.)	4	0.			0
Maple (Cary's)	1 T. (.6 oz.)	4	0.			0
Maple (Slim-ette)	1 T. (.5 oz.)		0.			
PANCREAS, raw (USDA):						
Beef, lean only	4 oz.	76	8.3			
Beef, medium-fat	4 oz.		28.4			
Calf	4 oz.		10.0			
Hog or hog sweetbread	4 oz.	50	22.6			
PAPAW, fresh (USDA):						
Whole	1 lb. (weighed with rind & seeds)		3.1			0
Flesh only	4 oz.		1.0			0
PAPAYA, fresh (USDA):						
Whole	1 lb. (weighed with skin & seeds)	9	.3			0
Flesh only	4 oz.	3	.1			0
Cubed	1 cup (6.4 oz.)	5	.2			0
PAPRIKA (Spice Islands)	1 tsp.	<2				(0)
PARISIAN-STYLE VEGETA-BLES, frozen (Birds Eye)	⅓ of 10-oz. pkg.	137	6.5			0

(USDA): United States Department of Agriculture
*Prepared as Package Directs

233

Food and Description	Measure or Quantity	Sodium (mg.)	Total	Fats in grams Satu-rated	Unsatu-rated	Choles-terol (mg.)
PARSLEY:						
Fresh (USDA):						
Whole	½ lb.	102	1.4			0
Chopped	1 T. (4 grams)	2	<.1			0
Dry (Spice Islands)	1 tsp.	5				
PARSNIP (USDA):						
Raw, whole	1 lb. (weighed					
	unpared)	46	1.9			0
Boiled without salt, drained,						
cut in pieces	½ cup (3.7 oz.)	8	.5			0
PARV-A-ZERT (SugarLo):						
Chocolate	⅓ pint (3.5 oz.)	59	10.6	9.	2.	19
Coffee, strawberry or vanilla	⅓ pint (3.5 oz.)	36	10.6	9.	2.	19
PASSION FRUIT, fresh (USDA):						
Whole	1 lb. (weighed					
	with shell)	66	1.7			0
Pulp & seeds	4 oz.	32	.8			0
PASTINAS, dry (USDA):						
Carrot	1 oz.		.5			
Egg[1]	1 oz.	1	1.2	Tr.	<1.	
Spinach	1 oz.		.5			
PASTRY SHELL (See also **PIE CRUST**):						
Home recipe, made with lard, baked (USDA)[2]	1 shell (1.5 oz.)	260	14.2	6.	9.	
Home recipe, made with vegetable shortening, baked (USDA)[3]	1 shell (1.5 oz.)	260	14.2	3.	11.	0
(Stella D'oro)	1 shell (1 oz.)		7.8			
Pot, bland (Stella D'oro)	1 shell (1.6 oz.)		11.3			
Pot pie, bland (Keebler)	4" shell (1.7 oz.)	137	11.5			
Tart, sweet (Keebler)	3" shell (1 oz.)	5	9.1			
Frozen (Pepperidge Farm)	1 shell (1.8 oz.)	210	18.1			
Frozen, tart (Pepperidge Farm)	1 pie tart (3 oz.)	176	15.5			

(USDA): United States Department of Agriculture
*Prepared as Package Directs
[1]Principal source of fat: egg.
[2]Principal source of fat: lard.
[3]Principal source of fat: vegetable shortening.

Food and Description	Measure or Quantity	Sodium (mg.)	—Fats in grams—			Choles-terol (mg.)
			Total	Satu-rated	Unsatu-rated	

PATE, canned:

Food and Description	Measure or Quantity	Sodium	Total			
De foie gras (USDA)	1 oz.		12.4			
De foie gras (USDA)	1 T. (.5 oz.)		6.6			
Liver (Hormel)	1 oz.		6.3			
Liver (Sell's)	1 T. (.5 oz.)	105	3.8			

PDQ:

Chocolate	1 T. (.6 oz.)		.3			
Egg Nog	2 heaping tsps. (1 oz.)		.4			
Strawberry	1T. (.5 oz.)		Tr.			

PEA, GREEN:
Raw (USDA):

In pod	1 lb. (weighed in pod)	3	.7			0
Shelled	1 lb.	9	1.8			0
Shelled	½ cup (2.4 oz.)	1	.3			0
Boiled without salt, drained (USDA)	½ cup (2.9 oz.)	<1	.3			0
Canned, regular pack:						
Drained solids (Del Monte)	½ cup (3 oz.)	198	.4			0
Seasoned, drained solids (Del Monte)	½ cup (3 oz.)	204	.4			0
Alaska, Early or June, solids & liq. (USDA)	½ cup (4.4 oz.)	293	.4			0
Alaska, Early or June, drained solids (USDA)	½ cup (3 oz.)	203	.3			0
Alaska, Early or June, drained liq. (USDA)	4 oz.	268	Tr.			0
Early, solids & liq., *April Showers*	½ of 8½-oz. can	374	.4			(0)
Early, solids & liq. (Le Sueur)	½ of 8.5-oz. can	374	.2			
Early, solids & liq. (Stokely-Van Camp)	½ cup (4.1 oz.)		.4			(0)
Early June, with onions (Green Giant)	¼ of 17-oz. can	518	.2			(0)
Sweet, solids & liq. (USDA)	½ cup (4.4 oz.)	293	.4			0
Sweet, drained solids (USDA)	½ cup (3 oz.)	203	.3			0
Sweet, drained liq. (USDA)	4 oz.	268	Tr.			0
Sweet, solids & liq. (Green Giant)	½ of 8.5-oz. can	337	.4			(0)

(USDA): United States Department of Agriculture
*Prepared as Package Directs

Food and Description	Measure or Quantity	Sodium (mg.)	Total	Satu-rated	Unsatu-rated	Choles-terol (mg.)
			— Fats in grams —			
Sweet, honey pod, solids & liq. (Stokely-Van Camp)	½ cup (4 oz.)		.4			(0)
Canned, dietetic pack:						
Alaska, Early or June, solids & liq. (USDA)	4 oz.	3	.3			0
Alaska, Early or June, drained solids (USDA)	4 oz.	3	.5			0
Alaska, Early or June, drained liq. (USDA)	4 oz.	3	Tr.			0
Sweet, solids & liq. (USDA)	4 oz.	3	.3			0
Sweet, drained solids (USDA)	4 oz.	3	.5			0
Sweet, drained liq. (USDA)	4 oz.	3	Tr.			0
Solids & liq. (Diet Delight)	½ cup (4.4 oz.)	10	.2			(0)
(S and W) _Nutradiet_, unseasoned	4 oz.	11	.1			(0)
Sweet, solids & liq. (Blue Boy)	4 oz.	8	.2			(0)
Sweet, solids & liq. (Tillie Lewis)	½ cup (4.4 oz.)	<10	.4			0
Frozen:						
Not thawed (USDA)	1 cup (2.5 oz.)	93	.2			0
Boiled, drained (USDA)	½ cup (3 oz.)	97	.3			0
Sweet (Birds Eye)	½ cup (3.3 oz.)	152	.3			0
Tender tiny (Birds Eye)	½ cup (3.3 oz.)	121	.3			0
In butter sauce, baby peas, (Le Sueur)	⅓ of 10-oz. pkg.	416	2.4			
In butter sauce, sweet (Green Giant)	⅓ of 10-oz. pkg.	402	2.8			
With cream sauce (Birds Eye)	⅓ of pkg. (2.7 oz.)	399	6.5			Tr.
With cream sauce (Green Giant)	⅓ of 10-oz. pkg.	151	.8			
With sliced mushroom (Birds Eye)	⅓ of 10-oz. pkg.	258	.3			0
PEA, MATURE SEED, dry:						
Raw:						
Whole (USDA)	1 lb.	159	5.9			0
Whole (USDA)	1 cup (7.1 oz.)	70	2.6			0
Split, without seed coat (USDA)	1 lb.	181	4.5			0
Split, without seed coat (USDA)	1 cup (7.2 oz.)	81	2.0			0
Cooked without salt, split, without seed coat, drained (USDA)	½ cup (3.4 oz.)	13	.3			0

(USDA): United States Department of Agriculture
*Prepared as Package Directs

Food and Description	Measure or Quantity	Sodium (mg.)	— Fats in grams —		Choles- terol (mg.)
			Total	Satu- rated	Unsatu- rated

Food and Description	Measure or Quantity	Sodium (mg.)	Total	Satu-rated	Unsatu-rated	Choles-terol (mg.)
PEA POD, edible-podded or Chinese (USDA):						
Raw	1 lb. (weighed untrimmed)		.9			0
Boiled, drained	4 oz.		.2			0
PEA & CARROT:						
Canned, regular pack, drained (Del Monte)	½ cup (3 oz.)	308	.4			0
Canned, dietetic pack, solids & liq.:						
(Blue Boy)	4 oz.	22	.2			(0)
(Diet Delight)	½ cup (4.2 oz.)	10	.2			(0)
(S and W) *Nutradiet*	4 oz.	24	.2			(0)
Frozen:						
Not thawed (USDA)	4 oz.	104	.3			0
Boiled, drained (USDA)	½ cup (3.1 oz.)	73	.3			0
(Birds Eye)	½ cup (3.3 oz.)	86	.3			0
In cream sauce (Green Giant)	⅓ of 10-oz. pkg.	236	.8			
PEA & CELERY, frozen (Birds Eye)	½ cup (3.3 oz.)	490	.2			0
PEA & ONION, frozen:						
(Birds Eye)	½ cup (3.3 oz.)	428	.3			0
In butter sauce (Green Giant)	⅓ of 10-oz. pkg.	340	2.6			
PEA & POTATO, with cream sauce, frozen (Birds Eye)	⅓ of pkg. (2.7 oz.)	406	7.1			Tr.
PEA SOUP, GREEN:						
Canned, condensed:						
(USDA)	8 oz. (by wt.)	1665	4.1			
*Prepared with equal volume water (USDA)	1 cup (8.6 oz.)	899	2.2			
*Prepared with equal volume milk (USDA)	1 cup (8.6 oz.)	963	6.4			
*(Campbell)	1 cup	939	1.8	1.	<1.	
Canned, low sodium:						
(Campbell)	7½-oz. can	40	2.3			
*(Claybourne)	8 oz.	23	.1			
Dry mix:						
(USDA)	1 oz.	669	1.2			

(USDA): United States Department of Agriculture
*Prepared as Package Directs

Food and Description	Measure or Quantity	Sodium (mg.)	— Fats in grams —			Choles- terol (mg.)
			Total	Satu- rated	Unsatu- rated	
*(USDA)	1 cup (8.5 oz.)	787	1.5			
*(Lipton)	1 cup	1078	1.7			
(Lipton) *Cup-a-Soup*	1.2-oz. pkg.	694	1.4	Tr.	1.	
Frozen, condensed:						
With ham (USDA)	8 oz. (by wt.)	1701	5.2			
*With ham, prepared with						
equal volume water (USDA)	8 oz. (by wt.)	850	2.7			
PEA SOUP, SPLIT:						
Canned, regular pack:						
Condensed (USDA)	8 oz. (by wt.)	1790	5.9			
*Prepared with equal volume						
water (USDA)	1 cup (8.6 oz.)	941	3.2			
*(Manischewitz)	1 cup		3.1			
*With ham (Campbell)	1 cup	864	3.0	1.	2.	
*With ham (Heinz)	1 cup (8¾ oz.)	953	2.4			
With smoked ham (Heinz)						
Great American	1 cup (9 oz.)	1132	5.2			
*Canned, dietetic (Slim-ette)	8 oz. (by wt.)	24	.5			
Canned, low sodium (Tillie						
Lewis)	1 cup (8 oz.)	25	.5			
PEACH:						
Fresh, without skin (USDA):						
Whole	1 lb. (weighed unpeeled)	4	.4			0
Whole	4-oz. peach (2½″ dia.)	1	.1			0
Diced	½ cup (4.7 oz.)	1	.1			0
Slices	½ cup (3 oz.)	1	.1			0
Canned, regular pack, solids & liq.:						
Juice pack (USDA)	4 oz.	2	.1			0
Light syrup (USDA)	4 oz.	2	.1			0
Heavy syrup (USDA)	2 med. halves & 2 T. syrup (4.1 oz.)	2	.1			0
Heavy syrup, halves (USDA)	½ cup (4.5 oz.)	3	.1			0
Heavy syrup, slices (USDA)	½ cup (4.4 oz.)	3	.1			0
Heavy syrup (Del Monte) cling	½ cup (4.6 oz.)	15	.2			0
Heavy syrup (Del Monte) freestone	½ cup (4.6 oz.)	16	.2			0
Heavy syrup (Hunt's)	½ cup (4.5 oz.)	3	.1			0
Extra heavy syrup (USDA)	4 oz.	2	.1			0

(USDA): United States Department of Agriculture
*Prepared as Package Directs

Food and Description	Measure or Quantity	Sodium (mg.)	Fats in grams — Total	Satu- rated	Unsatu- rated	Choles- terol (mg.)
Spiced, heavy syrup						
(Del Monte)	½ cup (4.5 oz.)	4	Tr.			0
(Stokely-Van Camp)	½ cup (4 oz.)		.1			(0)
Canned, dietetic or unsweetened pack:						
Water pack, solids & liq.						
(USDA)	½ cup (4.3 oz.)	2	.1			0
(Blue Boy) slices, solids & liq.	4 oz.	1	Tr.			(0)
(Diet Delight) cling, halves						
or slices	½ cup (4.4 oz.)	5	<.1			(0)
(Diet Delight) freestone, halves						
or slices	½ cup (4.4 oz.)	5	<.1			(0)
(S and W) *Nutradiet*, cling,						
halves, low calorie, undrained	2 halves (3.5 oz.)	2	.1			(0)
(S and W) *Nutradiet*, cling,						
halves, unsweetened	2 halves (3.5 oz.)	2	.1			(0)
(S and W) *Nutradiet*, cling,						
slices, low calorie, undrained	4 oz.	2	.1			(0)
(S and W) *Nutradiet*, cling,						
slices, unsweetened	4 oz.	2	<.1			(0)
(S and W) *Nutradiet*, freestone,						
halves, low calorie, undrained	4 halves (3.5 oz.)	5	.1			(0)
(S and W) *Nutradiet*, freestone,						
slices, low calorie, solids & liq.	4 oz.	2	.1			(0)
(Tillie Lewis) cling, solids &						
liq.	½ cup (4.3 oz.)	<10	.1			0
(Tillie Lewis) Elberta, solids &						
liq.	½ cup (4.3 oz.)	10	.1			0
Dehydrated, sulfured, nugget or pieces:						
Uncooked (USDA)	1 oz.	6	.3			0
Cooked with added sugar,						
solids & liq. (USDA)	½ cup (5-6 halves & 3 T. liq., 5.4 oz.)	8	.3			0
Dried:						
Uncooked (USDA)	1 lb.	73	3.2			0
Uncooked (USDA)	½ cup (3.1 oz.)	14	.6			0
Cooked, unsweetened (USDA)	½ cup (5-6 halves & 3 T. liq., 4.8 oz.)	7	.3			0
Cooked, with added sugar						
(USDA)	½ cup (5-6 halves & 3 T. liq., 5.4 oz.)	6	.3			0
Uncooked (Del Monte)	½ cup (3.1 oz.)	4	.7			0

(USDA): United States Department of Agriculture
*Prepared as Package Directs

Food and Description	Measure or Quantity	Sodium (mg.)	Total	Satu- rated	Unsatu- rated	Choles- terol (mg.)
			— Fats in grams —			
Frozen:						
Not thawed, slices, sweetened:						
(USDA)	12-oz. pkg.	7	.3			0
(USDA)	16-oz. can	9	.5			0
(USDA)	½ cup (4.2 oz.)	2	.1			0
Quick thaw (Birds Eye)	½ cup (5 oz.)	1	.1			0
PEACH BUTTER (Smucker's)	1 T. (.6 oz.)	2	Tr.			(0)
PEACH DUMPLING, frozen						
(Pepperidge Farm)	1 dumpling (3.3 oz.)	220	16.4			
PEACH NECTAR, canned:						
(USDA)	1 cup (8.8 oz.)	2	Tr.			0
(Del Monte)	1 cup (8.7 oz.)	15	Tr.			0
PEACH PIE:						
Home recipe, 2-crust (USDA):						
Made with lard[1]	⅙ of 9″ pie (5.6 oz.)	423	17.5	6.	11.	
Made with vegetable shortening[2]	⅙ of 9″ pie (5.6 oz.)	423	16.9	6.	11.	
(Tastykake)	4-oz. pie		15.0			
Frozen:						
(Banquet)	5-oz. serving		13.8			
(Morton)	⅙ of 20-oz. pie	244	10.8			
(Morton)	⅙ of 24-oz. pie	265	10.9			
(Morton)	⅛ of 46-oz. pie	311	16.5			
(Mrs. Smith's)	⅙ of 8″ pie (4.2 oz.)	283	14.2			
(Mrs. Smith's) old fashion	⅙ of 9″ pie (5.8 oz.)	470	22.7			
(Mrs. Smith's)	⅛ of 10″ pie (5.6 oz.)	381	18.4			
PEACH PIE FILLING:						
(Comstock)	½ cup (5.1 oz.)	183	<.1			
(Lucky Leaf)	8 oz.	132	.2			
PEACH PRESERVE:						
Sweetened (Bama)	1 T. (.7 oz.)	20	<.1			(0)
Low calorie or dietetic:						
(Kraft)	1 oz.	42	<.1			(0)
(Tillie Lewis)	1 T. (.5 oz.)	3	Tr.			0

(USDA): United States Department of Agriculture
*Prepared as Package Directs
[1]Principal sources of fat: lard & butter.
[2]Principal sources of fat: vegetable shortening & butter.

Food and Description	Measure or Quantity	Sodium (mg.)	— Fats in grams —			Choles- terol (mg.)
			Total	Satu- rated	Unsatu- rated	
PEACH TURNOVER, frozen						
(Pepperidge Farm)	1 turnover (3.3 oz.)	254	19.9			
PEANUT:						
Raw (USDA):						
In shell	1 lb. (weighed in shell)	17	157.3	35.	122.	0
With skins	1 oz.	1	13.5	3.	11.	0
Without skins, whole, unsalted	1 oz.	1	13.7	3.	11.	0
Boiled (USDA)	1 oz.	1	8.9	2.	7.	0
Roasted:						
Whole, unsalted (USDA)	1 lb. (weighed in shell)	15	148.0	33.	115.	0
With skins, unsalted (USDA)	1 oz.	1	13.8	3.	11.	0
Without skins, salted (USDA)	1 oz.	119	14.1	3.	11.	0
Halves, salted (USDA)	½ cup (2.5 oz.)	301	35.9	8.	28.	0
Chopped, salted (USDA)	½ cup (2.4 oz.)	288	34.4	8.	27.	0
Chopped, salted (USDA)	1 T. (9 grams)	38	4.5	1.	3.	0
Dry (Franklin)	1 oz.	283	12.0			(0)
Dry (Frito-Lay)	1 oz.	340	14.5	3.	12.	0
Dry, salted (Flavor House)	1 oz.	119	14.1			(0)
Dry (Planters)	1 oz. (jar)	340	13.3	3.	11.	0
Dry (Skippy)	1 oz.	194	14.5	3.	12.	0
Oil (Planters) cocktail	¾-oz. bag	165	10.9			0
Oil (Planters) cocktail	1 oz. (can)	220	14.6			0
Salted (Nabisco) *Nab*	¾-oz. pkg.	155	10.6			(0)
Salted (Nabisco) *Nab*	1¼-oz. pkg.	258	17.7			(0)
Salted (Nabisco) *Nab*	1⅞-oz. pkg.	387	26.5			(0)
Toasted (Tom Houston)	2 T. (1.1 oz.)	125	14.9			(0)
Spanish, oil-roasted, *Freshnut*	1 oz.	160	14.7	3.	12.	0
Spanish, dry-roasted (Planters)	1 oz. (jar)	340	14.4			0
Spanish, oil-roasted (Planters)	1 oz. (can)	220	15.3			0
PEANUT BUTTER:						
(USDA):						
Small amounts of fat & salt added	½ cup (4.4 oz.)	765	62.2	11.	51.	0
Small amounts of fat & salt added	1 T. (.6 oz.)	97	7.9	1.	6.	0
Small amounts of fat, sweetener & salt added	½ cup (4.4 oz.)	764	62.4	11.	51.	0
Small amounts of fat, sweetener & salt added	1 T. (.6 oz.)	97	7.9	1.	6.	0
Moderate amounts of fat, sweetener & salt added	½ cup (4.4 oz.)	762	63.8	11.	52.	0

(USDA): United States Department of Agriculture
*Prepared as Package Directs

(241)

Food and Description	Measure or Quantity	Sodium (mg.)	Fats in grams — Total	Satu- rated	Unsatu- rated	Choles- terol (mg.)
Moderate amounts of fat, sweetener & salt added	1 T. (.6 oz.)	97	8.1	1.	7.	0
(Bama) crunchy	1 T. (.6 oz.)	70	7.7	1.	7.	(0)
(Bama) smooth	1 T. (.6 oz.)	87	8.4	1.	7.	(0)
(Jif)	1 T. (.6 oz.)	89	8.6	2.	7.	0
(Peter Pan)	1 T. (.6 oz.)	78	7.8			0
(Planters)	1 T. (.5 oz.)	95	8.1	3.	6.	0
(Skippy) creamy[1]	1 T. (.6 oz.)	77	8.2	2.	6.	0
(Skippy) chunk[1]	1 T. (.6 oz.)	77	8.5	2.	7.	0
(Smucker's) creamy or crunchy	1 T. (.5 oz.)	81	6.7	1.	6.	(0)
(Smucker's) old fashioned	1 T. (.5 oz.)	81	6.9	1.	6.	(0)
& jelly (Smucker's) *Goober*	1 T. (.5 oz.)	43	3.3	<1.	3.	(0)
With crackers (See **CRACKERS**)						
PEANUT SPREAD:						
(USDA)[2]	1 oz.	169	14.8	3.	12.	0
Diet (Peter Pan)	1 T. (.5 oz.)	3	8.2			0
PEAR:						
Fresh:						
Whole (USDA)	1 lb. (weighed with stems & core)	8	1.7			0
Whole (USDA)	6.4-oz. pear (3″ x 2½″)	3	.6			0
Quartered (USDA)	½ cup (3.4 oz.)	2	.4			0
Slices, including skin (USDA)	½ cup (2.9 oz.)	2	.3			0
Canned, regular pack, solids & liq.:						
Juice pack (USDA)	4 oz.	1	.3			0
Light syrup (USDA)	4 oz.	1	.2			0
Heavy syrup, halves (USDA)	½ cup (4 oz.)	1	.2			0
Heavy syrup, halves (USDA)	2 med. halves & 2 T. syrup (4.1 oz.)	1	.2			0
Heavy syrup (Del Monte)	½ cup (4 oz.)	7	.6			0
Heavy syrup (Hunt's)	½ cup (4.5 oz.)	1	.3			(0)
Extra heavy syrup (USDA)	4 oz.	1	.2			0
(Stokely-Van Camp)	½ cup (4 oz.)		.2			(0)
Canned, unsweetened or low calorie:						
Water pack, solids & liq. (USDA)	½ cup (4.3 oz.)	1	.2			0
Solids & liq. (Blue Boy) Bartlett	4 oz.	1	.1			(0)

(USDA): United States Department of Agriculture
*Prepared as Package Directs
[1]Principal sources of fat: peanuts & vegetable oil.
[2]Principal sources of fat: peanuts & vegetable shortening.

Food and Description	Measure or Quantity	Sodium (mg.)	—Fats in grams—			Choles-terol (mg.)
			Total	Satu-rated	Unsatu-rated	
Solids & liq. (Diet Delight) halves or quarters	½ cup (4.4 oz.)	4	<.1			(0)
(S and W) *Nutradiet*, halves, low calorie, undrained	4 halves (3.5 oz.)	3	.1			(0)
(S and W) *Nutradiet*, halves, unsweetened	2 halves (3.5 oz.)	4	<.1			(0)
(S and W) *Nutradiet*, quartered, low calorie, undrained	4 oz.	2	.1			(0)
(S and W) *Nutradiet*, quartered, unsweetened	4 oz.	2	<.1			(0)
Solids & liq. (Tillie Lewis) Bartlett	½ cup (4.3 oz.)	<10	.2			0
Dried:						
(USDA)	1 lb.	32	8.2			0
Uncooked (Del Monte)	½ cup (2.8 oz.)	3	.6			0
Cooked without added sugar, solids & liq. (USDA)	4 oz.	3	.9			0
Cooked with added sugar, solids & liq. (USDA)	4 oz.	3	.9			0
PEAR, CANDIED (USDA)	1 oz.		.2			(0)
PEAR NECTAR:						
Sweetened (USDA)	1 cup (8.5 oz.)	2	.5			0
Sweetened (Del Monte)	1 cup (8.7 oz.)	17	Tr.			0
(S and W) *Nutradiet*	4 oz. (by wt.)	2	.1			(0)
PEAR PRESERVE (Bama)	1 T. (.7 oz.)	1	<.1			(0)
PECAN:						
In shell (USDA)	1 lb. (weighed in shell)	Tr.	171.2	12.	159.	0
Shelled, unsalted (USDA):						
Whole	1 lb.	Tr.	323.0	23.	300.	0
Halves	½ cup (1.9 oz.)	Tr.	38.4	3.	36.	0
Halves	12-14 halves (.5 oz.)	Tr.	10.0	<1.	9.	0
Chopped	½ cup (1.8 oz.)	Tr.	37.0	3.	34.	0
Chopped	1 T. (7 grams)	Tr.	5.0	Tr.	5.	0
Dry, roasted, salted (Flavor House)	1 oz.	32	20.2			(0)
Dry, roasted (Planters)	1 oz.	340	20.2	2.	18.	0

(USDA): United States Department of Agriculture
*Prepared as Package Directs

Food and Description	Measure or Quantity	Sodium (mg.)	Total	— Fats in grams — Satu- rated	Unsatu- rated	Choles- terol (mg.)
PECAN PIE:						
Home recipe, 1-crust (USDA):						
Made with lard[1]	¹/₆ of 9″ pie (4.9 oz.)	305	31.6	6.	26.	
Made with vegetable						
shortening[2]	¹/₆ of 9″ pie (4.9 oz.)	305	31.6	4.	27.	
Frozen (Morton)	¹/₆ of 20-oz. pie	320	15.4			
Frozen (Morton)	⅛ of 46-oz. pie	565	23.5			
Frozen (Mrs. Smith's)	¹/₆ of 8″ pie (4 oz.)	338	21.3			
Frozen (Mrs. Smith's)	⅛ of 10″ pie (4.5 oz.)	365	23.4			
PEP, cereal (Kellogg's)	1 cup (1 oz.)	184	.4			(0)
PEPPER:						
Black:						
(USDA)	1 cup (3.9 oz.)	13	7.5			0
(USDA)	1 tsp. (2 grams)	<1	.2			0
Seasoned (French's)	1 tsp. (3 grams)	4	.2			(0)
Seasoned (Lawry's)	1 pkg. (1.6 oz.)		2.4			(0)
Seasoned (Lawry's)	1 tsp. (2 grams)		.1			(0)
Whole or ground (Spice Islands)	1 tsp.	<1				(0)
Cayenne (Spice Islands)	1 tsp.	<1				(0)
Lemon (Durkee)	1 tsp. (3 grams)	698	Tr.			(0)
& lemon seasoning (French's)	1 tsp. (3 grams)	800	Tr.			(0)
White, whole or ground						
(Spice Islands)	1 tsp.	<1				(0)
PEPPER, HOT CHILI:						
Green (USDA):						
Raw, whole	4 oz.		.2			0
Raw, without seeds	4 oz.		.2			0
Canned, chili sauce	1 oz.		<.1			0
Canned, pods, without seeds,						
solids & liq.	4 oz.		1.1			0
Red:						
Raw, whole (USDA)	4 oz. (weighed with seeds)		2.6			0
Raw, trimmed, pods only (USDA)	4 oz.	21	3.2			0
Canned, chili sauce (USDA)	1 oz.		.2			0
Canned, solids & liq. (Del						
Monte)	¼ cup	554	.3			0
Canned, drained (Ortega)	¼ cup (1.8 oz.)	42	.1			(0)

(USDA): United States Department of Agriculture
*Prepared as Package Directs
[1]Principal sources of fat: pecans, lard & egg.
[2]Principal sources of fat: pecans, vegetable shortening & egg.

Food and Description	Measure or Quantity	Sodium (mg.)	—Fats in grams—			Choles- terol (mg.)
			Total	Satu- rated	Unsatu- rated	
Dried:						
Pods (USDA)	1 oz.	106	2.6			0
Pods (Chili Products)	1 oz.		2.5			(0)
Powder with added seasoning (USDA)	1 T. (.5 oz.)	236	1.9			0
*PEPPER POT SOUP, canned (Campbell)	1 cup	1093	3.6	1.	2.	
PEPPER, STUFFED:						
Home recipe, with beef & crumbs (USDA)[1]	2¾" x 2½" pepper with 1⅛ cups stuffing (6.5 oz.)	5881	10.2	6.	5.	56
Frozen, with veal (Weight Watchers)	12-oz. dinner	1123	6.8			
PEPPER, SWEET:						
Green:						
Raw, whole (USDA)	1 lb. (weighed untrimmed)	48	.7			0
Raw, without stem & seeds: (USDA)	1 med. pepper (2.6 oz.)	8	.1			0
Chopped (USDA)	½ cup (2.6 oz.)	10	.2			0
Slices (USDA)	½ cup (1.4 oz.)	5	.1			0
Strips (USDA)	½ cup (1.7 oz.)	6	.1			0
Boiled without salt, drained, (USDA)	1 med. pepper (2.6 oz.)	7	.1			0
Boiled without salt, strips, drained (USDA)	½ cup (2.4 oz.)	6	.1			0
Red (USDA):						
Raw, whole	1 lb. (weighed with stems & seeds)		1.1			0
Raw, without stem & seeds	1 med. pepper (2.2 oz.)		.2			0
PEPPERMINT, dry (Spice Islands)	1 tsp.	3				(0)
PEPPERMINT PIE, pink, frozen (Kraft)	¼ of 13-oz. pie (3.2 oz.)	69	17.6			

(USDA): United States Department of Agriculture
*Prepared as Package Directs
[1]Principal sources of fat: beef, butter, bread & milk.

Food and Description	Measure or Quantity	Sodium (mg.)	—Fats in grams— Total	Satu-rated	Unsatu-rated	Choles-terol (mg.)
PEPPERONI (Hormel)	1 oz.	425	13.0			
PERCH:						
Raw (USDA):						
White, whole	1 lb. (weighed whole)		6.5			
White, meat only	4 oz.		4.5			
Yellow, whole	1 lb. (weighed whole)	120	1.6			
Yellow, meat only	4 oz.	77	1.0			
Frozen, breaded (Gorton)	⅓ of 11-oz. pkg.	63	1.1			
PERSIMMON (USDA):						
Japanese or Kaki, fresh:						
With seeds	1 lb. (weighed with skin, calyx & seeds)	22	1.5			0
With seeds	4.4-oz. persimmon	6	.4			0
Seedless	1 lb. (weighed with skin & calyx)	23	1.5			0
Seedless	4.4-oz. persimmon	6	.4			0
Native, fresh, whole	1 lb. (weighed with seeds & calyx)	4	1.5			0
Native, fresh, flesh only	4 oz.	1	.5			0
PETITE MARMITE SOUP,						
canned (Crosse & Blackwell)	6½ oz. (½ can)		.7			
PETTIJOHNS (Quaker)						
rolled whole wheat	⅔ cup (1 oz. dry)	<1	.6			(0)
PHEASANT, raw (USDA):						
Ready-to-cook	1 lb. (weighed with bones)		20.5			
Meat & skin	4 oz.		5.9			
Meat only	4 oz.		7.7			
Giblets	2 oz.		2.8			
PICKEREL, chain, raw (USDA):						
Whole	1 lb. (weighed whole)		1.2			
Meat only	4 oz.		.6			

(USDA): United States Department of Agriculture
*Prepared as Package Directs

Food and Description	Measure or Quantity	Sodium (mg.)	—Fats in grams—			Choles- terol (mg.)
			Total	Satu- rated	Unsatu- rated	

PICKLE:
  Chow chow (See **CHOW CHOW**)
  Cucumber, fresh or bread &
    butter:

Food and Description	Measure or Quantity	Sodium (mg.)	Total	Satu-rated	Unsatu-rated	Choles-terol (mg.)
(USDA)	½ cup (3 oz.)	572	.2			0
(USDA)	3 slices (¼" x 1½", .7 oz.)	141	<.1			0
(Aunt Jane's)	4 slices or sticks (1 oz.)	191	<.1			(0)
(Aunt Jane's)	1 spear (1 oz.)	280				(0)
(Del Monte)	3 med. pieces (.9 oz.)	345	<.1			0
(Fanning's)	14-fl.-oz. bottle	2231	.4			0
(Heinz)	3 slices	160	Tr.			(0)
Dill:						
(USDA)	4" x 1¾" pickle (4.8 oz.)	1928	.3			0
(USDA)	3¾" x 1¼" pickle (2.3 oz.)	928	.1			0
(Aunt Jane's)	1 pickle (2 oz.)	811	.1			(0)
(Del Monte)	1 large pickle (3.5 oz.)	1940	.i			0
(Heinz)	4" pickle	1207	.1			(0)
Processed (Heinz)	3" pickle	511	.1			(0)
(Smucker's) baby, fresh pack	2¾" pickle (.8 oz.)	283	Tr.			(0)
Dill, candied sticks (Smucker's)	4" pickle (.8 oz.)	182	Tr.			(0)
Dill, hamburger (Heinz)	3 slices	245	Tr.			(0)
Dill, hamburger (Smucker's)	3 slices (.4 oz.)	141	Tr.			(0)
Hot, mixed (Smucker's)	4 pieces (.7 oz.)	290	Tr.			(0)
Hot peppers (Smucker's)	4" pepper (1 oz.)	401	Tr.			(0)
Kosher dill (Smucker's)	3½" pickle (.5 oz.)	642	Tr.			(0)
Sour:						
Cucumber (USDA)	1¾" x 4" pickle (4.8 oz.)	1827	.3			0
(Aunt Jane's)	1 pickle (2 oz.)	769	.1			(0)
Cucumber (Del Monte)	1 large pickle (3.5 oz.)	1490	.2			0
(Heinz)	2" pickle	166	Tr.			(0)
Sweet:						
Cucumber (USDA):						
Whole	1 oz.		.1			0
Whole, gherkin	2½" x ¾" pickle (.5 oz.)		<.1			0
Chopped	½ cup (2.6 oz.)		.3			0

(USDA): United States Department of Agriculture
*Prepared as Package Directs

PICKLE (Continued)

Food and Description	Measure or Quantity	Sodium (mg.)	— Fats in grams — Total	Saturated	Unsaturated	Cholesterol (mg.)
Chopped	1 T. (9 grams)		<.1			0
(Aunt Jane's)	1 pickle (1.5 oz.)	420	.2			(0)
(Del Monte)	1 med. pickle (.4 oz.)	99	<.1			0
(Smucker's)	2½" pickle (.4 oz.)	119	Tr.			(0)
Candied (Borden)	1 pickle (1.5 oz.)	420	.2			(0)
Candied, midgets (Smucker's)	2" pickle (9 grams)	89	Tr.			(0)
Cherry (Del Monte)	½ cup (1.9 oz.)	1023	.6			0
Chips, fresh pack (Smucker's)	1 piece (5 grams)	42	Tr.			(0)
Gherkin (Heinz)	2" pickle	76	Tr.			(0)
Mixed (Heinz)	3 pieces	120	Tr.			(0)
Mixed (Smucker's)	1 piece (8 grams)	82	Tr.			(0)
Sticks, fresh pack (Smucker's)	4" stick (1.1 oz.)	246	Tr.			(0)
Wax, mild (Del Monte)	½ cup (2.1 oz.)	1008	.4			0

PICKLING SPICE (Spice Islands)   1 tsp.   1   (0)

PIE (See individual kinds)

PIECRUST (See also **PASTRY SHELL**):

Food and Description	Measure or Quantity	Sodium (mg.)	Total	Saturated	Unsaturated
Home recipe, baked, 9":					
Made with lard[1]	1 crust (6.3 oz.)	1100	60.1	23.	37.
Made with vegetable shortening[2]	1 crust (6.3 oz.)	1100	60.1	14.	46.
Frozen:					
(Mrs. Smith's)	8" shell (5 oz.)	1130	56.7		
(Mrs. Smith's) old fashioned	9" shell (7 oz.)	1580	79.4		
(Mrs. Smith's)	10" shell (8 oz.)	1810	87.9		

PIECRUST MIX:

Food and Description	Measure or Quantity	Sodium (mg.)	Total	Saturated	Unsaturated
Dry, pkg. or stick (USDA)[2]	10-oz. pkg. (2 crusts)	1968	92.9	20	73.
*Prepared with water, baked (USDA)[2]	4 oz.	922	33.0	8.	25.
*Double crust (Betty Crocker)	⅙ of 2 crusts	389	21.5		
*Graham cracker (Betty Crocker)	⅙ of crust	137	7.9		
*(Flako)	⅙ of 9" shell	147	7.0		

PIE FILLING (See individual kinds)

PIGEON (See **SQUAB**)

(USDA): United States Department of Agriculture
*Prepared as Package Directs
[1]Principal source of fat: lard.
[2]Principal source of fat: vegetable shortening.

Food and Description	Measure or Quantity	Sodium (mg.)	— Fats in grams —			Choles- terol (mg.)
			Total	Satu- rated	Unsatu- rated	

PIGEONPEA (USDA):
Raw, immature seeds in pods	1 lb.	9	1.1			0
Dry seeds	1 lb.	118	6.4			0

PIGNOLIA (See **PINE NUT**)

PIGS FEET, pickled:
(USDA)	4 oz.		16.8	6.	11.	
(Hormel)	1-pt. can (8.6 oz.)		37.3			

PIKE, raw (USDA):
Blue, whole	1 lb. (weighed whole)		1.8			
Blue, meat only	4 oz.		1.0			
Northern, whole	1 lb. (weighed whole)		1.3			
Northern, meat only	4 oz.		1.2			
Walleye, whole	1 lb. (weighed whole)	132	3.1			
Walleye, meat only	4 oz.	58	1.4			

PILI NUT (USDA):
In shell	1 lb. (weighed in shell)	2	58.0			0
Shelled	4 oz.	3	80.6			0

PIMIENTO, canned:
Solids & liq. (USDA)	1 med. pod (1.3 oz.)		.2			0
Whole pods, slices, pieces (Dromedary)	1 oz.	2	.2			(0)
Drained (Ortega)	¼ cup (1.7 oz.)	18	.1			(0)
Solids & liq. (Stokely-Van Camp)	½ cup (4.1 oz.)		.6			(0)

PINA COLADA (Party Tyme)
	½-oz. pkg.	20	0.			(0)

PINCH of HERBS:
(Lawry's)	1 pkg. (2.2 oz.)		10.3			(0)
(Lawry's)	1 tsp. (3 grams)		.4			(0)

PINEAPPLE:
Fresh:
Whole (USDA)	1 lb. (weighed untrimmed)	2	.5			0

(USDA): United States Department of Agriculture
*Prepared as Package Directs

Food and Description	Measure or Quantity	Sodium (mg.)	Total	Satu- rated	Unsatu- rated	Choles- terol (mg.)
			—Fats in grams—			
Diced (USDA)	½ cup (2.8 oz.)	<1	.2			0
Slices (USDA)	¾" x 3½" slice (3 oz.)	<1	.2			0
(Del Monte)	½ cup (2.5 oz.)	<1	.4			(0)
Canned, regular pack, solids & liq.:						
Juice pack:						
(USDA)	4 oz.	1	.1			0
(Del Monte)	½ cup (4.9 oz.)	26	.3			0
Chunks or crushed (Dole)	½ cup (includes 2½ T. juice, 3.7 oz.)	1	.1			0
Slices (Dole)	2 med. slices & 2½ T. juice (3.7 oz.)	1	.1			0
Light syrup (USDA)	4 oz.	1	.1			0
Heavy syrup:						
Crushed (USDA)	½ cup (4.6 oz.)	1	.1			0
Slices (USDA)	½ cup (4.9 oz.)	1	.1			0
Slices (USDA)	2 small or 1 large slice & 2 T. syrup (4.3 oz.)	1	.1			0
Tidbits (USDA)	½ cup (4.6 oz.)	1	.1			0
(Del Monte)	½ cup (5 oz.)	14	.2			0
Chunks (Dole)	10 pieces & 2½ T. syrup (4 oz.)	1	.1			0
Crushed or tidbits (Dole)	½ cup (includes 2½ T. syrup, 4 oz.)	1	.1			0
Slices (Dole)	2 med. slices & 2½ T. syrup (4 oz.)	1	.1			0
Slices, chunks or crushed (Stokely-Van Camp)	½ cup (4 oz.)		.1			(0)
Extra heavy syrup (USDA)	4 oz.	1	.1			0
Canned, unsweetened, low calorie or dietetic, solids & liq.:						
Water pack, except crushed (USDA)	4 oz.	1	.1			0
Chunks:						
(Diet Delight)	½ cup (4.4 oz.)	5	Tr.			(0)
(S and W) *Nutradiet*	4 oz.	3	Tr.			(0)
Crushed (Diet Delight)	½ cup (4.4 oz.)	9	Tr.			(0)
Slices:						
(Diet Delight)	½ cup (4.4 oz.)	5	Tr.			(0)
(S and W) *Nutradiet*, low calorie	2½ slices (3.5 oz.)	3	Tr.			(0)

(USDA): United States Department of Agriculture
*Prepared as Package Directs

Food and Description	Measure or Quantity	Sodium (mg.)	Fats in grams			Cholesterol (mg.)
			Total	Saturated	Unsaturated	
(S and W) *Nutradiet*, unsweetened	2½ slices (3.5 oz.)	2	.1			(0)
Tidbits:						
(Diet Delight)	½ cup (4.4 oz.)	5	Tr.			(0)
(S and W) *Nutradiet*, low calorie	4 oz.	3	Tr.			(0)
(S and W) *Nutradiet*, unsweetened	4 oz.	1	.1			(0)
(Tillie Lewis)	½ cup (4.4 oz.)	<10	.1			0
Frozen:						
Chunks, sweetened, not thawed (USDA)	½ cup (4.3 oz.)	2	.1			0
Chunks in heavy syrup (Dole)	11 chunks & 2½ T. syrup (4 oz.)	2	.1			0
PINEAPPLE-APRICOT JUICE DRINK (Del Monte)	1 cup (8.6 oz.)	25	.2			(0)
PINEAPPLE CAKE MIX:						
*(Betty Crocker) layer	¹/₁₂ of cake	271	5.6			
*(Duncan Hines)	¹/₁₂ of cake (2.7 oz.)	366	6.1			50
(Pillsbury)	1 oz.					
PINEAPPLE, CANDIED (USDA)	1 oz.		.1			0
PINEAPPLE-CHERRY JUICE DRINK, canned (Del Monte) *Merry*	½ cup (4.3 oz.)	2	Tr.			
PINEAPPLE & GRAPEFRUIT JUICE DRINK, canned:						
(USDA) 40% fruit juices	½ cup (4.4 oz.)	Tr.	Tr.			0
(Del Monte)	½ cup (4.3 oz.)	42	Tr.			0
(Del Monte) pink	½ cup (4.3 oz.)	42	Tr.			0
(Dole) regular or pink	6-fl.-oz. can	Tr.	Tr.			0
(Dole) regular or pink	½ cup (4.3 oz.)	Tr.	Tr.			0
(Hi-C)	½ cup (4.2 oz.)	40	Tr.			0
PINEAPPLE JUICE:						
Canned, unsweetened:						
(USDA)	½ cup (4.4 oz.)	1	.1			0
(Del Monte)	½ cup (4.3 oz.)	1				0
(Dole)	6-fl.-oz. can	2	.2			0

(USDA): United States Department of Agriculture
*Prepared as Package Directs

Food and Description	Measure or Quantity	Sodium (mg.)	—Fats in grams—			Choles-terol (mg.)
			Total	Satu-rated	Unsatu-rated	
(Dole)	½ cup (4.6 oz.)	1	.1			0
(Heinz)	5½-fl.-oz. can	3	.3			(0)
(S and W) *Nutradiet*	4 oz. (by wt.)	1	.1			(0)
(Stokely-Van Camp)	½ cup (3.7 oz.)		.1			(0)
Frozen, concentrate:						
Unsweetened, undiluted (USDA)	6-fl.-oz. can (7.6 oz.)	6	.2			0
*Unsweetened, diluted with						
3 parts water (USDA)	½ cup (4.4 oz.)	1	Tr.			0
Unsweetened, undiluted						
(Dole)	6-fl.-oz. can	7	.2			0

PINEAPPLE & ORANGE JUICE DRINK, canned:

(USDA) 40% fruit juices	½ cup (4.4 oz.)	Tr.	.1			0
(Del Monte)	½ cup (4.3 oz.)	22	Tr.			0

PINEAPPLE-PEAR JUICE DRINK, canned (Del Monte)

	½ cup (4.3 oz.)	26	Tr.			(0)

PINEAPPLE PIE:

Home recipe, 2-crust (USDA):						
Made with lard[1]	¹⁄₆ of 9″ pie (5.6 oz.)	271	16.9	5.	12.	
Made with vegetable shortening[2]	¹⁄₆ of 9″ pie (5.6 oz.)	428	16.9	5.	12.	
Chiffon, 1-crust:						
Made with lard[3]	¹⁄₆ of 9″ pie (3.8 oz.)	276	13.1	4.	9.	
Made with vegetable shortening[4]	¹⁄₆ of 9″ pie (3.8 oz.)	276	13.1	3.	10.	
Custard, 1-crust:						
Made with lard[5]	¹⁄₆ of 9″ pie (5.4 oz.)	283	13.2	5.	9.	
Made with vegetable shortening[6]	¹⁄₆ of 9″ pie (5.4 oz.)	283	13.2	3.	10.	
(Tastykake)	4-oz. pie		15.6			
With cheese (Tastykake)	4-oz. pie		20.3			
Frozen (Morton)	⅛ of 46-oz. pie	315	16.4			

PINEAPPLE PIE FILLING, canned:

(Comstock)	½ cup (5.4 oz.)	1	<.1			
(Lucky Leaf)	8 oz.	134	.2			

(USDA): United States Department of Agriculture
*Prepared as Package Directs
[1]Principal sources of fat: lard & butter.
[2]Principal sources of fat: vegetable shortening & butter.
[3]Principal sources of fat: lard & cream.
[4]Principal sources of fat: vegetable shortening & cream.
[5]Principal sources of fat: lard, egg & milk.
[6]Principal sources of fat: vegetable shortening, egg & milk.

Food and Description	Measure or Quantity	Sodium (mg.)	Fats in grams Total	Satu- rated	Unsatu- rated	Choles- terol (mg.)
*PINEAPPLE PUDDING MIX, CREAM, instant (Jell-O)	½ cup (5.3 oz.)	406	4.7			13
PINEAPPLE PRESERVE:						
Sweetened (Bama)	1 T. (.7 oz.)	1	<.1			(0)
Low calorie (Tillie Lewis)	1 T. (.5 oz.)	3	Tr.			0
PINEAPPLE SOFT DRINK, sweetened:						
(Hoffman)	6 fl. oz.	14	0.			0
(Kirsch)	6 fl. oz.	<1	0.			0
(Nedick's)	6 fl. oz.	14	0.			0
PINE NUT (USDA):						
Pignolias, shelled	4 oz.		53.8			0
Piñon, whole	4 oz. (weighed in shell)		39.8			0
Piñon, shelled	4 oz.		68.6			0
PINK PANTHER FLAKES, cereal (Post)	⅔ cup (1 oz.)	209	.1			0
PISTACHIO NUT:						
In shell (USDA)	4 oz. (weighed in shell)		30.4	3.	27.	0
In shell (USDA)	½ cup (2.3 oz.)		17.8	2.	16.	0
Shelled (USDA)	½ cup (2.2 oz.)		33.3	3.	30.	0
Shelled (USDA)	1 T. (8 grams)		4.2	Tr.	4.	0
Dry, roasted, salted (Flavor House)	1 oz.	32	15.2			(0)
*PISTACHIO NUT PUDDING MIX, instant (Royal)	½ cup (5.1 oz.)	350	4.8			14
PITANGA, fresh (USDA):						
Whole	1 lb. (weighed whole)		1.5			0
Flesh only	4 oz.		.5			0
PIZZA PIE:						
Home recipe, with cheese topping:						
(USDA)[1]	4 oz.	796	9.4	3.	6.	
(USDA)[1]	5½" sector (⅛ of 14" pie, 2.6 oz.)	527	6.2	2.	4.	

(USDA): United States Department of Agriculture
*Prepared as Package Directs
[1]Principal sources of fat: cheese, vegetable shortening & olive oil.

Food and Description	Measure or Quantity	Sodium (mg.)	Total	Fats in grams Satu- rated	Unsatu- rated	Choles- terol (mg.)
Home recipe, with sausage topping:						
(USDA)[1]	4 oz.	827	10.5	3.	7.	
(USDA)[1]	5½" sector (⅛ of 14" pie, 2.6 oz.)	547	7.0	2.	5.	
Chilled, partially baked (USDA)[2]	4 oz.	610	6.6	2.	4.	
Chilled, baked (USDA)[2]	4 oz.	718	7.7	2.	5.	
Frozen:						
Partially baked (USDA)[2]	4 oz.	686	7.5	2.	5.	
Baked (USDA)[2]	4 oz.	734	8.1	2.	6.	
Baked (USDA)[2]	⅛ of 14" pie (2.6 oz.)	485	5.3	2.	4.	
(Celeste) *Bambino*	10-oz. pie	1560	28.8			
With cheese (Buitoni)	4 oz.		8.9			
With cheese (Celeste)	20-oz. pie	3280	50.4			
With cheese (Chef Boy-Ar-Dee)	⅙ of 12½" pie (2.1 oz.)	161	4.2			
With cheese, little (Chef Boy-Ar-Dee)	1 pie (2½ oz.)	373	5.1			
With cheese (Jeno's)	13-oz. pie		28.4			
With cheese (Jeno's) Serv-A-Slice	1 slice (1.7 oz.)		4.6			
With cheese (Jeno's) snack tray	½-oz. pizza		.5			
With cheese (Kraft)	14-oz. pie	2596	29.4			
With cheese, *Pee Wee* (Kraft)	2½-oz. pie	484	6.5			
With hamburger (Jeno's)	13½-oz. pie		28.3			
With pepperoni (Buitoni)	4 oz.		10.5			
With pepperoni (Chef Boy-Ar-Dee)	⅙ of 14-oz. pie	271	6.0			
With pepperoni (Jeno's)	13¼-oz. pie		45.8			
With pepperoni (Jeno's) Serv-A-Slice	1 slice (1.8 oz.)		6.6			
With pepperoni (Jeno's) snack tray	½-oz. pizza		1.3			
With sausage (Buitoni)	4 oz.		10.1			
With sausage (Celeste)	23-oz. pie	3880	78.4			
With sausage (Celeste) *Bambino*	9-oz. pie	1340	27.6			
With sausage (Chef Boy-Ar-Dee)	⅙ of 13¼-oz. pie	159	5.1			
With sausage, little (Chef Boy-Ar-Dee)	2½-oz. pie	387	5.4			

(USDA): United States Department of Agriculture
*Prepared as Package Directs
[1]Principal sources of fat: sausage, vegetable shortening & olive oil.
[2]Principal sources of fat: cheese, vegetable shortening & olive oil.

Food and Description	Measure or Quantity	Sodium (mg.)	—Fats in grams—			Choles-terol (mg.)
			Total	Satu-rated	Unsatu-rated	
With sausage (Jeno's)	13½-oz. pie		28.3			
With sausage (Jeno's) Serv-A-Slice	1 slice (2 oz.)		5.2			
With sausage (Jeno's) snack tray	½-oz. pizza		1.1			
With sausage (Kraft)	14½-oz. pie	2852	47.7			
With sausage, *Pee Wee* (Kraft)	2½-oz. pie	365	9.2			
PIZZA PIE MIX:						
*With cheese (Chef Boy-Ar-Dee)	⅕ of 15½-oz. pie	548	5.4			
*With cheese (Kraft)	4 oz.	1293	10.8			
*With sausage (Chef Boy-Ar-Dee)	⅕ of 17-oz. pie	695	7.8			
*With sausage (Kraft)	4 oz.	1182	12.2			
PIZZA ROLL, frozen (Jeno's):						
Cheeseburger, 12 to pkg.	½-oz. roll		2.1			
Pepperoni, 12 to pkg.	½-oz. roll		1.8			
Sausage, 12 to pkg.	½-oz. roll		2.0			
Shrimp, 12 to pkg.	½-oz. roll		1.5			
Snack tray:						
Hamburger	½-oz. roll		2.0			
Pepperoni	½-oz. roll		1.8			
Sausage	½-oz. roll		2.0			
PIZZA SAUCE:						
Canned (Buitoni)	4 oz.		2.9			
Canned (Chef Boy-Ar-Dee)	½ of 10½-oz. can	987	8.5			
Canned (Contadina)	1 cup	1064	4.0			
Mix (French's)	1-oz. pkg.	1790	.1			
*Mix (French's)	2 T.	155	.1			
PIZZA SEASONING (French's)	1 tsp. (4 grams)	390	.1			(0)
PIZZARIA MIX (Hunt's) *Skillet*[1]	14.1-oz. pkg.	2844	11.1	7.	4.	
PLANTAIN, raw (USDA):						
Whole	1 lb. (weighed with skin)	16	1.3			0
Flesh only	4 oz.	6	.5			0
PLUM:						
Damson, fresh (USDA):						
Whole	1 lb. (weighed with pits)	8	Tr.			0

(USDA): United States Department of Agriculture
*Prepared as Package Directs
[1]Principal sources of fat: cheese & egg.

Food and Description	Measure or Quantity	Sodium (mg.)	Total	Satu- rated	Unsatu- rated	Choles- terol (mg.)
			— Fats in grams —			
Flesh only	4 oz.	2	Tr.			0
Japanese & hybrid, fresh (USDA):						
Whole	1 lb. (weighed with pits)	4	.9			0
Whole	2″ plum (2.1 oz.)	<1	.1			0
Diced	½ cup (2.3 oz.)	<1	.2			0
Halves	½ cup (3.1 oz.)	<1	.2			0
Slices	½ cup (3 oz.)	<1	.2			0
Prune-type, fresh (USDA):						
Whole	1 lb. (weighed with pits)	4	.9			0
Halves	½ cup (2.8 oz.)	<1	.2			0
Canned, purple, regular pack, solids & liq.:						
Light syrup (USDA)	4 oz.	1	.1			0
Heavy syrup (USDA)	½ cup (with pits, 4.5 oz.)	1	.1			0
Heavy syrup (USDA)	½ cup (without pits, 4.2 oz.)	1	.1			0
Heavy syrup (USDA)	3 plums without pits & 2 T. syrup (4.3 oz.)	1	.1			0
Heavy syrup (Del Monte)	½ cup (4.1 oz.)	6	.1			0
Extra heavy syrup (USDA)	4 oz.	1	.1			0
(Stokely-Van Camp)	½ cup (4.2 oz.)		.1			(0)
Canned, unsweetened or low calorie, solids & liq.:						
Greengage, water pack (USDA)	4 oz.	1	.1			0
Purple:						
Water pack (USDA)	4 oz.	2	.2			0
(Diet Delight)	½ cup (4.4 oz.)	4	<.1			(0)
(S and W) *Nutradiet*	4 oz.	2	.1			(0)
(Tillie Lewis)	½ cup (4.5 oz.)	<10	.2			(0)
PLUM PIE (Tastykake)	4-oz. pie		15.0			
PLUM PRESERVE or JAM, sweetened, Damson or red (Bama)	1 T. (.7 oz.)	2	<.1			(0)
PLUM PUDDING:						
(Crosse & Blackwell)	4 oz.		1.0			
(Richardson & Robbins)	½ cup (4 oz.)	173	1.4			
P.M., fruit juice drink (Mott's)	½ cup		.1			(0)

(USDA): United States Department of Agriculture
*Prepared as Package Directs

Food and Description	Measure or Quantity	Sodium (mg.)	Fats in grams — Total	Satu- rated	Unsatu- rated	Choles- terol (mg.)
POHA (See **GROUND-CHERRY**)						
POKE SHOOTS (USDA):						
Raw	1 lb.		1.8			0
Boiled, drained	4 oz.		.5			0
POLISH-STYLE SAUSAGE:						
(USDA)	1 oz.		7.3			
(Oscar Mayer) all meat	1 oz. (from 8-oz. link)	221	6.8			
Kolbase (Hormel)	1 oz.	315	6.9	3.	4.	17
POLLOCK (USDA):						
Raw, drawn	1 lb. (weighed with head, tail, fins & bones)	98	1.8			
Raw, meat only	4 oz.	5.4	1.0			
Creamed[1]	4 oz.	126	6.7			
POLYNESIAN-STYLE DINNER,						
frozen (Swanson)	11¾-oz. dinner	1490	20.0			
POMEGRANATE, raw (USDA):						
Whole	1 lb. (weighed whole)	8	.8			0
Pulp only	4 oz.	3	.3			0
POMPANO, raw (USDA):						
Whole	1 lb. (weighed whole)	119	24.1			
Meat only	4 oz.	53	10.8			
POPCORN:						
Unpopped (USDA)	1 oz.	<1	1.3	Tr.	1.	0
Popped (USDA):						
Plain, large kernel	1 oz.	<1	1.4	Tr.	1.	0
Plain, large kernel	1 cup (6 grams)	<1	.3	Tr.	Tr.	0
Butter & salt added[2]	1 oz.	550	6.2	3.	3.	
Butter & salt added[2]	1 cup (9 grams)	175	2.0	<1.	1.	
Coconut oil & salt added[3]	1 oz.	550	6.2	4.	2.	
Coconut oil & salt added[3]	1 cup (9 grams)	175	2.0	1.	<1.	

(USDA): United States Department of Agriculture
*Prepared as Package Directs
[1]Prepared with flour, butter & milk.
[2]Principal source of fat: butter.
[3]Principal source of fat: coconut oil.

257

Food and Description	Measure or Quantity	Sodium (mg.)	— Fats in grams — Total	Satu- rated	Unsatu- rated	Choles- terol (mg.)
Sugar-coated, without salt[1]	1 oz.	<1	1.0	Tr.	<1.	
Sugar-coated, without salt[1]	1 cup (1.2 oz.)	<1	1.2	Tr.	<1.	
(Jiffy Pop)	½ pkg. (2½ oz.)	936	11.6			
(Tom Houston)	1 cup (.5 oz.)	291	3.3			
Buttered (Jiffy Pop)	½ pkg. (2½ oz.)	936	12.2			
Buttered (Wise)	1-oz. bag	235	7.1			
Caramel-coated:						
Without peanuts (Old London)	1¾-oz. bag	372	1.6			
With peanuts (Old London)	1 cup (1.3 oz.)	258	1.6			
Cracker Jack[2]	¾-oz. bag		2.0			
Cracker Jack[2]	1⅜-oz. box		3.6			
Cheese (Old London)	¾-oz. bag	102	6.3			
Cheese (Wise)	⅝-oz. bag	167	5.2			
Seasoned (Old London)	1¼-oz. bag	69	8.6			

POPOVER:

Food and Description	Measure or Quantity	Sodium (mg.)	Total	Satu- rated	Unsatu- rated	Choles- terol (mg.)
Home recipe (USDA)[3]	1.4-oz. popover (2¾" dia. at top, ¼ cup batter)	88	3.7	1.	2.	59
*Mix (Flako)	2.3-oz. popover (⅙ of pkg.)	300	4.8			

POPPY SEED (Spice Islands)

	1 tsp.	<1				(0)

POPSICLE, fruit flavors

(Popsicle Industries)	3-fl.-oz. bar (3.4 oz.)	7	0.			(0)

POP-UP (See **TOASTER CAKE**)

PORGY, raw (USDA):

Whole	1 lb. (weighed whole)	117	6.3			
Meat only	4 oz.	71	3.9			

PORK, medium-fat:
Fresh (USDA):
  All lean cuts:
    Lean only:

Raw, diced	1 cup (8.2 oz.)	667	26.4	10.	17.	140
Raw, strips	1 cup (8.2 oz.)	664	26.3	10.	17.	140
Roasted, chopped	1 cup (5 oz.)	554	20.0	7.	13.	125

(USDA): United States Department of Agriculture
*Prepared as Package Directs
[1] Principal source of fat: coconut oil.
[2] Principal sources of fat: corn & peanut oil.
[3] Principal sources of fat: vegetable shortening, egg & milk.

Food and Description	Measure or Quantity	Sodium (mg.)	— Fats in grams —			Choles- terol (mg.)
			Total	Satu- rated	Unsatu- rated	
Boston butt:						
Raw	1 lb. (weighed with bone & skin)	260	104.1	37.	67.	264
Roasted, lean & fat	4 oz.	74	32.3	11.	21.	101
Roasted, lean only	4 oz.	74	16.2	6.	10.	100
Chop:						
Broiled, lean & fat	4-oz. chop (weighed with bone)	49	23.9	8.	16.	67
Broiled, lean & fat	3-oz. chop (weighed without bone)	55	26.9	9.	18.	76
Broiled, lean only	3-oz. chop (weighed without bone)	55	13.1	5.	8.	75
Fat, separable, cooked	1 oz.		23.6	9.	15.	
Ham (See also **HAM**):						
Raw	1 lb. (weighed with bone & skin)	320	102.6	37.	66.	239
Roasted, lean & fat	4 oz.	74	34.7	12.	22.	101
Roasted, lean only	4 oz.	74	11.3	5.	7.	100
Loin:						
Raw	1 lb. (weighed with bone)	260	89.0	32.	57.	221
Broiled, lean & fat	4 oz.	74	35.9	12.	24.	101
Broiled, lean only	4 oz.	74	17.5	7.	11.	100
Roasted, lean & fat	4 oz.	74	32.3	11.	21.	101
Roasted, lean only	4 oz.	74	16.1	6.	10.	100
Picnic:						
Raw	1 lb. (weighed with bone & skin)	260	92.2	33.	59.	231
Simmered, lean & fat	4 oz.	74	34.6	12.	22.	101
Simmered, lean only	4 oz.	74	11.1	5.	7.	100
Spareribs:						
Raw, with bone	1 lb. (weighed with bone)	775	89.7	32.	58.	169
Raw, without bone	1 lb. (weighed without bone)	1295	150.6	54.	97.	281
Braised, lean & fat	4 oz.	74	44.1	16.	28.	101
Cured, light commercial cure:						
Bacon (see **BACON**)						
Boston butt (USDA):						
Raw	1 lb. (weighed with bone & skin)		101.7	37.	65.	
Roasted, lean & fat	4 oz.		29.1	10.	19.	
Roasted, lean only	4 oz.	1055	15.6	7.	9.	

(USDA): United States Department of Agriculture
*Prepared as Package Directs

259

Food and Description	Measure or Quantity	Sodium (mg.)	Fats in grams — Total	Satu- rated	Unsatu- rated	Choles- terol (mg.)
Smoked, Tasty Meat (Wilson)	4 oz.	1174	25.4	11.	15.	70
Ham (See also **HAM**):						
Raw (USDA)	1 lb. (weighed with bone & skin)		89.7	32.	58.	
Raw, lean only, ground (USDA)	1 cup (6 oz.)	1870	14.4	5.	9.	
Roasted, lean & fat (USDA)	4 oz.		25.1	9.	16.	
Roasted, lean only:						
(USDA)	4 oz.	1055	10.0	3.	7.	
Chopped (USDA)	1 cup (4.9 oz.)	1283	12.1	4.	8.	
Diced (USDA)	1 cup (5.2 oz.)	1367	12.9	4.	9.	
Ground (USDA)	1 cup (3.8 oz.)	1014	9.6	3.	6.	
Fully cooked, bone-in (Hormel)	4 oz.	1361	15.3			
Fully cooked, boneless:						
Parti-Style (Armour Star)	4 oz.		8.3			
Cure 81 (Hormel)	4 oz.	953	11.9			
Curemaster (Hormel)	4 oz.	1361	5.0			
(Wilson) Certified, rolled	4 oz.	1145	15.8	7.	9.	70
(Wilson) Festival, smoked	4 oz.	1281	11.1	4.	7.	70
Picnic:						
Raw (USDA)	1 lb. (weighed with bone & skin)		87.8	32.	56.	
Raw (Wilson) smoked	4 oz.	1247	24.4	10.	15.	70
Roasted, lean & fat (USDA)	4 oz.		28.6	10.	18.	
Roasted, lean only (USDA)	4 oz.	1055	11.2	5.	7.	
Canned (Hormel)	4 oz. (3-lb. can)		14.5			
PORK, CANNED, chopped luncheon meat:						
(USDA)[1]	1 oz.	350	7.1	3.	5.	
Chopped (USDA)*[1]	1 cup (4.8 oz.)	1678	33.9	12.	22.	
Diced (USDA)[1]	1 cup (5 oz.)	1740	35.1	13.	22.	
(Hormel)	1 oz. (8-lb. can)		5.8			
PORK DINNER, loin of pork, frozen (Swanson)	10-oz. dinner	820	21.9	6.	15.	
PORK, FREEZE DRY, canned (Wilson) *Campsite:*						
Chop, dry	2-oz. can	874	17.1	6.	11.	146

(USDA): United States Department of Agriculture
*Prepared as Package Directs
[1]Principal source of fat: pork.

Food and Description	Measure or Quantity	Sodium (mg.)	—Fats in grams—			Choles- terol (mg.)
			Total	Satu- rated	Unsatu- rated	
*Chop, reconstituted	4 oz.	695	13.8	5.	9.	119
Patties, dry	2-oz. can	1413	18.7	7.	12.	146
*Patties, reconstituted	4 oz.	1174	15.2	6.	10.	119
PORK & GRAVY, 90% pork, canned(USDA)	4 oz.		20.2	7.	13.	
PORK LOIN (Oscar Mayer) thin sliced	1 slice (16 to 3 oz.)	58	.7			
PORK RINDS, fried, *Baken·ets* (See also other brand names)	1 oz.	490	7.4	3.	5.	6
PORK ROAST, canned (Wilson) *Tender Made*	1 oz.	127	2.2	<1.	1.	18
PORK SAUSAGE:						
Uncooked:						
Links or bulk (USDA)[1]	1 oz.	210	14.4	5.	9.	
(Armour Star)	1-oz. sausage		13.5			
Country style (Hormel)	1 oz.	270	9.7	4.	6.	31
Little Sizzlers (Hormel)	1 piece (.8 oz.)	292	10.5			
Midget (Hormel)	1 piece (.8 oz.)	286	7.4	3.	4.	23
Smoked (Hormel)	1 oz.	318	8.9	3.	5.	14
Bulk style (Oscar Mayer)	1 oz.	210	12.8			
Italian Brand (Oscar Mayer)	1 oz.	299	13.3			
Little Friers (Oscar Mayer):						
6-8 per lb.	1 link (2.3 oz.)	482	29.9			
14-18 per lb.	1 link (1 oz.)	210	13.0			
(Wilson)	1 oz.	207	13.6	5.	8.	18
Cooked:						
Links or bulk (USDA)[1]	1 oz.	272	12.5	5.	8.	
Links, 16 per lb. raw (USDA)[1]	1 link	125	5.7	2.	4.	
Bulk style (Oscar Mayer)	1 oz.	409	7.9			
Italian Brand (Oscar Mayer)	1 oz.	299	6.8			
Little Friers (Oscar Mayer):						
6-8 per lb. raw	1 link (1 oz. cooked)	396	8.7	3.	5.	17
14-18 per lb. raw	1 link (.4 oz. cooked)	164	3.6	1.	2.	7
Canned, solids & liq. (USDA)[1]	1 oz.		10.9	4.	7.	
Canned, drained (USDA)[1]	1 oz.		9.3	3.	6.	

(USDA): United States Department of Agriculture
*Prepared as Package Directs
[1]Principal source of fat: pork.

Food and Description	Measure or Quantity	Sodium (mg.)	—Fats in grams—			Cholesterol (mg.)
			Total	Saturated	Unsaturated	
PORK, SWEET & SOUR, frozen						
(Chun King)	7½-oz. serving (½ pkg.)		10.0			
PORT WINE:						
(Gold Seal) 19% alcohol	3 fl. oz. (3.3 oz.)	3	0.			(0)
(Great Western) Solera, 18% alcohol	3 fl. oz.	34	0.			0
(Great Western) Solera, tawny, 18% alcohol	3 fl. oz.	34	0.			0
POST TOASTIES, cereal (Post)	1 cup (1 oz.)	238	.1			0
POSTUM, instant	1 cup (6 oz.)	3	Tr.			0
POTATO:						
Raw, whole (USDA)	1 lb. (weighed unpared)	11	.4			0
Raw, pared (USDA):						
Chopped	1 cup (5.2 oz.)	4	.1			0
Diced	1 cup (5.5 oz.)	5	.2			0
Slices	1 cup (5.2 oz.)	4	.1			0
Cooked:						
Au gratin or scalloped, without cheese (USDA)[1]	½ cup (4.3 oz.)	433	4.8	2.	2.	7
Au gratin, with cheese (USDA)[2]	½ cup (4.3 oz.)	545	9.6	5.	5.	18
Baked, peeled after baking, no salt added (USDA)	2½" dia. potato (3.5 oz., 3 raw per lb.)	4	.1			0
Baked, peeled after baking, salt added (USDA)	2½" dia. potato (3.5 oz., 3 raw per lb.)	234	.1			0
Boiled, peeled after boiling, no salt added (USDA)	4.8-oz. potato (3 raw per lb.)	4	.1			0
Boiled, peeled after boiling, salt added (USDA)	4.8-oz. potato (3 raw per lb.)	321	.1			0

(USDA): United States Department of Agriculture
*Prepared as Package Directs
[1]Principal sources of fat: butter & milk.
[2]Principal sources of fat: cheese, butter & milk.

Food and Description	Measure or Quantity	Sodium (mg.)	Total	Fats in grams — Satu- rated	Unsatu- rated	Choles- terol (mg.)
Boiled, peeled before boiling, no salt added (USDA):						
Whole	4.3-oz. potato (3 raw per lb.)	2	.1			0
Diced	½ cup (2.8 oz.)	2	<.1			0
Mashed	½ cup (3.7 oz.)	2	.1			0
Riced	½ cup (4 oz.)	2	.1			0
Slices	½ cup (2.8 oz.)	2	<.1			0
Boiled, peeled before boiling, salt added (USDA):						
Whole	4.3-oz. potato (3 raw per lb.)	288	.1			0
Diced	½ cup (2.8 oz.)	184	<.1			0
Mashed	½ cup (3.7 oz.)	245	.1			0
Riced	½ cup (4 oz.)	269	.1			0
Slices	½ cup (2.8 oz.)	189	<.1			0
French-fried in deep fat, no salt added (USDA)[1]	10 pieces (2″ x ½″ x ½″, 2 oz.)	3	7.5	2.	6.	
French-fried in deep fat, salt added (USDA)[1]	10 pieces (2″ x ½″ x ½″, 2 oz.)	135	7.5	2.	6.	
French-fried (McDonald's)	1 serving (2.4 oz.)	117	10.4			
Hash-browned (USDA)[2]	½ cup (3.4 oz.)	281	11.4	3.	8.	
Mashed, milk added (USDA)	½ cup (3.5 oz.)	295	.7			
Mashed, milk & butter added (USDA)[3]	½ cup (3.5 oz.)	324	4.2	2.	2.	
Pan-fried from raw (USDA)	½ cup (3 oz.)	190	12.1	3.	10.	
Scalloped (See Au gratin)						
Canned:						
Solids & liq., no added salt (USDA)	1 cup (8.8 oz.)	2	.5			0
Solids & liq., added salt (USDA)	1 cup (8.8 oz.)	590	.5			0
White (Butter Kernel)	3-4 small potatoes (4.1 oz.)		2.3			(0)
Solids & liq. (Stokely-Van Camp)	½ cup (4.1 oz.)		.2			(0)
Whole, new, solids & liq. (Del Monte)	½ cup (2.6 oz.)	382	Tr.			0

(USDA): United States Department of Agriculture
*Prepared as Package Directs
[1]Principal source of fat: cottonseed oil.
[2]Principal source of fat: vegetable shortening.
[3]Principal sources of fat: butter & milk.

Food and Description	Measure or Quantity	Sodium (mg.)	— Fats in grams — Total	Satu- rated	Unsatu- rated	Choles- terol (mg.)
Whole, new, drained solids (Del Monte)	½ cup (2.5 oz.)	261	<.1			0
Dehydrated, mashed:						
Flakes, without milk (USDA):						
Dry	½ cup (.8 oz.)	20	.1			0
*Prepared with water, milk & fat[1]	½ cup (3.8 oz.)	247	3.4	2.	1.	
Granules, without milk (USDA):						
Dry	½ cup (3.5 oz.)	84	.6			0
*Prepared with water, milk & butter[1]	½ cup (3.7 oz.)	256	3.8	2.	2.	
Granules with milk (USDA):						
Dry	½ cup (3.5 oz.)	82	1.1			
*Prepared with water & fat[2]	½ cup (3.7 oz.)	246	2.3	1.	1.	
Frozen:						
Au gratin (Stouffer's)	11½-oz. pkg.		11.5			
Diced for hash-browning, not thawed (USDA)	4 oz.	9	Tr.			
Diced, hash-browned (USDA)[3]	4 oz.	339	13.0	3.	10.	
French-fried:						
Not thawed, no salt added (USDA)[3]	9-oz. pkg.	8	16.6	5.	12.	
Not thawed, salt added (USDA)[3]	9-oz. pkg.	602	16.6	5.	12.	
Heated, no added salt (USDA)[3]	10 pieces (2″ x ½″ x ½″, 2 oz.)	2	4.8	2.	3.	
Heated, salt added (USDA)[3]	10 pieces (2″ x ½″ x ½″, 2 oz.)	135	4.8	2.	3.	
Mashed, not thawed (USDA)	4 oz.	90	.1			
Mashed, heated (USDA)[1]	4 oz.	407	3.2	2.	<1.	
Stuffed, baked (Holloway House):						
With cheese	1 potato (6 oz.)	842	13.6			
With some cream & chives	1 potato (6 oz.)	646	13.6			
POTATO CHIP:						
(USDA)[3,4]	1 oz.	284	11.3	3.	8.	0
(USDA)[3,4]	10 2″ chips or 7 3″ chips (.7 oz.)	200	8.0	2.	6.	0

(USDA): United States Department of Agriculture
*Prepared as Package Directs
[1]Principal source of fat: butter & milk.
[2]Principal source of fat: butter.
[3]Principal source of fat: cottonseed oil.
[4]Sodium content is variable & may be as high as 284 mg. per oz.

Food and Description	Measure or Quantity	Sodium (mg.)	— Fats in grams —			Cholesterol (mg.)
			Total	Saturated	Unsaturated	
Lay's	1 oz.	230	11.1	3.	8.	0
(Nalley's)	1 oz.		10.5			
(Pringle's)	10 chips (.5 oz.)	126	4.8	1.	3.	0
(Pringle's)	1 oz.	255	9.6	3.	7.	0
Ruffles	1 oz.	200	9.4	2.	7.	0
(Tom Houston)	10 chips (.7 oz.)	200	8.0			
(Wise)	1-oz. bag	104	10.3			
(Wonder)	1 oz.	220	11.1			
Barbecue, *Lay's*	1 oz.	380	10.8	3.	8.	0
Barbecue (Wise)	1-oz. bag	149	9.7			
Barbecue (Wonder)	1 oz.	199	10.6			
Onion-garlic (Wise)	1-oz. bag	149	10.2			
Ridgies (Wise)	1-oz. bag	119	10.1			

POTATO MIX:

Food and Description	Measure or Quantity	Sodium (mg.)	Total			
*Au gratin (Betty Crocker)	½ cup	468	7.0			
Au gratin (French's)	5½-oz. pkg.	2880	13.0			
*Au gratin (French's)	½ cup	480	2.2			
Buds (Betty Crocker)	½ cup	401	6.4			
Mashed, country style (French's)	2⅔-oz. pkg.	60	.5			
*Mashed, country style (French's)	½ cup	390	7.0			
Mashed, granules (French's)	3¼-oz. pkg.	77	.6			
*Mashed, granules, salt added (French's)	½ cup	306	4.8			
*Scalloped (Betty Crocker)	½ cup	668	5.8			
Scalloped (French's)	5⅝-oz. pkg.	2160	1.1			
*Scalloped (French's)	½ cup	383	2.1			
Whipped (Borden)	¼ cup (.5 oz.)	<.1				

POTATO PANCAKE MIX:

Food and Description	Measure or Quantity	Sodium (mg.)	Total			
(French's)	3-oz. pkg.	1440	.5			
*(French's)	3 small pancakes (¼ pkg.)	375	1.4			

POTATO SALAD:

Food and Description	Measure or Quantity	Sodium (mg.)	Total	Saturated	Unsaturated	Cholesterol (mg.)
Home recipe, with cooked salad dressing & seasonings (USDA)[1]	4 oz.	599	3.2	1.	2.	
Home recipe, with mayonnaise & French dressing, hard-cooked eggs & seasonings (USDA)[2]	½ cup (4.4 oz.)	600	11.5	2.	9.	81
Canned (Nalley's)	4 oz.		9.9			

(USDA): United States Department of Agriculture
*Prepared as Package Directs
[1]Principal sources of fat: butter, milk & egg.
[2]Principal sources of fat: soybean oil, cottonseed oil, corn oil & eggs.

Food and Description	Measure or Quantity	Sodium (mg.)	— Fats in grams —			Choles- terol (mg.)
			Total	Satu- rated	Unsatu- rated	
POTATO SOUP, Cream of:						
*Canned (Campbell)	1 cup	1049	4.4	<1.	4.	
Frozen (USDA):						
Condensed[1]	8 oz. (by wt.)	2223	9.8	5.	5.	
*Prepared with equal volume water[1]	1 cup (8.5 oz.)	1176	5.3	2.	3.	
*Prepared with equal volume milk[2]	1 cup (8.6 oz.)	1264	9.6	5.	5.	
POTATO SOUP MIX:						
*(Lipton)	1 cup	1272	1.2	Tr.	<1.	
(Wyler's)	1 oz.		1.0			
POTATO STICK:						
(USDA)[3,4]	1 oz.	284	10.3	3.	8.	
(Durkee) O & C	1¾-oz. can	496	18.7			
Julienne (Wise)	1-oz. bag	128	9.3			
POUND CAKE:						
Home recipe, old fashioned, equal weights flour, sugar, eggs & butter (USDA)[5]	1.1-oz. slice (3½″ x 3″ x ½″)	33	7.9	4.	4.	
Home recipe, old fashioned, equal weights flour, sugar, eggs & vegetable shortening (USDA)[6]	1.1-oz. slice (3½″ x 3″ x ½″)	33	8.8	2.	7.	
Home recipe, traditional, made with butter (USDA)[7]	1.1-oz. slice (3½″ x 3″ x ½″)	53	4.8	2.	2.	
Home recipe, traditional, made with vegetable shortening (USDA)[8]	1.1-oz. slice (3½″ x 3″ x ½″	53	5.6	2.	4.	
Frozen (Morton)	1 oz.	136	5.9			

(USDA): United States Department of Agriculture
*Prepared as Package Directs
[1]Principal source of fat: cream.
[2]Principal sources of fat: cream & milk.
[3]Principal source of fat: cottonseed oil.
[4]Sodium content is variable & may be as high as 284 mg. per oz.
[5]Principal sources of fat: egg & butter.
[6]Principal sources of fat: egg & vegetable shortening.
[7]Principal sources of fat: butter, egg & milk.
[8]Principal sources of fat: vegetable shortening, egg & milk.

Food and Description	Measure or Quantity	Sodium (mg.)	— Fats in grams —			Choles- terol (mg.)
			Total	Satu- rated	Unsatu- rated	

POUND CAKE MIX:

*(Betty Crocker)	$^1/_{12}$ of cake	171	9.2			
*(Dromedary)	1″ slice (2.9 oz.)	388	14.9			

PRESERVE, sweetened (See also individual listings by flavor):

(USDA)	1 oz.	3	<.1			0
(USDA)	1 T. (.7 oz.)	2	Tr.			0
(Kraft)	1 oz.	<1	<.1			(0)
(Smucker's)	1 T. (.7 oz.)	<1	0.			(0)

PRETZEL:

(USDA)[1]	1 oz.	476	1.3			
Dutch, twist (USDA)[1]	1 pretzel (.6 oz.)	269	.7			
Stick, small (USDA)[1]	10 small sticks (2¼″, 3 grams)	50	.1			
Stick, regular (USDA)[1]	5 regular sticks (3⅛″, 3 grams)	50	.1			
Thin, twist (USDA)[1]	1 pretzel (6 grams)	101	.3			
(Nab) *Mister Salty Veri-Thin*	75 pieces (¾-oz. pkg.)	671	.5			
(Nab) *Mister Salty Veri-Thin*	150 pieces (1½-oz. pkg.)	1342	1.0			
(Nab) Pretzelette	16 pieces (1¼-oz. pkg.)	639	1.5			
(Nabisco) *Mister Salty,* Dutch	1 piece (.5 oz.)	239	.2			
(Nabisco) *Mister Salty,* pretzelette	1 piece (2 grams)	31	<.1			
(Nabisco) *Mister Salty*, 3-ring	1 piece (3 grams)	56	.1			
(Nabisco) *Mister Salty Veri-Thin*	1 piece (5 grams)	88	.2			
(Nabisco) *Mister Salty Veri-Thin*, stick	1 piece (<1 gram)	9	Tr.			
(Old London) nugget	2-oz. bag	1108	2.6			
(Old London) ring	1½-oz. bag	1059	1.9			
(Rold Gold) rod	1 oz.	390	1.1	Tr.	<1.	0
(Rold Gold) twist	1 oz.	300	.9	Tr.	<1.	0

PRICKLY PEAR, fresh (USDA):

Whole	1 lb. (weighed with rind & seeds)	4	.2			0
Flesh only	4 oz.	2	.1			0

(USDA): United States Department of Agriculture
*Prepared as Package Directs
[1]Sodium content is variable. For example, very thin pretzel sticks contain about twice the average amount listed.

Food and Description	Measure or Quantity	Sodium (mg.)	— Fats in grams —			Choles- terol (mg.)
			Total	Satu- rated	Unsatu- rated	
PRODUCT 19, cereal (Kellogg's)	1 cup (1 oz.)	271	.4			(0)
PRUNE:						
Dried, "softenized," uncooked:						
Small (USDA)	1 prune (5 grams)	<1	Tr.			0
Medium, whole with pits (USDA)	1 cup (6.6 oz.)	13	1.0			0
Medium (USDA)	1 prune (7 grams)	<1	Tr.			0
Large (USDA)	1 prune (9 grams)	<1	Tr.			0
Pitted, chopped (USDA)	1 cup (5.3 oz.)	12	.9			0
Pitted, ground (USDA)	1 cup (9.7 oz.)	22	1.6			0
Dried, moist-pak (Del Monte)	1 cup (8 oz.)	5	.9			0
Dried, ready-to-eat (Del Monte)	1 cup (6.6 oz.)	2	4.7			0
Dried, "softenized," cooked, unsweetened (USDA)	1 cup (17-18 med. with ⅓ cup liq., 9.5 oz.)	10	.7			0
Dried, "softenized," cooked with sugar (USDA)	1 cup (16-18 prunes & ⅓ cup liq., 11.1 oz.)	9	.6			0
Dehydrated (USDA):						
Nugget-type & pieces	8 oz.	25	1.1			0
Nugget-type & pieces, cooked with sugar, solids & liq.	1 cup (8.9 oz.)	10	.5			0
Canned:						
Cooked (Sunsweet)	1 cup		.4			(0)
Stewed (Del Monte)	1 cup (9.4 oz.)	3	1.1			0
Stewed, pitted (Del Monte)	1 cup (9.2 oz.)	14	1.3			0
PRUNE JUICE:						
(USDA)	½ cup (4.5 oz.)	2	.1			0
(Bennett's)	½ cup (4.5 oz.)	3	.1			0
(Del Monte)	½ cup (4.3 oz.)	10	Tr.			0
(Heinz)	5½-fl.-oz. can	13	.2			(0)
(Mott's) Super	4 oz. (by wt.)		.1			(0)
RealPrune	½ cup (4.5 oz.)	33	.3			(0)
(Sunsweet)	½ cup		<.1			(0)
& apple (Sunsweet)	4 oz. (by wt.)		.1			(0)
With lemon (Sunsweet)	½ cup		<.1			(0)

(USDA): United States Department of Agriculture
*Prepared as Package Directs

Food and Description	Measure or Quantity	Sodium (mg.)	—Fats in grams—			Choles-terol (mg.)
			Total	Satu-rated	Unsatu-rated	
PRUNE WHIP, home recipe (USDA)	1 cup (4.8 oz.)	221	.3			
PUDDING or PUDDING MIX (See individual kinds)						
PUFF (See **CRACKER** or individual kinds of hors d'oeuvres, such as **CHICKEN PUFF**)						
PUFFA PUFFA RICE, cereal	1 cup (1 oz.)	25	3.0			(0)
PUFFED OAT CEREAL (USDA):						
Added nutrients	1 oz.	359	1.6			0
Sugar-coated, added nutrients	1 oz.	167	1.0			0
PUFFED RICE CEREAL:						
(USDA) added nutrients, unsalted	1 cup (.5 oz.)	<1	<.1			0
(USDA) honey & added nutrients	1 oz.	200	.2			0
(USDA) honey or cocoa fat, added nutrients	1 oz.	101	1.1			0
(Checker)	½ oz.	2	<.1			(0)
(Quaker)	1¼ cups (½ oz.)	<1	.1			(0)
(Sunland)	½ oz.	2	<.1			(0)
(Whiffs)	½ oz.	2	<.1			(0)
PUMPKIN:						
Fresh, whole (USDA)	1 lb. (weighed with rind & seeds)	3	.3			0
Flesh only (USDA)	4 oz.	1	.1			0
Canned:						
Salted (USDA)	½ cup (4.3 oz.)	288	.4			0
Unsalted (USDA)	½ cup (4.3 oz.)	2	.4			0
(Del Monte)	½ cup (4.3 oz.)	1	.1			0
(Stokely-Van Camp)	½ cup (4.1 oz.)		.4			(0)
PUMPKIN PIE:						
Home recipe, 1-crust (USDA):						
Made with lard[1]	⅙ of 9″ pie (5.4 oz.)	325	17.0	6.	11.	93

(USDA): United States Department of Agriculture
*Prepared as Package Directs
[1]Principal sources of fat: lard & butter.

Food and Description	Measure or Quantity	Sodium (mg.)	— Fats in grams —			Cholesterol (mg.)
			Total	Saturated	Unsaturated	
Made with vegetable shortening[1]	1/6 of 9″ pie (5.4 oz.)	325	17.0	5.	12.	
(Tastykake)	4-oz. pie		16.0			
Frozen:						
(Banquet)	5-oz. serving		10.7			
(Morton)	1/6 of 20-oz. pie	218	5.9			
(Morton)	1/8 of 46-oz. pie	393	27.3			
(Mrs. Smith's)	1/6 of 8″ pie (4 oz.)	245	9.4			
(Mrs. Smith's)	1/8 of 10″ pie (5.6 oz.)	310	12.0			
PUMPKIN PIE FILLING:						
(Comstock)	1/2 cup (5.4 oz.)	298	.3			
(Del Monte)	1 cup (9 oz.)	512	.5			
PUMPKIN PIE SPICE						
(Spice Islands)	1 tsp	<1				(0)
PUMPKIN SEED, dry (USDA):						
Whole	4 oz. (weighed in hull)		39.2	7.	32.	0
Hulled	4 oz.		53.0	9.	44.	0
PUNCH DRINK canned (Hi-C)	6 fl. oz. (6.3 oz.)	<1				0
PURPLE PASSION, soft drink						
(Canada Dry) bottle or can	6 fl. oz.	13+	0.			(0)
PURSLANE, including stems (USDA):						
Raw	1 lb.		1.8			0
Boiled, drained	4 oz.		.3			0
PUSSYCAT MIX (Bar-Tender's)	1 serving (2/3 oz.)	21	.2			(0)

Q

QUAIL, raw (USDA):						
Ready-to-cook	1 lb. (weighed with bones)		27.8			
Meat & skin only	4 oz.	45	7.9			
Giblets	2 oz.		3.5			

(USDA): United States Department of Agriculture
*Prepared as Package Directs
[1]Principal sources of fat: vegetable shortening & butter.

Food and Description	Measure or Quantity	Sodium (mg.)	—Fats in grams—			Choles- terol (mg.)
			Total	Satu- rated	Unsatu- rated	
QUAKE, cereal (Quaker)	1 cup (1 oz.)	116	2.3			0
QUANGAROOS, cereal (Quaker)	1 cup (1 oz.)	146	1.1			0
QUIK (See individual kinds)						
QUINCE, fresh (USDA):						
Untrimmed	1 lb. (weighed with skin & seeds)	11	.3			0
Flesh only	4 oz.	5	.1			0
QUININE SOFT DRINK or TONIC WATER:						
Sweetened:						
(Canada Dry) bottle	6 fl. oz.	0+	0.			0
(Canada Dry) can	6 fl. oz.	<1+	0.			0
(Dr. Brown's)	6 fl. oz.	3	0.			0
(Fanta)	6 fl. oz.	5	0.			0
(Hoffman)	6 fl. oz.	3	0.			0
(Kirsch)	6 fl. oz.	<1	0.			0
(Schweppes)	6 fl. oz.	10	0.			0
(Shasta)	6 fl. oz.	10	0.			0
(Yukon Club)	6 fl. oz.	3	0.			0
Low calorie (No-Cal)	6 fl. oz.	12	0.			0
QUISP, cereal (Quaker)	1¹/₆ cups (1 oz.)	215	2.7			0

R

Food and Description	Measure or Quantity	Sodium (mg.)	Total	Satu- rated	Unsatu- rated	Choles- terol (mg.)
RABBIT (USDA):						
Domesticated:						
Ready-to-cook	1 lb. (weighed with bones)	154	29.	11.	18.	
Raw meat only	4 oz.	49	9.1			74
Stewed, flesh only	4 oz.	46	11.5			103
Stewed, flesh only, chopped or diced	1 cup (4.9 oz.)	57	14.1			127
Wild, ready-to-cook	1 lb. (weighed with bones)		18.1			
Wild, raw, meat only	4 oz.		5.7			
RACOON, roasted, meat only (USDA)	4 oz.		16.4			

USDA): United States Department of Agriculture
ᵉPrepared as Package Directs

(271)

Food and Description	Measure or Quantity	Sodium (mg.)	— Fats in grams —			Cholesterol (mg.)
			Total	Saturated	Unsaturated	

RADISH (USDA):
Common, raw:

Food and Description	Measure or Quantity	Sodium (mg.)	Total	Saturated	Unsaturated	Cholesterol (mg.)
Untrimmed, without tops	½ lb. (weighed untrimmed)	36	.2			0
Trimmed, whole	4 small radishes (1.4 oz.)	7	<.1			0
Trimmed, whole	1 cup (4.7 oz.)	24	.1			0
Trimmed, sliced	½ cup (2 oz.)	10	<.1			0
Oriental, raw, without tops	½ lb. (weighed unpared)		.2			0
Oriental, raw, trimmed & pared	4 oz.		.1			0

RAISIN:
Dried:

Food and Description	Measure or Quantity	Sodium (mg.)	Total	Saturated	Unsaturated	Cholesterol (mg.)
Whole (USDA)	4 oz.	31	.2			0
Whole (USDA)	1 pkg. (.5 oz.)	4	<.1			0
Whole, pressed down (USDA)	½ cup (2.9 oz.)	22	.2			0
Whole, pressed down (USDA)	1 T. (.4 oz.)	3	<.1			0
Chopped (USDA)	½ cup (2.9 oz.)	22	.2			0
Ground (USDA)	½ cup (4.7 oz.)	36	.3			0
Cinnamon-coated (Del Monte)	½ cup (2.5 oz.)	16	.4			(0)
Seeded, Muscat (Del Monte)	½ cup (2.5 oz.)	42	.6			(0)
Seeded, Muscat (Sun-Maid)	15-oz. pkg.		<.1			0
Seeded, Muscat (Sun-Maid)	1 oz.		.1			0
Seedless, California Thompson (Del Monte)	½ cup (2.5 oz.)	17	.6			(0)
Seedless, California Thompson (Sun-Maid)	½ cup (2.8 oz.)	14	.2			0
Seedless, California Thompson (Sun-Maid)	1 T. (.4 oz.)	2	<.1			0
Seedless, golden (Del Monte)	½ cup (2.5 oz.)	23	.2			0
Seedless, golden (Sun-Maid)	15-oz. pkg.	85	.8			0
Cooked, added sugar, solids & liq. (USDA)	½ cup (4.3 oz.)	16	.1			0

RAISIN PIE:
Home recipe, 2-crust (USDA):

Food and Description	Measure or Quantity	Sodium (mg.)	Total	Saturated	Unsaturated	Cholesterol (mg.)
Made with lard[1]	⅙ of 9″ pie (5.6 oz.)	450	16.9	6.	11.	
Made with vegetable shortening[2]	⅙ of 9″ pie (5.6 oz.)	450	16.9	5.	12.	
(Tastykake)	4-oz. pie		14.6			

RAISIN PIE FILLING:

Food and Description	Measure or Quantity	Sodium (mg.)	Total	Saturated	Unsaturated	Cholesterol (mg.)
(Comstock)	½ cup (5.4 oz.)	84	<.1			
(Lucky Leaf)	8 oz.	242	1.6			

(USDA): United States Department of Agriculture
*Prepared as Package Directs
[1]Principal sources of fat: lard & butter.
[2]Principal sources of fat: vegetable shortening & butter.

Food and Description	Measure or Quantity	Sodium (mg.)	—Fats in grams— Total	Satu- rated	Unsatu- rated	Choles- terol (mg.)
RAJA FISH (See **SKATE**)						
RASPBERRY:						
Black:						
Fresh:						
(USDA)	½ lb. (weighed with caps & stems)	2	3.1			0
(USDA) without caps & stems	½ cup (2.4 oz.)	<1	.9			0
Canned, water pack, unsweet- ened, solids & liq. (USDA)	4 oz.	1	1.2			0
Red:						
Fresh:						
(USDA)	½ lb. (weighed with caps & stems)	2	1.1			0
(USDA) without caps & stems	½ cup (2.5 oz.)	<1	.4			0
Canned, water pack, un- sweetened, or low calorie:						
Solids & liq. (USDA)	4 oz.	1	.1			0
Solids & liq. (Blue Boy)	4 oz.	1	1.0			(0)
Frozen, sweetened:						
Not thawed (USDA)	10-oz. pkg.	3	.6			0
Not thawed (USDA)	½ cup (4.4 oz.)	1	.2			0
Quick-thaw (Birds Eye)	½ cup (5 oz.)	1	.4			0
RASPBERRY PIE FILLING:						
Red (Comstock)	½ cup (5.3 oz.)	166	.2			
Red (Lucky Leaf)	8 oz.	260	.4			
RASPBERRY PRESERVE or JAM:						
Sweetened, black or red (Bama)	1 T. (.7 oz.)	2	<.1			(0)
Low calorie or dietetic:						
(Diet Delight)	1 T. (.6 oz.)	9	Tr.			(0)
Black (Kraft)	1 oz.	5	<.1			(0)
(S and W) *Nutradiet*	1 T. (.5 oz.)		<.1			(0)
Low sugar, black (Slenderella)	1 T. (.7 oz.)	16	<.1			(0)
RASPBERRY RENNET MIX:						
Powder:						
Dry (Junket)	1 oz.	11	Tr.			

(USDA): United States Department of Agriculture
*Prepared as Package Directs

273

Food and Description	Measure or Quantity	Sodium (mg.)	Total	Fats in grams — Satu-rated	Unsatu-rated	Choles-terol (mg.)
*(Junket)	4 oz.	56	3.8			
Tablet:						
Dry (Junket)	1 tablet (<1 gram)	197	Tr.			
*& sugar (Junket)	4 oz.	98	3.9			
RASPBERRY SOFT DRINK:						
Sweetened:						
(Clicquot Club)	6 fl. oz.	11	0.			0
(Cott)	6 fl. oz.	11	0.			0
(Dr. Brown's) black	6 fl. oz.	14	0.			0
(Hoffman) black	6 fl. oz.	14	0.			0
(Kirsch) black	6 fl. oz.	<1	0.			0
(Mission)	6 fl. oz.	11	0.			0
(Shasta) Wild	6 fl. oz.	22	0.			0
(Yukon Club) black	6 fl. oz.	14	0.			0
Low calorie:						
(Clicquot Club)	6 fl. oz.	45	0.			0
(Cott)	6 fl. oz.	45	0.			0
(Dr. Brown's) black	6 fl. oz.	20	0.			0
(Hoffman) black	6 fl. oz.	20	0.			0
(Key Food) black	6 fl. oz.	20	0.			0
(Mission)	6 fl. oz.	45	0.			0
(No-Cal) black	6 fl. oz.	12	0.			0
(Shasta) Wild	6 fl. oz.	37	0.			0
(Waldbaum) black	6 fl. oz.	20	0.			0
RASPBERRY SYRUP, low calorie:						
(No-Cal)	1 tsp. (5 grams)	<1	0.			(0)
RASPBERRY TURNOVER,						
frozen (Pepperidge Farm)	1 turnover (3.3 oz.)	258	20.0			
RAVIOLI:						
Canned:						
Beef or meat:						
(Buitoni)	8 oz.		4.5			
In brine (Buitoni)	8 oz.		3.8			
(Chef Boy-Ar-Dee)	1/5 of 40-oz. can	1349	6.4			
(Nalley's)	8 oz.		10.2			
(Prince)	3.7-oz. can		4.8			
Cheese:						
(Buitoni)	8 oz.		7.4			
In brine (Buitoni)	8 oz.		3.3			

(USDA): United States Department of Agriculture
*Prepared as Package Directs

Food and Description	Measure or Quantity	Sodium (mg.)	—Fats in grams— Total	Satu- rated	Unsatu- rated	Choles- terol (mg.)
(Chef Boy-Ar-Dee)	½ of 15-oz. can	1347	10.9			
(Prince)	3.7-oz. can		3.7			
Chicken (Nalley's)	8 oz.		10.0			
Frozen:						
Beef (Celeste)	7 ravioli (4 oz.)	300	5.7			
Beef, dinner (Celeste)	½ of 15-oz. pkg.	660	6.4			
Beef (Kraft)	12½-oz. pkg.	1628	14.9			
Cheese (Buitoni)	4 oz.		7.1			
Cheese (Celeste)	7 ravioli (4 oz.)	250	6.3			
Cheese, dinner (Celeste)	½ of 15-oz. pkg.	665	6.2			
Cheese (Kraft)	12½-oz. pkg.	1848	16.6			
Meat, without sauce (Buitoni)	4 oz.		5.9			
Meat, with sauce (Buitoni)	4 oz.		4.4			
Raviolettes (Buitoni)	4 oz.		4.0			

REDFISH (See **DRUM, RED & OCEAN PERCH,** Atlantic)

RED & GRAY SNAPPER, raw:

Whole (USDA)	1 lb. (weighed whole)	158	2.1			
Meat only (USDA)	4 oz.	76	1.0			

REDHORSE, SILVER, raw (USDA):

Drawn	1 lb. (weighed eviscerated)		4.8			
Meat only	4 oz.		2.6			

RED POP, soft drink (No-Cal) — 6 fl. oz. — 12 — 0. — (0)

REINDEER, raw, lean only (USDA) — 4 oz. — 4.3

RELISH:

Barbecue (Crosse & Blackwell)	1 T. (.7 oz.)	220	0.			
Barbecue (Heinz)	1 T.	139	.1			
Corn (Crosse & Blackwell)	1 T. (.6 oz.)	340	0.			
Hamburger (Crosse & Blackwell)	1 T. (.6 oz.)	220	0.			
Hamburger (Del Monte)	1 T. (.9 oz.)	402	<.1			0
Hamburger (Heinz)	1 T.	146	Tr.			
Hot dog (Crosse & Blackwell)	1 T. (.7 oz.)		0.			
Hot dog (Del Monte)	1 T. (.9 oz.)	422	.2			0

(USDA): United States Department of Agriculture
*Prepared as Package Directs

Food and Description	Measure or Quantity	Sodium (mg.)	— Fats in grams —			Cholesterol (mg.)
			Total	Satu-rated	Unsatu-rated	
Hot dog (Heinz)	1 T.	111	.1			
Hot pepper (Crosse & Blackwell)	1 T. (.7 oz.)		0.			
India (Crosse & Blackwell)	1 T. (.7 oz.)	220	0.			
India (Heinz)	1 T.	115	.1			
Piccalilli (Crosse & Blackwell)	1 T. (.7 oz.)		0.			
Piccalilli (Heinz)	1 T.	123	.1			
Sour (USDA)	½ cup (4.3 oz.)		1.1			0
Sour (USDA)	1 T. (.5 oz.)		.1			0
Sweet:						
(USDA) finely chopped	½ cup (4.3 oz.)	869	.7			0
(USDA) finely chopped	1 T. (.5 oz.)	107	<.1			0
(Aunt Jane's)	1 rounded tsp. (.4 oz.)	71	<.1			(0)
(Crosse & Blackwell)	1 T. (.7 oz.)		0.			
(Del Monte)	1 T. (.9 oz.)	355	.2			0
(Heinz)	1 T.	181	.1			
(Smucker's)	1 T. (.6 oz.)	158	Tr.			

RENNIN CUSTARD PRODUCTS
(See individual flavors)

RHINE WINE:

(Gold Seal) 12% alcohol,	3 fl. oz. (3.1 oz.)	3	0.			(0)
(Great Western) 12.5% alcohol, regular	3 fl. oz.	25	0.			0
(Great Western) 12.5% alcohol, Dutchess	3 fl. oz.	27	0.			0

RHUBARB:
Fresh (USDA):

Partly trimmed	1 lb. (weighed with part leaves, ends & trimmings)	7	.3			0
Trimmed	4 oz.	2	.1			0
Diced	½ cup (2.2 oz.)	1	<.1			0
Cooked, sweetened, solids & liq. (USDA)	½ cup (4.2 oz.)	2	.1			0
Frozen, sweetened:						
Not thawed (USDA)	½ cup (3.9 oz.)	4	.2			0
Cooked, added sugar, solids & liq. (USDA)	½ cup (4.4 oz.)	4	.2			0
(Birds Eye)	½ cup (4 oz.)	2	.1			0

(USDA): United States Department of Agriculture
*Prepared as Package Directs

Food and Description	Measure or Quantity	Sodium (mg.)	Total	—Fats in grams— Satu- rated	Unsatu- rated	Choles- terol (mg.)
RHUBARB PIE, home recipe,						
2-crust (USDA)	1/6 of 9″ pie (5.6 oz.)	427	16.9	5.	12.	
RICE:						
Brown:						
Raw (USDA)	½ cup (3.7 oz.)	9	2.0			0
Raw (USDA)	1 oz.	3	.5			0
Cooked:						
With salt added:						
(USDA)	4 oz.	320	.7			0
(Carolina)	4 oz.	320	.7			(0)
(River Brand)	4 oz.	320	.7			(0)
(Water Maid)	4 oz.	320	.7			(0)
Parboiled (Uncle Ben's) with no added butter or salt	⅔ cup (4.2 oz.)	6	1.1			0
Parboiled (Uncle Ben's) with added butter	⅔ cup (4.3 oz.)	32	3.2			
Frozen, in beef stock (Green Giant)	⅓ of 12-oz. pkg.	782	2.8			
White:						
Instant or precooked:						
Dry, long-grain (USDA)	½ cup (1.9 oz.)	<1	.1			0
Dry, long-grain (USDA)	1 oz.	<1	<.1			0
Cooked:						
With salt added:						
Long-grain (USDA)	⅔ cup (3.3 oz.)	254	Tr.			0
(Carolina)	⅔ cup (3.3 oz.)	254	Tr.			(0)
Without salt:						
(Minute Rice) no added butter	⅔ cup (4 oz.)	1	Tr.			0
Long-grain (Uncle Ben's Quick) no added butter	⅔ cup (4 oz.)	9	<.1			0
Long-grain (Uncle Ben's Quick) with added butter	⅔ cup (4.1 oz.)	37	2.5			
Parboiled:						
Dry, long-grain (USDA)	1 oz.	3	<.1			0
Cooked:						
With added salt:						
Long-grain (USDA)	⅔ cup (4.1 oz.)	419	.1			0
(Aunt Caroline)	⅔ cup (4.1 oz.)	419	.1			(0)

(USDA): United States Department of Agriculture
*Prepared as Package Directs

Food and Description	Measure or Quantity	Sodium (mg.)	Total	Fats in grams Satu- rated	Unsatu- rated	Choles- terol (mg.)
No added salt, long-grain (Uncle Ben's Converted), no added butter	⅔ cup (4.3 oz.)	3	.2			0
Regular:						
Raw (USDA)	½ cup (3.5 oz.)	5	.4			0
Cooked with salt:						
(USDA)	⅔ cup (4.8 oz.)	512	.1			0
Extra long-grain (Carolina)	⅔ cup (4.8 oz.)	512	.1			(0)
Long-grain (Mahatma)	⅔ cup (4.8 oz.)	512	.1			(0)
(River Brand) fluffy	⅔ cup (4.8 oz.)	512	.1			(0)
(Water Maid)	⅔ cup (4.8 oz.)	512	.1			(0)
White & wild, frozen (Green Giant)	⅓ of 12-oz. pkg.	522	1.1			
Wild (See **WILD RICE**)						
RICE BRAN (USDA)	1 oz.	Tr.	4.5			0
RICE CEREAL (USDA):						
With casein & other added nutrients	1 oz.	170	<.1			0
Wheat gluten & other added nutrients	1 oz.	227	<.1			0
RICE CHEX, cereal (Ralston)	1⅛ cups (1 oz.)	261	.1			(0)
RICE FLAKES, cereal, added nutrients (USDA)	1 cup (1.1 oz.)	296	<.1			0
RICE, FRIED:						
Frozen, with almonds (Green Giant)	⅓ of 12-oz. pkg.	743	6.8			
Seasoning Mix (Durkee)	1-oz. pkg.	1931	1.1			
*Seasoning Mix (Durkee)	2 cups (1-oz. pkg.)	3195	1.4			
RICE KRISPIES, cereal (Kellogg's)	1 cup (1 oz.)	261	.1			(0)
RICE MIX:						
Beef:						
(Uncle Ben's)	6-oz. pkg.	3995	3.4			0
*(Uncle Ben's) no added butter or salt	½ cup (4.2 oz.)	692	.6			Tr.
*(Uncle Ben's) with added butter, no added salt	½ cup (4.3 oz.)	715	2.5			

(USDA): United States Department of Agriculture
*Prepared as Package Directs

Food and Description	Measure or Quantity	Sodium (mg.)	—Fats in grams— Total	Satu- rated	Unsatu- rated	Choles- terol (mg.)
*(Village Inn)	½ cup		1.7			
Brown & wild:						
(Uncle Ben's)	6-oz. pkg.	1980	5.1			0
*(Uncle Ben's) no added butter or salt	½ cup (4.3 oz.)	319	.8			0
*(Uncle Ben's) with added butter, no added salt	½ cup (4.3 oz.)	344	2.9			
Chicken:						
(Uncle Ben's)	6-oz. pkg.	2542	5.4			0
*(Uncle Ben's) no added butter or salt	½ cup (3.6 oz.)	416	.9			
*(Uncle Ben's) with added butter, no added salt	½ cup (3.8 oz.)	460	4.5			
*(Village Inn)	½ cup		1.7			
Curried or curry:						
(Uncle Ben's)	6-oz. pkg.	3179	1.2			0
*(Uncle Ben's) no added butter or salt	½ cup (4.2 oz.)	541	.2			0
*(Uncle Ben's) with added butter, no added salt	½ cup (4.2 oz.)	564	2.1			
*(Village Inn)	½ cup		1.7			
*Drumstick (Minute Rice)	½ cup (4.1 oz.)	634	5.7			23
*Herb (Village Inn)	½ cup		1.7			
*Keriyaki dinner (Betty Crocker)	1 cup	223	20.3			
Long-grain & wild:						
(Uncle Ben's)	6-oz. pkg.	2926	1.5			0
*(Uncle Ben's) no added butter or salt	½ cup (4 oz.)	482	.2			
*(Uncle Ben's) with added butter, no added salt	½ cup (4.1 oz.)	505	2.1			
*(Village Inn)	½ cup		1.7			
*Milanese (Betty Crocker)	½ cup	797	6.4			
*Oriental, dinner (Jeno's) Add 'n Heat	40-oz. pkg.		81.6			
Pilaf:						
(Uncle Ben's)	6-oz. pkg.	2703	2.2			0
*(Uncle Ben's) no added butter or salt	½ cup (3.3 oz.)	435	.4			
*(Uncle Ben's) with added butter, no added salt	½ cup (3.5 oz.)	479	3.9			
*Provence (Betty Crocker)	½ cup	895	5.7			
*Rib Roast (Minute Rice)	½ cup (4.1 oz.)	508	4.0			12

(USDA): United States Department of Agriculture
*Prepared as Package Directs

Food and Description	Measure or Quantity	Sodium (mg.)	—Fats in grams—			Choles- terol (mg.)
			Total	Satu- rated	Unsatu- rated	
Spanish (See also **RICE, SPANISH**):						
*(Minute Rice)	½ cup (5.6 oz.)	866	3.7			12
(Uncle Ben's)	5½-oz. pkg.	4365	.9			0
*(Uncle Ben's) no added butter or salt	½ cup (4.6 oz.)	888	.2			0
*(Uncle Ben's) with added butter, no added salt	½ cup (4.6 oz.)	916	2.5			
*(Village Inn)	½ cup		1.7			
*Yellow (Village Inn)	½ cup		1.7			
RICE & PEAS with MUSH-ROOMS:						
Frozen (Birds Eye)	⅓ of pkg. (2.3 oz.)	536	.1			0
Frozen (Green Giant)	⅓ of 12-oz. pkg.	437	2.3			
RICE PILAF, frozen (Green Giant)	⅓ of 12-oz. pkg.	607	1.1			
RICE POLISH (USDA)	1 oz.	Tr.	3.6			0
RICE PUDDING:						
Home recipe, with raisins (USDA)[1]	½ cup (4.7 oz.)	94	4.1	3.	2.	15
Canned:						
(Betty Crocker)	½ cup	166	4.2			
(Hunt's)[2]	5-oz. can	192	11.5	2.	9.	
RICE, SPANISH:						
Home recipe (USDA)	4 oz.	358	1.9			
Canned:						
(Heinz)	8¾-oz. can	1545	4.5			
(Nalley's)	4 oz.		1.6			
(Van Camp)	½ cup (.9 oz.)		1.8			
Frozen (Green Giant)	⅓ of 12-oz. pkg.	476	.6			
RICE, SPANISH, SEASONING MIX (Lawry's)	1½-oz. pkg.		1.3			
RICE VERDI, frozen (Green Giant)	⅓ of 12-oz. pkg.	544	2.3			
ROAST 'n BOAST (General Foods):						
For beef	1½-oz. pkg.	2678	.7			0
For chicken	1⅜-oz. pkg.	3440	.3			0

(USDA): United States Department of Agriculture
*Prepared as Package Directs
[1]Principal source of fat: milk.
[2]Principal source of fat: soybean oil.

Food and Description	Measure or Quantity	Sodium (mg.)	Fats in grams Total	Satu- rated	Unsatu- rated	Choles- terol (mg.)
For pork	1¾-oz. pkg.	5399	.4			0
For stew	1½-oz. pkg.	4106	.7			0
ROCKFISH (USDA):						
Raw, meat only	1 lb.	272	8.2			
Oven-steamed, with onion	4 oz.	77	2.8			
ROE (USDA):						
Raw, carp, cod, haddock, herring, pike or shad	4 oz.		2.6			
Raw, salmon, sturgeon, turbot	4 oz.		11.8			401
Baked or broiled,[1] cod & shad	4 oz.	83	3.2			
Canned, cod, haddock or herring, solids & liq.	4 oz.		3.2			
ROLL & BUN:						
Barbeque (Arnold)	1 bun (1.6 oz.)		2.7			
Brown & serve:						
Unbrowned (USDA)[2]	1 oz.	145	1.9	Tr.	2.	
Browned (USDA)[2]	1 oz.	159	2.2	Tr.	2.	
(Wonder)	1 roll (1 oz.)	134	1.5			
Butter crescent (Pepperidge Farm)	1 roll (1.2 oz.)	203	8.0			
Butterfly (Pepperidge Farm)	1 roll (.6 oz.)	87	2.1			
Cinnamon nut (Pepperidge Farm)	1 bun (1 oz.)	76	4.8			
Cloverleaf, home recipe (USDA)[3]	1 roll (1.2 oz.)	98	3.0	<1.	2.	
Club (Pepperidge Farm)	1 roll (1.6 oz.)	267	.7			
Deli Twist (Arnold)	1 roll (1.2 oz.)		3.3			
Diet size (Arnold)	1 roll (.5 oz.)		1.1			
Dinner (Arnold) 12 or 24 to pkg.	1 roll (¾ oz.)		2.2			
Dinner (Pepperidge Farm)	1 roll (.7 oz.)	90	1.6			
Dutch Egg, sandwich (Arnold)	1 bun (1.7 oz.)		3.9			
Finger:						
(Arnold) handipan	1 roll (.7 oz.)		1.7			
Egg (Arnold) family	1 roll (.7 oz.)		1.7			
Frankfurter:						
(USDA)[2]	1 roll (1.4 oz.)	202	2.2	Tr.	2.	
(Arnold)	1 roll (1.4 oz.)		2.6			
New England (Arnold)	1 roll (1.6 oz.)		2.8			
(Pepperidge Farm)	1 roll (1.4 oz.)	205	2.3			
(Wonder)	1 bun (1.5 oz.)	232	2.2			

(USDA): United States Department of Agriculture
*Prepared as Package Directs
[1] Prepared with butter or margarine & lemon juice or vinegar.
[2] Principal source of fat: vegetable shortening.
[3] Principal sources of fat: vegetable shortening, milk & egg.

Food and Description	Measure or Quantity	Sodium (mg.)	Fats in grams — Total	Satu- rated	Unsatu- rated	Choles- terol (mg.)
French:						
Triple (Pepperidge Farm)	1 roll (3.5 oz.)	589	1.5			
Twin (Pepperidge Farm)	1 roll (5.2 oz.)	867	2.1			
Golden Twist (Pepperidge Farm)	1 roll (1.2 oz.)	178	6.7			
Hamburger:						
(USDA)[1]	1 roll (1.4 oz.)	202	2.2	Tr.	2.	
(Pepperidge Farm)	1 roll (1.4 oz.)	201	2.2			
(Wonder)	1 bun (1.5 oz.)	232	2.2			
Hard, round or rectangular (USDA)	1 roll (1.8 oz.)	312	1.6	<1.	1.	
Hearth (Pepperidge Farm)	1 roll (.8 oz.)	118	.9			
Honey, frozen (Morton)	1 serving (2.2 oz.)	56	6.5			
Kaiser, brown & serve (Arnold)	1 roll (1.7 oz.)		1.8			
Old Fashioned (Pepperidge Farm)	1 roll (.6 oz.)	90	2.2			
Parker (Arnold) handipan	1 roll (.7 oz.)		1.7			
Party Pan (Pepperidge Farm):						
Finger	1 roll (.7 oz.)	84	1.5			
Round	1 roll (.4 oz.)	51	.9			
Pecan, coffee (Pepperidge Farm)	1 bun (1.7 oz.)	186	11.6			
Plain (USDA)[1]	1 roll (1 oz.)	143	1.6	Tr.	1.	
Raisin (USDA)[1]	1 oz.	109	.8	Tr.	<1.	
Sandwich, soft (Arnold)	1 roll (1.5 oz.)		4.0			
Sesame crisp (Pepperidge Farm):						
Midwest	1 roll (.8 oz.)	112	1.4			
East	1 roll (.9 oz.)	119	1.5			
Soft (Arnold) handipan	1 roll (.7 oz.)		1.6			
Sweet (USDA)[2]	1 bun (1.5 oz.)	167	3.9	<1.	3.	
Whole-wheat (USDA)[1]	1 roll (1.3 oz.)	214	1.1	Tr.	<1.	
ROLL DOUGH:						
Frozen, unraised (USDA)[1]	1 oz.	137	1.4	Tr.	1.	
Frozen, baked (USDA)[1]	1 oz.	159	1.5	Tr.	1.	
ROLL MIX:						
Dry (USDA)[1]	1 oz.	117	1.7	Tr.	1.	
*Prepared with water (USDA)[1]	1 oz.	89	1.3	Tr.	1.	
(Pillsbury) hot	1 oz.					
***ROMAN MEAL CEREAL,** dry*	¾ cup (1.3 oz.)	2	.7	Tr.	<1.	0
ROOT BEER SOFT DRINK:						
Sweetened:						
(Canada Dry) *Rooti*	6 fl. oz.	13+	0.			0

(USDA): United States Department of Agriculture
*Prepared as Package Directs
[1]Principal source of fat: vegetable shortening.
[2]Principal sources of fat: vegetable shortening, milk & egg.

Food and Description	Measure or Quantity	Sodium (mg.)	Fats in grams — Total	Satu- rated	Unsatu- rated	Choles- terol (mg.)
(Clicquot Club)	6 fl. oz.	12	0.			0
(Cott)	6 fl. oz.	12	0.			0
(Fanta)	6 fl. oz.	7	0.			0
(Dr. Brown's)	6 fl. oz.	3	0.			0
(Hires)	6 fl. oz.	2	0.			0
(Hoffman)	6 fl. oz.	3	0.			0
(Key Food)	6 fl. oz.	3	0.			0
(Kirsch)	6 fl. oz.	<1	0.			0
(Mission)	6 fl. oz.	12	0.			0
(Nedick's)	6 fl. oz.	3	0.			0
(Nehi)	6 fl. oz. (6.6 oz.)	0+	0.			0
(Shasta) draft	6 fl. oz.	22	0.			0
(Waldbaum)	6 fl. oz.	3	0.			0
(Yukon Club)	6 fl. oz.	3	0.			0
Low calorie:						
(Canada Dry)	6 fl. oz.	12+	0.			0
(Clicquot Club)	6 fl. oz.	45	0.			0
(Cott)	6 fl. oz.	45	0.			0
(Hoffman)	6 fl. oz.	35	0.			0
(Mission)	6 fl. oz.	45	0.			0
(No-Cal)	6 fl. oz.	11	0.			0
(Shasta) draft	6 fl. oz.	37	0.			0
(Yukon Club)	6 fl. oz.	64	0.			0
ROSE APPLE, raw (USDA):						
Whole	1 lb. (weighed with caps & seeds)		.9			0
Flesh only	4 oz.		.3			0
ROSEMARY (Spice Islands)	1 tsp.	<1				(0)
ROSE WINE:						
(Great Western) 12.5% alcohol	3 fl. oz.	38	0.			0
(Great Western) Isabella, 12.5% alcohol	3 fl. oz.	<1	0.			0
RUSK:						
(USDA)[1]	1 piece (.5 oz.)	35	1.2	Tr.	1.	
Holland (Nabisco)	1 piece (.4 oz.)	34	.6			
RUTABAGA:						
Raw, without tops (USDA)	1 lb. (weighed with skin)	19	.4			0

(USDA): United States Department of Agriculture
*Prepared as Package Directs
[1]Principal sources of fat: vegetable shortening, egg & milk.

Food and Description	Measure or Quantity	Sodium (mg.)	Fats in grams — Total	Satu- rated	Unsatu- rated	Choles- terol (mg.)
Raw, diced (USDA)	½ cup (2.5 oz.)	4	<.1			0
Boiled without salt, diced, drained (USDA)	½ cup (3 oz.)	3	<.1			0
Boiled without salt, mashed (USDA)	½ cup (4.3 oz.)	5	.1			0
Canned (King Pharr)	½ cup					
RYE, whole grain (USDA)	1 oz.	<1	.5			0

RYE FLOUR (See **FLOUR**)

RYE WHISKEY (See **DISTILLED LIQUOR**)

RYE-KRISP (See **CRACKER**)

RY-KING (See **BREAD**)

S

SABLEFISH, raw (USDA):						
Whole	1 lb. (weighed whole)	107	28.4			
Meat only	4 oz.	64	16.9			
SAFFLOWER SEED (USDA):						
Kernels, dry, in hull	½ lb. (weighed in hull)		68.8	6.	63.	
Kernels, dry, hulled	1 oz.		16.9	1.	15.	
Meal, partially defatted	1 oz.		2.3	Tr.	2.	
SAFFRON (Spice Islands)	1 tsp.	<1				(0)
SAGE (Spice Islands)	1 tsp.	<1				(0)
SAINT JOHN'S-BREAD FLOUR (See **FLOUR,** Carob)						
SALAD DRESSING (See also **SALAD DRESSING, LOW CALORIE**):						
Avocado, refrigerated (Marzetti)	1 T. (.5 oz.)		8.2			10
(Bama)	1 T. (.5 oz.)	315	5.1			

(USDA): United States Department of Agriculture
*Prepared as Package Directs

Food and Description	Measure or Quantity	Sodium (mg.)	—Fats in grams— Total	Satu- rated	Unsatu- rated	Choles- terol (mg.)
Bennett's	1 T. (.5 oz.)	127	4.7			
Blendaise (Marzetti)	1 T. (.5 oz.)		5.5			7
Bleu or blue cheese:						
(USDA)[1]	1 oz.	310	14.8	3.	12.	
(USDA)[1]	1 T. (.5 oz.)	164	7.8	2.	6.	
(Bernstein's) Danish	1 T. (.5 oz.)	115	4.4	<1.	4.	
(Kraft) Imperial	1 T. (.5 oz.)	138	7.1			
(Kraft) refrigerated	1 T. (.5 oz.)	171	7.6			
(Kraft) Roka	1 T. (.5 oz.)	177	5.4			
(Lawry's)	1 T. (.5 oz.)		5.8			
(Marzetti)	1 T. (.5 oz.)		7.0			7
(Marzetti) refrigerated	1 T. (.5 oz.)		8.6			9
(Wish-Bone) chunky	1 T. (.5 oz.)	149	7.7	1.	6.	1
Boiled, home recipe (USDA)[2]	1 T. (.6 oz.)	116	1.6	<1.	<1.	12
Caesar (Kraft) Golden	1 T. (.5 oz.)	170	6.7			
Caesar (Kraft) Imperial	1 T. (.5 oz.)	164	8.2			
Caesar (Lawry's)	1 T. (.5 oz.)		7.4			
Canadian (Lawry's)	1 T. (.5 oz.)		7.5			
Coleslaw (Bernstein's)	1 T. (.5 oz.)	188	4.4	<1.	4.	
Coleslaw (Kraft)	1 T. (.5 oz.)	187	5.5			
French:						
Home recipe with corn oil (USDA)[3]	1 T. (.6 oz.)	105	11.2	1.	10.	
Home recipe with cottonseed oil (USDA)[4]	1 T. (.6 oz.)	105	11.2	3.	8.	
Commercial (USDA)[5]	1 T. (.6 oz.)	219	6.2	1.	5.	
(Bennett's)	1 T. (.5 oz.)	217	7.2			0
(Bernstein's)	1 T. (.5 oz.)	224	5.4	<1.	5.	
(Bernstein's) New Orleans	1 T. (.5 oz.)	292	5.4	<1.	5.	
(Hellmann's) Family[6]	1 T. (.6 oz.)	272	6.0	<1.	5.	2
(Kraft)	1 T. (.5 oz.)	231	6.5			
(Kraft) Casino	1 T. (.5 oz.)	180	5.5			
(Kraft) Catalina	1 T. (.5 oz.)	172	5.2			
(Kraft) *Miracle*	1 T. (.5 oz.)	303	5.3			
(Lawry's)	1 T. (.5 oz.)		5.8			
(Lawry's) San Francisco	1 T. (.5 oz.)		5.4			
(Marzetti) Blue, refrigerated	1 T. (.5 oz.)		6.1			6
(Marzetti) Country	1 T. (.5 oz.)		6.3			5

(USDA): United States Department of Agriculture
*Prepared as Package Directs
[1]Principal sources of fat: soybean oil, cottonseed oil, corn oil & cheese.
[2]Principal sources of fat: butter, milk & egg.
[3]Principal source of fat: corn oil.
[4]Principal source of fat: cottonseed oil.
[5]Principal sources of fat: soybean oil, cottonseed oil & corn oil.
[6]Principal sources of fat: vegetable oil & egg yolk.

Food and Description	Measure or Quantity	Sodium (mg.)	Fats in grams — Total	Satu- rated	Unsatu- rated	Choles- terol (mg.)
(Nalley's)	.5 oz.		.5			
(Wish-Bone)	1 T. (.6 oz.)	163	1.1	Tr.	<1.	
(Wish-Bone) Deluxe	1 T. (.6 oz.)	83	5.5	<1.	5.	
(Wish-Bone) Garlic	1 T. (.6 oz.)	317	6.0	<1.	5.	
Fruit (Kraft)	1 T. (.5 oz.)	103	4.6			
Garlic:						
French (Hellmann's) *Old Home-stead*[1]	1 T. (.6 oz.)	240	6.1	1.	5.	5
(Marzetti) creamy	1 T. (.5 oz.)		7.2			4
(Marzetti) creamy, refrigerated	1 T. (.5 oz.)		9.0			9
(Wish-Bone) creamy	1 T. (.5 oz.)	174	8.1	1.	7.	
German Style (Marzetti)	1 T. (.5 oz.)		5.0			0
Green Goddess:						
(Bernstein's)	1 T. (.5 oz.)	225	4.4	<1.	4.	
(Kraft)	1 T. (.5 oz.)	138	8.2			
(Kraft) Imperial	1 T. (.5 oz.)	64	9.4			
(Lawry's)	1 T. (.5 oz.)		6.1			
(Wish-Bone)	1 T. (.5 oz.)	150	7.0	1.	6.	<1
Green onion (Kraft)	1 T. (.5 oz.)	152	7.6			
Hawaiian (Lawry's)	1 T. (.6 oz.)		5.5			
Herb & garlic (Kraft)	1 T. (.5 oz.)	147	9.7			
Italian:						
(USDA)[2]	1 T. (.5 oz.)	314	9.0	2.	8.	
(Bernstein's)	1 T. (.5 oz.)	184	6.4	1.	5.	
(Hellmann's) True[3]	1 T. (.5 oz.)	315	9.0	1.	8.	0
(Kraft)	1 T. (.5 oz.)	222	9.7			
(Lawry's)	1 T. (.5 oz.)		8.1			
(Lawry's) with cheese	1 T. (.5 oz.)		4.5			
(Marzetti) creamy	1 T. (.5 oz.)		7.3			4
(Marzetti) Sunny	1 T. (.5 oz.)		8.8			0
(Wish-Bone)	1 T. (.5 oz.)	362	8.1	1.	7.	
(Wish-Bone) Rosé	1 T. (.5 oz.)	317	6.3	<1.	5.	0
Mayonnaise (See **MAYONNAISE**)						
Mayonnaise-type salad dressing (USDA)[4]	1 T. (.5 oz.)	88	6.3	1.	5.	8
Mayonnaise, imitation (Healthlife)	.5 oz.	71	6.4	<1.	6.	0
Miracle Whip (Kraft)	1 T. (.5 oz.)	90	7.0			
Oil & vinegar (Kraft)	1 T. (.5 oz.)	231	7.1			
Onion, California (Wish-Bone)	1 T. (.5 oz.)	164	8.0	1.	7.	

(USDA): United States Department of Agriculture
*Prepared as Package Directs
[1]Principal sources of fat: vegetable oil & egg yolk.
[2]Principal sources of fat: soybean oil, cottonseed oil & corn oil.
[3]Principal source of fat: vegetable oil.
[4]Principal sources of fat: soybean oil, cottonseed oil, corn oil & egg.

Food and Description	Measure or Quantity	Sodium (mg.)	— Fats in grams —			Cholesterol (mg.)
			Total	Saturated	Unsaturated	
Potato salad (Marzetti)	1 T. (.5 oz.)		5.8			9
Ranch Style (Marzetti)	1 T. (.5 oz.)		5.3			0
Red wine vinegar & oil (Lawry's)	1 T. (.6 oz.)		4.6			
Rich 'n' Tangy (Dutch Pantry)	1 T. (.6 oz.)	188	6.0	<1.	5.	0
Romano Caesar (Marzetti)	1 T. (.5 oz.)		7.2			<1
Roquefort:						
(USDA)[1]	1 T. (.5 oz.)	164	7.8	2.	6.	
(Bernstein's)	1 T. (.5 oz.)	142	5.0	<1.	4.	
(Kraft) refrigerated	1 T. (.5 oz.)	172	5.7			
(Kraft) refrigerated, Imperial	1 T. (.5 oz.)	145	7.1			
(Marzetti) refrigerated	1 T. (.5 oz.)		8.3			9
Russian:						
(USDA)[2]	1 T. (.5 oz.)	130	7.6	1.	6.	
(Kraft)	1 T. (.5 oz.)	126	4.3			
(Kraft) creamy	1 T. (.5 oz.)	56	6.7			
(Marzetti) creamy	1 T. (.5 oz.)		7.4			6
(Wish-Bone)	1 T. (.5 oz.)	167	2.9	Tr.	2.	
(Saffola)	1 T. (.5 oz.)	88	4.8	Tr.	4.	3
Salad Bowl (Kraft)	1 T. (.5 oz.)	99	5.0			
Salad 'n Sandwich (Kraft)	1 T. (.5 oz.)	92	4.7			
Salad Secret (Kraft)	1 T. (.5 oz.)	245	5.3			
Sherry (Lawry's)	1 T. (.5 oz.)		5.3			
Slaw (Marzetti) regular or refrigerated[3]	1 T. (.5 oz.)		6.8			14
Spin Blend (Best Foods)	1 T. (.5 oz.)	100	5.0	<1.	4.	5
Spin Blend (Hellmann's)[3]	1 T. (.5 oz.)	100	5.0	<1.	4.	5
Sweet & Saucy (Marzetti)	1 T. (.5 oz.)		6.7			0
Sweet 'n' Sour (Dutch Pantry)	1 T. (.6 oz.)	63	7.1	2.	5.	0
Sweet & Sour (Kraft)	1 T. (.5 oz.)	64	.2			
Tahitian Isle (Wish-Bone)	1 T. (.5 oz.)	167	2.9	Tr.	2.	
Tang (Nalley's)			4.5			
Tart & Creamy (Bama)	1 T. (.5 oz.)	196	8.2			
Thousand Island:						
(USDA)[2]	1 T. (.6 oz.)	112	8.0	1.	7.	
(Bernstein's)	1 T. (.5 oz.)	127	4.6	<1.	4.	
(Best Foods)	1 T. (.5 oz.)	180	5.4	1.	4.	Tr.
(Kraft)	1 T. (.5 oz.)	99	7.1			
(Kraft) Imperial	1 T. (.5 oz.)	60	7.9			
(Kraft) pourable	1 T. (.5 oz.)	127	5.2			
(Kraft) refrigerated	1 T. (.5 oz.)	78	7.3			

(USDA): United States Department of Agriculture
*Prepared as Package Directs
[1]Principal sources of fat: soybean oil, cottonseed oil, corn oil & cheese.
[2]Principal sources of fat: soybean oil, cottonseed oil, corn oil & egg.
[3]Principal sources of fat: vegetable oil & egg yolk.

Food and Description	Measure or Quantity	Sodium (mg.)	Fats in grams — Total	Satu- rated	Unsatu- rated	Choles- terol (mg.)
(Lawry's)	1 T. (.5 oz.)		6.4			
(Marzetti)	1 T. (.5 oz.)		6.7			5
(Marzetti) refrigerated	1 T. (.5 oz.)		6.9			6
Tomato 'n' Spice (Dutch Pantry)	1 T. (.6 oz.)	200	6.1	<1.	5.	0
Vinaigrette (Bernstein's)	1 T. (.5 oz.)	180	4.2	<1.	4.	
(Wish-Bone)	1 T. (.5 oz.)	138	6.7	1.	6.	5

SALAD DRESSING, DIETETIC or LOW CALORIE:

Bleu or blue cheese:

Food and Description	Measure or Quantity	Sodium (mg.)	Total	Satu- rated	Unsatu- rated	Choles- terol (mg.)
Low fat, 6% fat (USDA)[1]	1 T. (.6 oz.)	177	.9	Tr.	<1.	
Low fat, 1% fat (USDA)	1 T. (.5 oz.)	170	.2			
(Frenchette) chunky	1 T. (.5 oz.)	286	1.4			4
(Kraft)	1 T. (.5 oz.)	286	.9			
(Marzetti)	1 T. (.5 oz.)		1.5			4
(Slim-ette)	1 T. (.5 oz.)		1.0			
Caesar (Frenchette)	1 T. (.5 oz.)		2.7			4
Catalina (Kraft)	1 T. (.5 oz.)	110	.5			
Cheese (Tillie Lewis)	1 T. (.5 oz.)		1.1			
Chef Style (Kraft)	1 T. (.5 oz.)	107	.6			
Chef's (Slim-ette)	1 T. (.5 oz.)		1.5			
Chef's (Tillie Lewis)	1 T. (.5 oz.)		Tr.			
Coleslaw (Kraft)	1 T. (.5 oz.)	165	1.6			
Diet Mayo 7 (Bennett's)	1 T. (.5 oz.)	164	1.5			
French:						
Low fat, 6% fat (USDA)[2]	1 T. (.6 oz.)	126	.7	Tr.	<1.	
Low fat, 1% fat, with artificial sweetener (USDA)	1 T. (.5 oz.)	118	<.1			
Medium fat, with artificial sweetener (USDA)[2]	1 T. (.5 oz.)	118	2.5	Tr.	2.	
(Bennett's)	1 T. (.5 oz.)	148	1.0			0
(Frenchette)	1 T. (.5 oz.)	138	<.1			0
(Kraft)	1 T. (.5 oz.)	267	1.5			
(Marzetti)	1 T. (.5 oz.)		Tr.			0
(Tillie Lewis)	1 T. (.5 oz.)		Tr.			
Gourmet (Frenchette)	1 T. (.5 oz.)	202	1.3			0
Green Goddess (Frenchette)	1 T. (.5 oz.)	254	1.5			6
Green Goddess (Slim-ette)	1 T. (.5 oz.)		1.0			
Italian:						
(USDA)[2]	1 T. (.5 oz.)	118	.7	Tr.	<1	
(Bennett's)	1 T. (.5 oz.)	288	.4			0

(USDA): United States Department of Agriculture
*Prepared as Package Directs
[1]Principal source of fat: cheese.
[2]Principal sources of fat: soybean oil, cottonseed oil & corn oil.

Food and Description	Measure or Quantity	Sodium (mg.)	—Fats in grams—			Choles- terol (mg.)
			Total	Satu- rated	Unsatu- rated	
(Bernstein's)	1 T. (.4 oz.)	209	<.1	Tr.	Tr.	
(Bernstein's) with cheese	1 T. (.4 oz.)	209	.2	Tr.	Tr.	
Italianette (Frenchette)	1 T. (.5 oz.)	300	.2			0
(Kraft)	1 T. (.5 oz.)	173	.8			
(Marzetti)	1 T. (.5 oz.)		.2			0
(Slim-ette)	1 T. (.5 oz.)		.6			
(Tillie Lewis)	1 T. (.5 oz.)		Tr.			
(Wish-Bone)	1 T. (.5 oz.)	193	1.5	Tr.	1.	
Mayonnaise, imitation:						
(USDA)[1]	1 T. (.6 oz.)	19	2.0	Tr.	2.	
May-Lo-Naise (Tillie Lewis)	1 T. (.5 oz.)		.9			2
Mayonette Gold (Frenchette)	1 T. (.5 oz.)	203	2.8			10
Remoulade (Tillie Lewis)	1 T. (.5 oz.)		.9			
Russian (Wish-Bone)	1 T. (.6 oz.)	162	.6	Tr.	Tr.	
Slaw (Frenchette)	1 T. (.5 oz.)		1.8			13
Slaw (Marzetti)	1 T. (.5 oz.)		1.8			13
Supreme (McCormick)	1 oz.		6.0			
Thousand Island:						
(USDA)[1]	1 T. (.5 oz.)	105	2.1	Tr.	2.	
(Frenchette)	1 T. (.5 oz.)	158	1.1			
(Kraft)	1 T. (.5 oz.)	133	2.1			
(Marzetti)	1 T. (.5 oz.)		1.0			7
(Wish-Bone)	1 T. (.5 oz.)	173	1.6	Tr.	1.	5
Vinaigrette (Bernstein's)	1 T. (.4 oz.)	132	<.1	Tr.	Tr.	
Whipped (Tillie Lewis)	1 T. (.5 oz.)		.9			2

SALAD DRESSING MIX, regular
& low calorie:

Food and Description	Measure or Quantity	Sodium (mg.)	Total	Satu- rated	Unsatu- rated	Choles- terol (mg.)
Bacon (Lawry's)	1 pkg. (.8 oz.)		1.0			
Bleu or blue cheese:						
*(Good Seasons)	1 T. (.5 oz.)	170	9.3			Tr.
*(Good Seasons) thick, creamy	1 T. (.5 oz.)	106	9.2			7
(Lawry's)	1 pkg. (.7 oz.)		5.0			
Caesar garlic cheese (Lawry's)	1 pkg. (.8 oz.)		2.4			
*Cheese garlic (Good Seasons)	1 T. (.5 oz.)	166	9.2			Tr.
*French, old fashion (Good Seasons)	1 T. (.5 oz.)	270	9.2			0
*French (Good Seasons) thick, creamy	1 T. (.5 oz.)	203	9.1			5
French, old fashion (Lawry's)	1 pkg. (.8 oz.)		<.1			
*French, Riviera (Good Seasons)	1 T. (.6 oz.)	285	9.2			0
*Garlic (Good Seasons)	1 T. (.5 oz.)	166	9.2			Tr.

(USDA): United States Department of Agriculture
*Prepared as Package Directs
[1]Principal sources of fat: soybean oil, cottonseed oil, corn oil & egg.

Food and Description	Measure or Quantity	Sodium (mg.)	Fats in grams Total	Satu- rated	Unsatu- rated	Choles- terol (mg.)
Green Goddess (Lawry's)	1 pkg. (.8 oz.)		.9			
Italian:						
*(Good Seasons)	1 T. (.5 oz.)	166	9.2			Tr.
*(Good Seasons) cheese	1 T. (.5 oz.)	170	9.3			Tr.
*(Good Seasons) mild	1 T. (.5 oz.)	170	9.3			Tr.
*(Good Seasons) thick, creamy	1 T. (.5 oz.)	183	9.2			6
(Lawry's)	1 pkg. (.6 oz.)		<.1			
(Lawry's) cheese	1 pkg. (.8 oz.)		2.3			
*Low calorie (Good Seasons)	1 tsp.	55	Tr.			0
*Onion (Good Seasons)	1 T. (.5 oz.)	166	9.2			Tr.
*Thousand Island (Good Seasons) thick, creamy	1 T. (.6 oz.)	98	7.7			5
SALAD HERBS (Spice Islands)	1 tsp.	<1				(0)
SALAD SEASONING:						
(Durkee)	1 tsp. (4 grams)	1151	.5			
Salad Mate (Durkee)	1 tsp. (4 grams)	1296	.6			
With cheese (Durkee)	1 tsp. (3 grams)	786	.7			
Salad Lift (French's)	1 tsp. (4 grams)	640	.1			
SALAMI:						
Dry (USDA)	1 oz.		10.8			
Cooked (USDA)	1 oz.		7.3			
Cotto, all meat (Oscar Mayer)	.8-oz. slice (10 per ½ lb.)	217	4.4	2.	3.	9
Cotto, pure beef (Oscar Mayer)	.8-oz. slice	251	4.4			
Dilusso Genoa (Hormel)	1 oz.	454	11.1			
For beer (Oscar Mayer)	.8-oz. slice	251	4.1			
Hard (Hormel) dairy	1 oz.	425	10.3	3.	5.	18
Hard, all meat (Oscar Mayer)	1 slice (.4 oz.)	177	3.6			
Machiaeh Brand, pure beef, cooked (Oscar Mayer)	.8-oz. slice	251	4.8			
SALISBURY STEAK:						
Frozen:						
(Banquet) buffet	2-lb. pkg.		103.1			
(Banquet) cooking bag	5-oz. bag		16.2			
(Morton House) & mushroom gravy	4¹⁄₆-oz. serving	512	11.0	5.	6.	38
(Swanson) *Hungry Man*	17-oz. dinner	1999	56.5			
(Swanson) with potato	6-oz. pkg.	775	18.9			

(USDA): United States Department of Agriculture
*Prepared as Package Directs

Food and Description	Measure or Quantity	Sodium (mg.)	—Fats in grams— Total	Satu- rated	Unsatu- rated	Choles- terol (mg.)
Dinner (Banquet):						
Meat compartment	6.3 oz.		17.2			
Potato compartment	2.8 oz.		.4			
Peas & carrots compartment	1.9 oz.		.8			
Complete dinner	11-oz. dinner		18.4			
Dinner (Morton)	11-oz. dinner	1000	22.5			
Dinner (Morton) 3-course	1-lb. 1-oz. dinner	1570	27.3			
Dinner (Swanson) 3-course	16-oz. dinner	1653	24.5			
SALMON:						
Atlantic (USDA):						
Raw, whole	1 lb. (weighed whole)		39.5			
Raw, meat only	4 oz.		15.2			
Canned, solids & liq., including bones	4 oz.		13.8			
Chinook or King (USDA):						
Raw, steak	1 lb. (weighed with bones)	180	62.3	19.	43.	
Raw, meat only	4 oz.	51	17.7	6.	12.	
Canned, solids & liq., including bones, no salt added	4 oz.	51	15.9	4.	11.	
Chum, raw, meat only (USDA)	4 oz.	60				
Chum, canned, solids & liq., including bones, no salt added (USDA)	4 oz.	60	5.9			
Coho, raw, meat only (USDA)	4 oz.	54				
Coho, raw, meat only, dipped in brine (USDA)	4 oz.	244				
Coho, canned, solids & liq., no salt added (USDA)	4 oz.	54	8.1			
Coho, canned, solids & liq., salt added (USDA)	4 oz.	398	8.1			
Pink or Humpback:						
Raw, steak (USDA)	1 lb. (weighed with bones)	255	14.8	4.	11.	
Raw, meat only (USDA)	4 oz.	73	4.2	1.	3.	
Raw, meat only, dipped in in brine (USDA)	4 oz.	536	4.2	1.	3.	
Canned, solids & liq.:						
No salt added (USDA)	4 oz.	73	6.7	2.	4.	
Salt added (USDA)	4 oz.	439	6.7	2.	4.	
(Del Monte)	7¾-oz. can	1371	9.2			
(Del Monte)	1 cup (8 oz.)	1414	9.5			

(USDA): United States Department of Agriculture
*Prepared as Package Directs

(291)

Food and Description	Measure or Quantity	Sodium (mg.)	Fats in grams Total	Satu- rated	Unsatu- rated	Choles- terol (mg.)
Sockeye or Red or Blueback:						
Raw, meat only (USDA)	4 oz.	54				40
Raw, steak (USDA)	1 lb.	192				141
Canned, solids & liq.:						
Including bones, no salt added (USDA)	4 oz.	54	10.5			40
Including bones, salt added (USDA)	4 oz.	592	10.5			40
(Del Monte)	7¾-oz. can	1217	16.7			
(Del Monte)	1 cup (8 oz.)	1255	17.2			
Unspecified kind of salmon, baked or broiled with vegetable shortening:						
(USDA)	4-oz. steak (approx. 4″ x 3″ x ½″, 4.2 oz.)	139	8.9			53
(USDA)	6¾″ x 2½″ x 1″ (5.1 oz.)	168	10.7			68
SALMON RICE LOAF, home recipe (USDA)	4 oz.		5.1			
SALMON, SMOKED:						
(USDA)	4 oz.		10.5			
Lox, drained (Vita)	4-oz. jar		6.7			
Nova, drained (Vita)	4-oz. can		14.6			
SALSIFY (USDA):						
Raw, without tops, freshly harvested	1 lb. (weighed untrimmed)		2.4			0
Raw, without tops, after storage	1 lb. (weighed untrimmed)		2.4			0
Boiled, drained, freshly harvested	4 oz.		.7			0
Boiled, drained, after storage	4 oz.		.7			0
SALT:						
Butter-flavored, imitation:						
(Durkee)	1 tsp. (5 grams)	1362	.3			
(French's)	1 tsp. (4 grams)	1090	.9			
Garlic (French's)	1 tsp. (6 grams)	1850	.1			
Garlic, parslied (French's)	1 tsp. (4 grams)	1050	.1			
Hickory smoke (French's)	1 tsp. (4 grams)	1170	Tr.			
Garlic (Lawry's)	2.9-oz. pkg.		.3			

(USDA): United States Department of Agriculture
*Prepared as Package Directs

Food and Description	Measure or Quantity	Sodium (mg.)	— Fats in grams —			Choles- terol (mg.)
			Total	Satu- rated	Unsatu- rated	
Garlic (Lawry's)	1 tsp. (4 grams)		Tr.			
Lite Salt (Morton)	1 tsp. (6 grams)	1188	0.			
Onion (French's)	1 tsp. (5 grams)	1620	.1			
Onion (Lawry's)	3-oz. pkg.		1.0			
Onion (Lawry's)	1 tsp. (3 grams)		<.1			
Seasoning (French's)	1 tsp. (5 grams)	1620	.1			
Seasoned (Lawry's)	3-oz. pkg.		.5			
Seasoned (Lawry's)	1 tsp. (5 grams)		<.1			
Substitute (Adolph's)	1 tsp. (4 grams)	<1	0.			0
Substitute (Morton)	1 tsp. (6 grams)	<1	0.			
Substitute, seasoned (Adolph's)	1 tsp. (4 grams)	<1	<.1	Tr.	Tr.	0
Substitute, seasoned (Morton)	1 tsp. (6 grams)	<1	0.			
Table (USDA)	1 tsp. (6 grams)	2325	0.			0
Table (Morton)	1 tsp. (6 grams)	2544	0.			(0)

SALT PORK, raw (USDA):
With skin	1 lb. (weighed with skin)	5278	370.0	141.	229.	
Without skin	1 oz.	344	24.1	9.	15.	

SALT STICK (See **BREAD STICK**)

SAND DAB, raw (USDA):
Whole	1 lb. (weighed whole)	117	1.2			
Meat only	4 oz.	88	.9			

SANDWICH SPREAD:
(USDA)	1 cup (8.7 oz.)	1540	89.1			
(USDA)	1 T. (.5 oz.)	94	5.4			
(USDA) low calorie	1 T. (.5 oz.)	94	1.4			
(Bama)	1 T. (.5 oz.)	238	4.2			
(Bennett's)[1]	1 T. (.5 oz.)	139	3.2			5
(Best Foods)	1 T. (.5 oz.)	185	5.8	<1.	5.	2
(Hellmann's)[1]	1 T. (.5 oz.)	185	5.8	<1.	5.	2
(Kraft)	1 oz.	183	9.4			
(Kraft) *Salad Bowl*	1 oz.	183	8.9			
(Nalley's)	1 oz.		7.9			
(Oscar Mayer)	1 oz.	243	3.4			
Chicken salad (Carnation)[2]	1/5 can (1.5 oz.)	164	7.0	<1.	6.	12

(USDA): United States Department of Agriculture
*Prepared as Package Directs
[1]Principal source of fat: oil.
[2]Principal sources of fat: chicken & corn oil.

293

Food and Description	Measure or Quantity	Sodium (mg.)	— Fats in grams —			Choles- terol (mg.)
			Total	Satu- rated	Unsatu- rated	
Corned beef (Carnation)[1]	¹/₃ can (1.5 oz.)	277	6.5	1.	5.	13
Ham & cheese (Carnation)[2]	¹/₃ can (1.5 oz.)	280	4.7	2.	3.	7
Ham salad (Carnation)[3]	¹/₃ can (1.5 oz.)	265	6.3	1.	5.	8
Pimento (Kraft)	1 oz.	231	11.3			
Tuna salad (Carnation)[4]	¹/₃ can (1.5 oz.)	141	5.3	<1.	5.	14
Turkey salad (Carnation)[5]	¹/₃ can (1.5 oz.)	156	6.0	<1.	5.	13
SANGRIA MIX (Party Tyme)	½-oz. pkg.	<1	0.			(0)
SAPODILLA, fresh (USDA):						
Whole	1 lb. (weighed with skin & seeds)	44	4.0			0
Flesh only	4 oz.	14	1.2			0
SAPOTE or MARMALADE PLUM, fresh (USDA):						
Whole	1 lb. (weighed with skin & seeds)		2.1			0
Flesh only	4 oz.		.7			0
SARDINE:						
Atlantic, canned in oil (USDA):						
Solids & liq.	3¾-oz. can	541	25.9			127
Drained solids	3¾-oz. can	757	10.2			129
Atlantic, canned in tomato sauce, solids & liq. (Del Monte)	1½ large sardines	321	6.8			·
Norwegian, canned:						
(Snow)	1 oz.		6.3			
In mustard sauce (Underwood)	3¾-oz. can	837	14.2			
In oil, drained (Underwood)	3¾-oz. can	182	16.0			
In tomato sauce (Underwood)	3¾-oz. can	422	9.8			
Pacific (USDA):						
Raw	4 oz.		9.8			
Canned, in brine or mustard, solids & liq.	4 oz.	862	13.6			
Canned in tomato sauce, solids & liq.	4 oz.	454	13.8			

(USDA): United States Department of Agriculture
*Prepared as Package Directs
[1]Principal sources of fat: beef & corn oil.
[2]Principal sources of fat: ham, milk & cottonseed oil.
[3]Principal sources of fat: ham & corn oil.
[4]Principal sources of fat: tuna & corn oil.
[5]Principal sources of fat: turkey & corn oil.

Food and Description	Measure or Quantity	Sodium (mg.)	—Fats in grams—			Choles- terol (mg.)
			Total	Satu- rated	Unsatu- rated	

SARSAPARILLA SOFT DRINK,
sweetened:

(Hoffman)	6 fl. oz.	3	0.			0
(Yukon Club)	6 fl. oz.	3	0.			0

SAUCE, regular & dietetic:

A1	1 T. (.6 oz.)	278	Tr.			
Barbecue:						
(USDA)	½ cup (4.4 oz.)	1019	8.6	1.	7.	
(USDA)	1 T. (.6 oz.)	130	1.1	Tr.	<1.	
(Contadina) oven	1 fl. oz. (1.2 oz.)	105	Tr.			0
(French's)	1 T.	190	Tr.			
(French's) mild	1 T.	165	Tr.			
(French's) smoky	1 T.	210	Tr.			
(General Foods) hickory smoke, *Open Pit*	1 T. (.6 oz.)	258	.2			0
(General Foods) hot 'n spicy, *Open Pit*	1 T. (.6 oz.)	211	.2			0
(General Foods) original, *Open Pit*	1 T. (.6 oz.)	224	.2			0
(General Foods) original flavor, with onions, *Open Pit*	1 T. (6. oz.)	226	.2			0
(Heinz) with onions, regular	1 T.	196	.1			
(Heinz) with onions, hickory smoke	1 T.	197	.1			
(Kraft)	1 oz.	466	.5			
(Kraft) garlic	1 oz.	484	.3			
(Kraft) hickory smoke	1 oz.	466	.5			
(Kraft) hot	1 oz.	572	.6			
(Kraft) mustard-flavored	1 oz.	399	.6			
(Kraft) onion	1 oz.	371	.7			
Cheese (Kraft) *Deluxe Dinner*	1 oz.	510	5.8			
Chili (See **CHILI SAUCE**)						
Creole (Contadina)[1]	1 fl. oz. (1.1 oz.)	138	.3	Tr.	Tr.	0
Escoffier Sauce Diable	1 T. (.6 oz.)	49	.2			
Escoffier Sauce Robert	1 T. (.6 oz.)	62	Tr.			
Famous (Durkee)	1 T. (6½-oz. bottle, .5 oz.)	435	6.4	2.	5.	
Famous (Durkee)	1 T. (10-oz. bottle, .5 oz.)	435	6.1	1.	5.	
57 (Heinz)	1 T.	265	.2			
Hard (Crosse & Blackwell)	1 T. (.5 oz.)		3.4			

(USDA): United States Department of Agriculture
*Prepared as Package Directs
[1]Principal sources of fat: soybean oil & olive oil.

295

Food and Description	Measure or Quantity	Sodium (mg.)	—Fats in grams—			Cholesterol (mg.)
			Total	Saturated	Unsaturated	
Horseradish (Marzetti)	1 T. (.5 oz.)		5.4			8
H.P. Steak Sauce (Lea & Perrins)	1 T. (1 oz.)	280	Tr.	0.	Tr.	0
Marinara (Buitoni)	4 oz.		2.5			
Marinara (Chef Boy-Ar-Dee)	¼ of 15-oz. can	912	2.0			
Meat loaf (Contadina)[1]	1 fl. oz. (1.1 oz.)	237	Tr.	Tr.	Tr.	Tr.
Mushroom (Contadina)[2]	1 fl. oz. (1.1 oz.)	138	1.2	1.	0.	0
Mushroom steak, *Dawn Fresh*	5¾-oz. can	823	0.			
Newburg, canned (Snow)	4 oz.		9.1			
Savory (Heinz)	1T.	162	.1			
Seafood (Bernstein's)	1 T. (.5 oz.)	172	Tr.			
Seafood cocktail (Crosse & Blackwell)	1 T. (.6 oz.)	204	0.			
Seafood cocktail (Del Monte)	1 T. (.6 oz.)	261	<.1			
Sloppy Joe (Contadina)	1 fl. oz. (1.1 oz.)	212	Tr.			
Sloppy Joe, chili (Contadina)	1 fl. oz. (1.1 oz.)	188	Tr.			
Sloppy Joe, pizza (Contadina)[1]	1 fl. oz. (1.1 oz.)	204	.2	Tr.	Tr.	<1
Soy (USDA)	1 oz.	2077	.4			
Spaghetti (See **SPAGHETTI SAUCE**)						
Steak (Crosse & Blackwell)	1 T. (.7 oz.)	276				
Steak (Marzetti)	1 T. (.5 oz.)		.2			0
Stroganoff (Contadina)	1 fl. oz. (1.1 oz.)	148	.7			
Sweet 'n sour (Contadina)[3]	1 fl. oz. (1.2 oz.)	88	.6	Tr.	<1.	0
Sweet & sour (Kraft)	1 oz.	128	.5			
Swiss steak (Contadina)	1 fl. oz. (1.1 oz.)	133	.1			0
Tartar:						
(USDA) regular	1 T. (.5 oz.)	99	8.1			7
(USDA) low calorie	1 T. (.5 oz.)	99	3.1			
(Bama)	1 T. (.5 oz.)	162	9.6			
(Bennett's)	1 T. (.5 oz.)	103	8.5			12
(Best Foods)[4]	1 T. (.5 oz.)	182	8.0	1.	7.	4
(Hellmann's)[4]	1 T. (.5 oz.)	182	8.0	1.	7.	4
(Kraft)	1 oz.	401	15.6			
(Marzetti)	1 T. (.5 oz.)		7.3			6
(Mrs. Paul's)	4.2-oz. pkg.		70.2			
Tomato (See **TOMATO SAUCE**)						
White, home recipe (USDA):						
Thin[5]	1 cup (8.8 oz.)	878	21.8	12.	9.	36
Medium[5]	1 cup (9 oz.)	966	31.9	18.	14.	36
Thick[5]	1 cup (8.7 oz.)	986	38.5	22.	16.	30

(USDA): United States Department of Agriculture
*Prepared as Package Directs
[1]Principal source of fat: milk.
[2]Principal source of fat: modified coconut oil.
[3]Principal source of fat: soybean oil.
[4]Principal source of fat: oils.
[5]Principal sources of fat: butter & milk.

Food and Description	Measure or Quantity	Sodium (mg.)	—Fats in grams—			Choles-terol (mg.)
			Total	Satu-rated	Unsatu-rated	
Worcestershire:						
(Crosse & Blackwell)	1 T.	265	0.			
(French's)	1 T.	150	Tr.			
(Heinz)	1 T.	234	Tr.			
(Lea & Perrins)	1 T. (.6 oz.)	175	Tr.	0.	Tr.	0
SAUCE MIX:						
*A la King (Durkee)	1 cup (1½-oz. pkg.)	1799	17.5			
*Barbecue (Kraft)	1 oz.	550	.2			
*Bordelaise (Betty Crocker)	¼ cup	280	2.0			
Cheese:						
*(Betty Crocker)	¼ cup	678	9.2			
*(Durkee)	1 cup (1.1-oz. pkg.)	1978	20.7			
(French's)	1¼-oz. pkg.	1150	9.3			
*(French's)	¼ cup	313	4.6			
*Cheddar (Kraft)	1 oz.	107	3.9			
(McCormick)	1¼-oz. pkg.	685	12.6			
*McCormick)	2-oz. serving	200	5.0			
Hollandaise:						
*(Betty Crocker)	¼ cup	308	7.2			
*(Durkee)	⅔ cup (1¼-oz. pkg.)	63	18.0			
(French's)	1⅛-oz. pkg.	1000	16.3			
*(French's)	1 T.	83	1.4			
*(Kraft)	1 oz.	49	4.8			
*(McCormick)	2-oz. serving	140	6.2			
Miracle (Mrs. Paul's)	½-oz. pkg.		<.1			
*Mushroom (Betty Crocker)	¼ cup	236	1.7			
*Newburg (Betty Crocker)	¼ cup	399	3.8			
Sloppy Joe (See **SLOPPY JOE MIX**)						
Sour Cream:						
*(Durkee)	⅔ cup (1.1-oz. pkg.)	954	16.0			
(French's)	1¼-oz. pkg.	300	15.0			
*(French's)	1 T.	34	1.8			
*(Kraft)	1 oz.	397	4.3			
*(McCormick)	2-oz. serving	131	.5			
Spaghetti (See **SPAGHETTI SAUCE MIX**)						
Stroganoff (French's)	1¾-oz. pkg.	1300	9.2			
*Stroganoff (French's)	⅓ cup	387	5.5			
*Sweet-sour (Durkee)	1 cup (2-oz. pkg.)	1053	5.7			
Tartar (Lawry's)	.6-oz. pkg.		1.4			
*Teri-yaki (Durkee)	⅔ cup (1¼-oz. pkg.)	3484	.1			

(USDA): United States Department of Agriculture
*Prepared as Package Directs

Food and Description	Measure or Quantity	Sodium (mg.)	Total	Satu- rated	Unsatu- rated	Choles- terol (mg.)
			— Fats in grams —			
White:						
*(Durkee)	1 cup (1-oz. pkg.)	799	19.7			
*(Kraft)	1 oz.	94	3.3			
*Supreme (McCormick)	2-oz. serving	268	1.5			
SAUERKRAUT, canned:						
Solids & liq. (USDA) 1.9% salt	1 cup (8.3 oz.)	1755	.5			0
Drained solids (USDA)	1 cup (5 oz.)		.4			0
Solids & liq. (Del Monte)	1 cup (8 oz.)	1655	.4			0
Solids & liq. (Stokely-Van Camp)	1 cup (7.8 oz.)		.4			
SAUERKRAUT JUICE, canned						
(USDA) 2% salt	½ cup (4.3 oz.)	952	Tr.			0
SAUGER, raw (USDA):						
Whole	1 lb. (weighed whole)		13.2			
Meat only	4 oz.		.9			
SAUSAGE (See also individual kinds):						
Breakfast (Hormel)	8-oz. can		80.6			
Breakfast, smoked, all meat						
(Oscar Mayer)	1 link (7 to 5-oz. pkg.)	172	6.0			
Brown & serve:						
Before browning (USDA)	1 oz.		10.2			
After browning (USDA)	1 oz.		10.7			
(Hormel)	1 piece (.8 oz.)	209	7.1	2.	4.	12
After browning (Swift)	1 link (.8 oz.)	193				7.9
New England Brand, all meat						
(Oscar Mayer)	.8-oz. slice	260	2.1			
In sauce, canned (Prince)	3.7-oz. can		14.2			
SAUTERNES:						
(Barton & Guestier) French white						
Bordeaux, 13% alcohol	3 fl. oz.	3	0.			(0)
(Gold Seal) dry, 12% alcohol	3 fl. oz.	3	0.			(0)
(Gold Seal) semi-soft, 12% alcohol	3 fl. oz. (3.2 oz.)	3	0.			(0)
(Great Western) Aurora, 12.5% alcohol	3 fl. oz.	34	0.			0
SAVORY (Spice Islands)	1 tsp.	<1				(0)

(USDA): United States Department of Agriculture
*Prepared as Package Directs

Food and Description	Measure or Quantity	Sodium (mg.)	—Fats in grams—			Choles- terol (mg.)
			Total	Satu- rated	Unsatu- rated	
SCALLION (See **ONION, GREEN**)						
SCALLOP:						
Raw, muscle only[1] (USDA)	4 oz.	289	.2			40
Steamed (USDA)	4 oz.	301	1.6			60
Frozen:						
Breaded, fried, reheated (USDA)	4 oz.		9.5			
Breaded, fried (Mrs. Paul's)	7-oz. pkg.		14.0			
Crisps (Gorton)	½ of 7-oz. pkg.	175	7.0			
SCHAV SOUP (Manischewitz)	8 oz. (by wt.)		.2			
***SCOTCH BROTH,** canned (Campbell)	1 cup	1009	2.7	1.	1.	
SCOTCH-COMFORT	1 fl. oz.	Tr.	0.			(0)
SCOTCH WHISKY (See **DISTILLED LIQUOR**)						
SCRAPPLE:						
(USDA)	4 oz.		15.4			
(Oscar Mayer)	4 oz.		14.7			
SCREWDRIVER:						
Canned (National Distillers)						
Duet, 12½% alcohol	8-fl.-oz. can	Tr.	0.			0
Dry mix (Bar-Tender's)	1 serving (⅝ oz.)	57	.2			(0)
SCUP (See **PORGY**)						
SEABASS, WHITE, raw, meat only (USDA)	4 oz.		.6			
SEAFOOD CHOWDER, New England (Snow)	8 oz.		5.6			
SEAFOOD PLATTER, breaded, fried, with potato puffs, frozen (Mrs. Paul's)	9-oz. pkg.		21.4			
SEAFOOD SEASONING (French's)	1 tsp. (5 grams)	1410	Tr.			(0)

(USDA): United States Department of Agriculture
*Prepared as Package Directs
[1]Frozen scallops, possibly brined.

Food and Description	Measure or Quantity	Sodium (mg.)	— Fats in grams —			Cholesterol (mg.)
			Total	Saturated	Unsaturated	
SENEGALESE SOUP (Crosse & Blackwell)	½ can (6½oz.)		2.0			
SESAME SEED:						
Dry, whole (USDA)	1 oz.	17	13.9	2.	12.	0
Dry, hulled (USDA)	1 oz.		15.1	2.	13.	0
Hulled (Spice Islands)	1 tsp.	2				(0)
Liquid, Tahini (A. Sahadi)	1 T. (8 grams)		5.1			
SEVEN-UP, soft drink:						
Regular	6 fl. oz.	16+	0.			0
Low calorie	6 fl. oz. (6.3 oz.)	19+	0.			0
SHAD (USDA):						
Raw, whole	1 lb. (weighed whole)	118	21.8			
Raw, meat only	4 oz.	61	11.3			
Cooked, home recipe:						
Baked with butter or margarine & bacon slices	4 oz.	90	12.8			
Creole, made with tomatoes, onion, green pepper, butter & flour	4 oz.	83	9.9			
Canned, solids & liq.	4 oz.		10.0			
SHAD, GIZZARD, raw (USDA):						
Whole	1 lb. (weighed whole)		21.0			
Meat only	4 oz.		15.9			
SHAKE 'N BAKE, seasoned mixes:						
Chicken-coating, regular	2⅜-oz. pkg.	2557	10.2			0
Chicken-coating, Italian	2⅜-oz. pkg.	2557	10.2			3
Fish-coating	2-oz. pkg.	2486	7.8			0
Hamburger-coating	2-oz. pkg.	4889	.8			0
Pork-coating	2⅜-oz. pkg.	2863	4.3			0
SHALLOT, raw (USDA):						
With skin	1 oz.	3	<.1			0
With skin removed	1 oz.	3	<.1			0

SHEEFISH (See **INCONNU**)

(USDA): United States Department of Agriculture
*Prepared as Package Directs

Food and Description	Measure or Quantity	Sodium (mg.)	— Fats in grams —			Choles- terol (mg.)
			Total	Satu- rated	Unsatu- rated	
SHEEPSHEAD, Atlantic, raw (USDA):						
Whole	1 lb. (weighed whole)	142	3.9			
Meat only	4 oz.	115	3.2			
SHERBET (See also individual brands):						
Orange (USDA)	1 cup (6.8 oz.)	19	2.3			
Orange (USDA)	¼ pint (3.4 oz.)	10	1.2			
Any flavor (Borden)	¼ pint (3 oz.)		1.3			
Orange (Sealtest)	¼ pint (3.1 oz.)	29	1.0			
SHERRY:						
(Great Western) Solera, 18% alcohol	3 fl. oz.	34	0.			0
Cocktail (Gold Seal) 19% alcohol	3 fl. oz. (3.1 oz.)	3	0.			(0)
Cooking (Great Western) 18% alcohol	3 fl. oz.	31	0.			0
Cream (Gold Seal) 19% alcohol	3 fl. oz. (3.3 oz.)	3	0.			(0)
Cream (Great Western) 18% alcohol	3 fl. oz.	32	0.			0
Dry (Great Western) Solera, 18% alcohol	3 fl. oz.	34	0.			0
SHORTENING (See **FATS**)						
SHREDDED OATS, cereal (USDA)[1]	1 oz.	173	.6	Tr.	<1.	0
SHREDDED WHEAT, cereal:						
(USDA) plain, without salt	1 cup (1.2 oz.)	1	.7	Tr.	<1.	0
(USDA) with malt, salt & sugar	1 cup (2.1 oz.)	418	1.7	Tr.	2.	0
(Kellogg's) cinnamon or sugar- frosted, *Mini-Wheats*	4 biscuits (1 oz.)	4	.3			(0)
(Nabisco)	1 biscuit (.9 oz.)	<1	.5			(0)
(Nabisco) *Spoon Size*	⅔ cup (1 oz.)	<1	.5			(0)
(Nabisco) *Spoon Size*	1 piece (1 gram)	Tr.	<.1			(0)
(Quaker)	2 biscuits (1⅓ oz.)	2	.7			(0)
SHRIMP:						
Raw (USDA):						
Whole	1 lb. (weighed in shell)	438	2.5			470

(USDA): United States Department of Agriculture
*Prepared as Package Directs
[1]Includes protein & other added nutrients.

Food and Description	Measure or Quantity	Sodium (mg.)	Fats in grams — Total	Satu- rated	Unsatu- rated	Choles- terol (mg.)
Meat only	4 oz.	159	.9			170
Canned, dry pack or drained						
(USDA)	1 cup (22 large or 76 small, 4.5 oz.')		1.4			192
Cooked, french-fried[1] (USDA)	4 oz.	211	12.2			
Frozen:						
Raw:						
Breaded, not more than 50% breading (USDA)	4 oz.		.8			
Breaded (Gorton)	¼ of 1-lb. pkg.	80	.8			
Cooked (Sau-Sea)	4 oz.		.5			
Cooked (Weight Watchers)	½ pkg. (4 oz.)		1.0			
Fried (Mrs. Paul's)	4 oz.		13.1			
Scampi (Gorton)	½ of 7½-oz. pkg.	415	23.0			
SHRIMP CAKE:						
Frozen, fried, breaded (Mrs. Paul's)	1 cake (3 oz.)		1.9			
Frozen, thins (Mrs. Paul's)	10-oz. pkg.		29.5			
SHRIMP COCKTAIL:						
(Sau-Sea)	4-oz. jar		.6			
(Sea Snack)	4-oz. jar	3	1.7			
SHRIMP DINNER, frozen:						
(Morton)	7¾-oz. dinner	480	16.4			
(Swanson)	8-oz. dinner	1035	13.1			
SHRIMP PASTE, canned (USDA)	1 oz.		2.7			
SHRIMP PUFF, frozen (Durkee)	1 piece (.5 oz.)		4.3			
SHRIMP SOUP, Cream of:						
*Canned (Campbell)	1 cup (8 oz.)	988	10.4			
Canned (Crosse & Blackwell)	½ can (6½ oz.)		5.3			
Frozen:						
Condensed (USDA)	8 oz. (by wt.)	1950	22.5			
*Prepared with equal volume water (USDA)	1 cup (8.5 oz.)	1032	12.0			
*Prepared with equal volume milk (USDA)	1 cup (8.6 oz.)	1117	16.4			
SIMBA, soft drink	6 fl. oz.	12	Tr.			0

(USDA): United States Department of Agriculture
*Prepared as Package Directs
[1]Dipped in egg, bread crumbs & flour or in batter.

Food and Description	Measure or Quantity	Sodium (mg.)	Total	Fats in grams Satu-rated	Unsatu-rated	Choles-terol (mg.)
SKATE, raw, meat only (USDA)	4 oz.		.8			
SLENDER (Carnation):						
Dry:						
Chocolate[1]	1 pkg. (1 oz.)	Tr.	.5	Tr.	Tr.	Tr.
Chocolate malt[2]	1 pkg. (1 oz.)	Tr.	.6	Tr.	Tr.	1
Chocolate marshmallow[1]	1 pkg. (1 oz.)	Tr.	.7	Tr.	Tr.	Tr.
Coffee[3]	1 pkg. (1 oz.)	Tr.	.2	Tr.	Tr.	Tr.
Dutch chocolate[1]	1 pkg. (1 oz.)	Tr.	.9	<1.	Tr.	Tr.
Milk chocolate[1]	1 pkg. (1 oz.)	Tr.	.5	Tr.	Tr.	Tr.
Strawberry, wild[3]	1 pkg. (1 oz.)	Tr.	.2	Tr.	Tr.	Tr.
Vanilla, French	1 pkg. (1 oz.)	Tr.	.2	Tr.	Tr.	Tr.
Liquid:						
Butterscotch	10-fl.-oz. can	Tr.	5.0	<1.	Tr.	<1
Chocolate[4]	10-fl.-oz. can	Tr.	5.0	<1.	Tr.	Tr.
Chocolate fudge[4]	10-fl.-oz. can	Tr.	5.0	<1.	Tr.	Tr.
Chocolate malt[5]	10-fl.-oz. can	Tr.	5.0	1.	4.	<1
Chocolate marshmallow[4]	10-fl.-oz. can	Tr.	5.0	1.	4.	Tr.
Coffee	10-fl.-oz. can	Tr.	5.0	1.	4.	<1
Eggnog[6]	10-fl.-oz. can	Tr.	5.0	1.	4.	<1
Milk chocolate[4]	10-fl.-oz. can	Tr.	5.0	1.	4.	Tr.
Vanilla[6]	10-fl.-oz. can	Tr.	5.0	1.	4.	<1
SLIM JIM:						
Sausage	1 piece (½ oz.)	252	7.1			
Polish sausage, all beef	1 piece (1¾ oz.)	461	9.1			
SLOPPY JOE:						
Frozen (Banquet) cooking bag	5-oz. bag		16.4			
Mix, including seasoning mix:						
(Durkee)	1½-oz. pkg.	3512	.2			
*With tomato paste & meat (Durkee)	3 cups (1½-oz. pkg.)	3866	97.0			
(French's)	1½-oz. pkg.	3300	.3			
*(Kraft)	1 oz.	101	2.8			
(Lawry's)	1½-oz. pkg.		1.5			
*(Wyler's)	¾ cup		.5			

(USDA): United States Department of Agriculture
*Prepared as Package Directs
[1]Principal sources of fat: milk, cocoa & lecithin.
[2]Principal sources of fat: wort solids, milk, cocoa & lecithin.
[3]Principal sources of fat: milk & lecithin.
[4]Principal sources of fat: milk, cocoa & corn oil.
[5]Principal sources of fat: wort solids, milk, cocoa & corn oil.
[6]Principal sources of fat: milk & corn oil.

Food and Description	Measure or Quantity	Sodium (mg.)	—Fats in grams—			Cholesterol (mg.)
			Total	Saturated	Unsaturated	

Food and Description	Measure or Quantity	Sodium (mg.)	Total	Satu-rated	Unsatu-rated	Choles-terol (mg.)
SMELT, Atlantic, jack & bay (USDA):						
Raw, whole	1 lb. (weighed whole)					
Raw, meat only	4 oz.		2.4			
Canned, solids & liq.	4 oz.		15.3			
SMOKIE SAUSAGE:						
(Hormel)	1 piece (.8 oz.)	261	6.6	2.	4.	14
(Oscar Mayer):						
8 links per ¾ lb.	1 link (1.5 oz.)	360	11.6	4.	7.	18
7 links per 5 oz.	1 link (.7 oz.)	168	5.4	2.	3.	8
Cheese	1.5-oz. link	369	11.6			
Little Smokies, all meat	1 link (16 per 5 oz.)	76	2.7			
Smoky Snax	1 link (4 oz.)	1106	29.5			
(Wilson)	1 oz.	293	7.3			17
SMOKY SNAX SPREAD (Oscar Mayer)	1 oz.	287	9.4			
SNACK (See CRACKER, POP-CORN, POTATO CHIPS, etc.)						
SNAIL, raw:						
(USDA)	4 oz.		1.6			
Giant African (USDA)	4 oz.		1.6			
SNAPPER (See RED SNAPPER)						
SNO BALL (Hostess) 2 to pkg.	1 cake (1.5 oz.)	117	3.5			
SODA or SOFT DRINK (See individual kinds listed by flavor or brand name)						
SOFT SWIRL (Jell-O):						
*All flavors except chocolate	½ cup (3.8 oz.)	301	5.8			9
*Chocolate	½ cup (4 oz.)	302	8.0			9
SOLE:						
Raw, whole (USDA)	1 lb. (weighed whole)	117	1.2			
Raw, meat only (USDA)	4 oz.	88	.9			

(USDA): United States Department of Agriculture
*Prepared as Package Directs

Food and Description	Measure or Quantity	Sodium (mg.)	—Fats in grams—			Choles- terol (mg.)
			Total	Satu- rated	Unsatu- rated	
Frozen:						
(Gorton)	⅓ of 1-lb. pkg.	117	1.2			
Dinner (Weight Watchers)	18-oz. dinner		3.6			
& cauliflower, luncheon (Weight Watchers)	9½-oz. luncheon		5.4			
In lemon butter (Gorton)	⅓ of 9-oz. pkg.	173	11.0			
SORGHUM (USDA):						
Grain	1 oz.		.9	Tr.	<1.	0
SORREL (See **DOCK**)						
SOUP (See individual listing by kind)						
SOUP BASE:						
Beef (Wyler's) no salt added	1 tsp. (6 grams)	12	1.4			
Chicken (Wyler's) no salt added	1 tsp. (6 grams)	4	1.4			
SOURSOP, raw (USDA):						
Whole	1 lb. (weighed with skin & seeds)	43	.9			0
Flesh only	4 oz.	16	.3			0
SOUSE (USDA)[1]	1 oz.		3.8	1.	2.	
SOUTHERN COMFORT	1 fl. oz.	Tr.	0.			(0)
SOYBEAN (USDA):						
Young seeds:						
Raw	1 lb. (weighed in pods)		12.3	2.	10.	0
Cooked without salt, drained	4 oz.		5.8	1.	5.	0
Canned, solids & liq.	4 oz.	268	3.6	1.	2.	0
Canned, drained solids	4 oz.	268	5.7	1.	5.	0
Mature seeds, dry:						
Raw	1 lb.	23	80.3	12.	68.	0
Raw	1 cup (7.4 oz.)	10	37.2	6.	31.	0
Cooked without salt	4 oz.	2	6.5	1.	5.	0
Roasted:						
Unsalted, *Soy Town*	1 oz.	6	10.5	1.	9.	0
Sea-salted, *Soy Town*	1 oz.	249	10.5	1.	9.	0

(USDA): United States Department of Agriculture
*Prepared as Package Directs
[1]Principal source of fat: pork.

Food and Description	Measure or Quantity	Sodium (mg.)	Total	Saturated	Unsaturated	Cholesterol (mg.)
				—Fats in grams—		Choles-terol
SOYBEAN CURD or TOFU						
(USDA)	4.2-oz. cake (2¾" x 2½" x 1")	8	5.0	1.	4.	0
SOYBEAN FLOUR (See **FLOUR**)						
SOYBEAN GRITS, high fat						
(USDA)	1 cup (4.9 oz.)	1	16.7	3.	14.	0
SOYBEAN MILK (USDA):						
Fluid	4 oz.		1.7			0
Powder	1 oz.		5.8	<1.	5.	0
Sweetened:						
Liquid concentrate[1]	4 oz. (by wt.)	49	8.	1.	7.	0
Dry powder[2]	1 oz.	<1	6.6	3.	4.	0
SOYBEAN PROTEIN (USDA)	1 oz.	60	<.1			0
SOYBEAN PROTEINATE (USDA)	1 oz.	340	<.1			0
SOYBEAN SPROUT (See **BEAN SPROUT**)						
SOYNUT, unsalted, *Soy Ahoy*	1 oz.	6	10.2	1.	9.	0
SOY SAUCE (See **SAUCE**)						
SOY SPREAD, *Soy Town*	1 oz.	113	14.2			0
SPAGHETTI (USDA):						
Dry	1 oz.	<1	.3			0
Dry, broken	1 cup (2.5 oz.)	1	.9			0
Cooked:						
8-10 minutee' "al dente"	1 cup (5.1 oz.)	1	.7			0
8-10 minutes, "al dente"	4 oz.	1	.6			0
14-20 minutes, tender	1 cup (4.9 oz.)	1	.6			0
14-20 minutes, tender	4 oz.	1	.5			0
SPAGHETTI DINNER:						
*With meat balls (Chef Boy-Ar-Dee)	8¾-oz. pkg.	1451	5.7			

(USDA): United States Department of Agriculture
*Prepared as Package Directs
[1]Principal source of fat: soybean.
[2]Principal sources of fat: soybean, coconut oil & olive oil.

SPAGHETTI with MEAT SAUCE

| Food and Description | Measure or Quantity | Sodium (mg.) | —Fats in grams— | | | Cholesterol (mg.) |
			Total	Saturated	Unsaturated	
*With meat sauce (Chef Boy-Ar-Dee)	7-oz. pkg.	1818	4.6			
*With meat sauce (Kraft) *Deluxe*	4 oz.	423	3.4			
*With mushroom sauce (Chef Boy-Ar-Dee)	7-oz. pkg.	1819	3.2			
Frozen, with meat balls:						
(Banquet):						
Spaghetti compartment	7.1 oz.		12.6			
Apple compartment	2.5 oz.		.2			
Peas compartment	1.9 oz.		1.2			
Complete dinner	11.5-oz. dinner		14.0			
(Morton)	11-oz. dinner	1329	10.7			
(Swanson)	12-oz. dinner	1138	9.8			

SPAGHETTI & FRANKFURTERS in TOMATO SAUCE, canned:

Food and Description	Measure or Quantity	Sodium (mg.)	Total	Saturated	Unsaturated	Cholesterol (mg.)
SpaghettiO's (Franco-American)	1 cup	816	12.7			
(Heinz)	8½-oz. can	1475	16.9			

SPAGHETTI & GROUND BEEF in TOMATO SAUCE, canned:

Food and Description	Measure or Quantity	Sodium (mg.)	Total	Saturated	Unsaturated	Cholesterol (mg.)
(Buitoni)	8 oz.		14.9			
(Chef Boy-Ar-Dee)	½ of 15-oz. can	1202	6.2			
(Franco-American)	1 cup	1150	13.9			
(Nalley's)	8 oz.		7.8			

SPAGHETTI & MEATBALLS in TOMATO SAUCE:

Food and Description	Measure or Quantity	Sodium (mg.)	Total	Saturated	Unsaturated	Cholesterol (mg.)
Home recipe (USDA)[1]	1 cup (8.7 oz.)	1009	11.7	5.	7.	75
Canned:						
(USDA)[1]	1 cup (8.8 oz.)	1220	10.2	5.	5.	39
(Buitoni)	8 oz.		13.1			
(Chef Boy-Ar-Dee)	⅕ of 40-oz. can	1061	8.2			
(Franco-American)	1 cup	1107	13.6			
SpaghettiO's (Franco-American)	1 cup	1411	8.9			
(Hormel)	15-oz. can		16.6			
(Van Camp)	1 cup (7.8 oz.)		9.0			
Frozen (Buitoni)	8 oz.		9.1			

SPAGHETTI with MEAT SAUCE:

Food and Description	Measure or Quantity	Sodium (mg.)	Total	Saturated	Unsaturated	Cholesterol (mg.)
Canned (Heinz)	8½-oz. can	1246	6.8			

(USDA): United States Department of Agriculture
*Prepared as Package Directs
[1]Principal sources of fat: olive oil, pork, beef, cheese, egg, bread crumbs & milk.

Food and Description	Measure or Quantity	Sodium (mg.)	—Fats in grams—			Choles- terol (mg.)
			Total	Satu- rated	Unsatu- rated	
Frozen:						
(Banquet) cooking bag	8-oz. bag		15.1			
(Banquet) buffet	2-lb. pkg.		73.4			
(Banquet) entrée	8-oz. pkg.		14.4			
(Kraft)	12½-oz. pkg.	1143	11.7			
(Morton)	8-oz. casserole	937	13.1			
(Morton)	20-oz. casserole	2342	32.8			
SPAGHETTI MIX:						
*American style (Kraft)	4 oz.	318	1.9			
*Italian style (Kraft)	4 oz.	338	2.2			
SPAGHETTI SAUCE:						
Clam, red (Buitoni)	4 oz.		7.3			
Clam, white (Buitoni)	4 oz.		9.6			
Italian (Contadina)[1]	4 fl. oz. (4.4 oz.)	492	2.8	Tr.	2.	0
Italian, frozen (Celeste)	½ cup	740	2.4			
Meat:						
(Buitoni)	4 oz.		8.4			
(Chef Boy-Ar-Dee)	¼ of 15-oz. can	953	4.3			
(Heinz)	½ cup	754	3.8			
(Prince)	½ cup (4.9 oz.)	1030	5.0			
Meatball (Chef Boy-Ar-Dee)	⅓ of 15-oz. can	940	9.2			
With ground meat (Chef Boy-Ar-Dee)	¹/₇ of 29-oz. jar	1088	8.0			
Meatless or plain:						
(Buitoni)	4 oz.		2.7			
(Chef Boy-Ar-Dee)	¼ of 16-oz. jar	829	1.9			
(Heinz)	½ cup	770	2.7			
(Prince)	½ cup (4.6 oz.)	970	3.6			
Mushroom:						
(Buitoni)	4 oz.		2.7			
(Chef Boy-Ar-Dee)	¼ of 15-oz. can	974	1.8			
(Heinz)	½ cup	794	3.0			
Mushroom with meat (Heinz)	½ cup	822	3.9			
SPAGHETTI SAUCE MIX:						
*(Kraft)	4 oz.	572	2.4			
(McCormick)	1½-oz. pkg.					
*(McCormick)	4-oz. serving	878	3.0			
Italian (French's)	1½-oz. pkg.	3600	.4			
*Italian (French's)	⅝ cup	918	5.8			

(USDA): United States Department of Agriculture
*Prepared as Package Directs
[1]Principal source of fat: olive oil.

Food and Description	Measure or Quantity	Sodium (mg.)	Total	Fats in grams — Satu- rated	Unsatu- rated	Choles- terol (mg.)
*Prepared without oil (Spatini)	½ cup (4.2 oz.)	402	.2			0
*Prepared with oil (Spatini)	½ cup (4.2 oz.)	380	2.8			0
With meatballs (Lawry's)	3½-oz. pkg.		2.2			
With mushrooms (French's)	1⅜-oz. pkg.	3940	3.3			
*With mushrooms (French's)	⅝ cup	1003	3.8			
With mushrooms (Lawry's)	1½-oz. pkg.		3.3			
*With mushrooms & tomato paste (Durkee)	2⅔ cups (1.2-oz. pkg.)	3106	.8			
*Without meat (Durkee)	2½ cups (1½-oz. pkg.)	3937	1.0			
*(Wyler's)	¾ cup		.5			

SPAGHETTI WITH TOMATO SAUCE:

Twists (Buitoni)	8 oz.		1.7			
(Van Camp)	1 cup (7.8 oz.)		1.4			
With cheese:						
Home recipe (USDA)[1]	1 cup (8.8 oz.)	955	8.8	2.	6.	
Canned:						
(USDA)	1 cup (8.8 oz.)	955	1.5			
(Chef Boy-Ar-Dee)	⅕ of 40-oz. can	1283	1.4			
(Franco-American)	1 cup	925	1.6			
SpaghettiO's (Franco-American)	1 cup	1064	2.3			
Italian Style (Franco-American)	1 cup	982	2.0			
(Heinz)	8½-oz. can	1330	3.5			

SPAM (Hormel), canned:

Regular	3 oz.	1020	22.5	6.	12.	45
Spread	1 oz.	360	6.7	2.	4.	8
& cheese	3 oz.	1020	22.3			

SPANISH MACKEREL, raw (USDA):

Whole	1 lb. (weighed whole)	188	28.8			
Meat only	4 oz.	77	11.8			

SPANISH-STYLE VEGE-TABLES, frozen (Birds Eye) | ⅓ of 10-oz. pkg. | 339 | 6.6 | | | 0 |

SPEARMINT, dry (Spice Islands) | 1 tsp. | 2 | | | | (0) |

SPECIAL K, cereal (Kellogg's) | 1¼ cups (1 oz.) | 168 | .1 | | | |

(USDA): United States Department of Agriculture
*Prepared as Package Directs
[1]Principal sources of fat: olive oil & cheese.

Food and Description	Measure or Quantity	Sodium (mg.)	Fats in grams Total	Saturated	Unsaturated	Cholesterol j)mg.)
SPICE CAKE MIX:						
*(Duncan Hines)	$^1/_{12}$ of cake (2.7 oz.)	384	6.1			50
*Apple with raisins, layer (Betty Crocker)	$^1/_{12}$ ofcake	272	5.5			
*Layer (Betty Crocker)	$^1/_{12}$ of cake	261	5.6			
Honey:						
(USDA)[1]	1 oz.	106	4.0	1.	3.	
*Prepared with eggs, water, caramel icing (USDA)[2]	2 oz.	139	6.1	2.	4.	
SPICE PARISIENNE (Spice Islands)	1 tsp.	<1				(0)
SPINACH:						
Raw (USDA):						
Untrimmed	1 lb. (weighed with large stems & roots)	232	1.0			0
Trimmed or packaged	1 lb.	322	1.4			0
Trimmed, whole leaves	1 cup (1.2 oz.)	23	<.1			0
Trimmed, chopped	1 cup (1.8 oz.)	37	.2			0
Boiled without salt, whole leaves, drained (USDA)	1 cup (5.5 oz.)	78	.5			0
Canned, regular pack:						
Solids & liq. (USDA)	½ cup (4.1 oz.)	274	.5			0
Drained solids (USDA)	½ cup (4 oz.)	264	.7			0
Drained liq. (USDA)	4 oz.	268	0.			0
Drained solids (Del Monte)	½ cup (4 oz.)	389	.6			0
Solids & liq. (Stokely-Van Camp)	½ cup (3.9 oz.)		.5			(0)
Canned, dietetic pack, low-sodium:						
Solids & liq. (USDA)	4 oz.	39	.5			0
Drained solids (USDA)	4 oz.	36	.6			0
Drained liq. (USDA)	4 oz.	36	0.			0
Solids & liq. (Blue Boy)	4 oz.	39	.4			(0)
Frozen:						
Chopped:						
Not thawed (USDA)	4 oz.	65	.3			0
Boiled, drained (USDA)	4 oz.	59	.3			0
(Birds Eye)	⅓ of 10. oz. pkg.	86	.3			0
Deviled, with cheddar cheese, casserole (Green Giant)	⅓ of 10-oz. pkg.	387	3.3			
In cream sauce (Green Giant)	⅓ of 10-oz. pkg.	359	2.8			
Leaf:						
Not thawed (USDA)	4 oz.	60	.3			0

(USDA): United States Department of Agriculture
*Prepared as Package Directs
[1]Principal source of fat: vegetable shortening.
[2]Principal sources of fat: vegetable shortening & egg.

Food and Description	Measure or Quantity	Sodium (mg.)	Fats in grams Total	Satu- rated	Unsatu- rated	Choles- terol (mg.)
Boiled, drained (USDA)	4 oz.	56	.3			0
(Birds Eye)	⅓ of 10-oz. pkg.	82	.3			0
Creamed (Birds Eye)	⅓ of 10-oz. pkg.	277	3.8			Tr.
In butter sauce (Green Giant)	⅓ of 10-oz. pkg.	340	2.6			

SPINACH, NEW ZEALAND (See **NEW ZEALAND SPINACH**)

SPINACH SOUFFLE, frozen

(Stouffer's)	12-oz. pkg.	2206	32.0			

SPINY LOBSTER (See **CRAYFISH**)

SPLEEN, raw (USDA):

Beef & calf	4 oz.		3.4			
Hog	4 oz.		4.3			
Lamb	4 oz.		4.4			

SPONGE CAKE, home recipe (USDA)[1]

	¹/₁₂ of 10″ cake (2.3 oz.)	110	3.8	1.	2.	162

SPOT, fillets (USDA):

Raw[2]	1 lb.	277	72.1			
Baked, salt added	4 oz.	354	24.8			

SPRITE, soft drink

	6 fl. oz.	21	0.			0

SQUAB, pigeon, raw (USDA):

Dressed	1 lb. (weighed with feet, inedible viscera & bones)		45.1			
Meat & skin	4 oz.		27.0			
Meat only	4 oz.		8.5			
Light meat only, without skin	4 oz.		4.8			
Giblets	1 oz.		2.0			

SQUASH SEEDS, dry (USDA):

In hull	4 oz.		39.2	7.	32	0
Hulled	1 oz.		13.2	2.	11.	0

(USDA): United States Department of Agriculture
*Prepared as Package Directs
[1]Principal source of fat: egg.
[2]Based on samples caught in October. Fat content may vary greatly in samples caught at other seasons of year.

Food and Description	Measure or Quantity	Sodium (mg.)	Total	Satu-rated	Unsatu-rated	Choles-terol (mg.)
			—Fats in grams—			

SQUASH, SUMMER:
Fresh (USDA):
Crookneck & Straightneck, yellow:

Food and Description	Measure or Quantity	Sodium (mg.)	Total	Satu-rated	Unsatu-rated	Choles-terol (mg.)
Whole	1 lb. (weighed untrimmed)	4	.9			0
Boiled, drained, diced	½ cup (3.6 oz.)	1	.2			0
Boiled, drained, slices	½ cup (3.1 oz.)	<1	.2			0
Scallop, white & pale green:						
Whole	1 lb. (weighed untrimmed)	4	.4			0
Boiled, drained, mashed	½ cup (4.2 oz.)	1	.1			0
Zucchini & Cocozelle, green:						
Whole	1 lb. (weighed untrimmed)	4	.4			0
Boiled, drained slices	½ cup (2.7 oz.)	<1	2.1			0
Canned, zucchini in tomato sauce (Del Monte)	½ cup (4.1 oz.)	416	.1			
Frozen:						
Not thawed (USDA)	4 oz.	3	.1			0
Boiled, drained (USDA)	4 oz.	3	.1			0
Fried, breaded, zucchini (Mrs. Paul's)	9-oz. pkg.		30.1			
Parmesan, zucchini (Mrs. Paul's)	12-oz. pkg.		13.5			
Summer squash, slices (Birds Eye)	½ cup (3.3 oz.)	3	.1			0
Zucchini (Birds Eye)	⅓ of 10.-oz. pkg.	1	.1			0

SQUASH, WINTER:
Fresh (USDA):
Acorn:

Food and Description	Measure or Quantity	Sodium (mg.)	Total	Satu-rated	Unsatu-rated	Choles-terol (mg.)
Whole	1 lb. (weighed with skin & seeds)	3	.3			0
Baked, flesh only, mashed	½ cup (3.6 oz.)	1	.1			0
Boiled, mashed	½ cup (4.1 oz.)	1	.1			0
Butternut:						
Whole	1 lb. (weighed with skin & seeds)	3	.3			0
Baked, flesh only	4 oz.	1	.1			0
Boiled, flesh only	4 oz.	1	.1			0
Hubbard:						
Whole	1 lb. (weighed with skin & seeds)	3	.9			0

(USDA): United States Department of Agriculture
*Prepared as Package Directs

Food and Description	Measure or Quantity	Sodium (mg.)	Fats in grams Total	Satu- rated	Unsatu- rated	Choles- terol (mg.)
Baked, flesh only	4 oz.	1	.5			0
Baked, flesh only, mashed	½ cup (3.6 oz.)	1	.4			0
Boiled, flesh only, diced	½ cup (4.2 oz.)	1	.4			0
Boiled, flesh only, mashed	½ cup (4.3 oz.)	1	.3			0
Frozen:						
Not thawed (USDA)	4 oz.	1	.3			0
Heated (USDA)	½ cup (4.2 oz.)	1	.4			0
(Birds Eye)	⅓ of 12-oz. pkg.	1	.3			0
SQUID, raw, meat only (USDA)	4 oz.		10.2			
STARCH (See **CORNSTARCH**)						
__START__, instant breakfast drink	½ cup (4.7 oz.)	45	.1			0
__STOCKPOT SOUP,__ canned (Campbell)	1 cup	925	3.9	<.1	3.	
STOMACH, PORK, scalded (USDA)	4 oz.		10.2			
STRAINED FOOD (See **BABY FOOD**)						
STRAWBERRY:						
Fresh, whole (USDA)	1 lb. (weighed with caps & stems)	4	2.2			0
Fresh, whole, capped (USDA)	1 cup (5.1 oz.)	1	.7			0
Canned, unsweetened or low calorie:						
Water pack, solids & liq. (USDA)	4 oz.	1	.1			0
Solids & liq. (Blue Boy)	4 oz.	1	.7			(0)
Low calorie, solids & liq. (S and W) *Nutradiet*	4 oz.	1	<.1			(0)
Frozen:						
Sweetened, whole, not thawed:						
(USDA)	16-oz. can	5	.9			0
(USDA)	½ cup (4.5 oz.)	1	.3			0
Sweetened, sliced, not thawed (USDA)	10-oz. pkg.	3	.6			0
Sweetened, sliced, not thawed (USDA)	½ cup (4.5 oz.)	1	.3			0

(USDA): United States Department of Agriculture
*Prepared as Package Directs

Food and Description	Measure or Quantity	Sodium (mg.)	— Fats in grams —			Choles- terol (mg.)
			Total	Satu- rated	Unsatu- rated	
Whole (Birds Eye)	¼ of 1-lb. pkg.	1	.2			0
Halves (Birds Eye)	½ cup (5.3 oz.)	2	.6			0
Quick-thaw (Birds Eye)	½ cup (5 oz.)	6	.3			0
*STRAWBERRY CAKE MIX (Duncan Hines)	$^1/_{12}$ of cake	240	6.1			50
STRAWBERRY DRINK MIX *Quik*	2 heaping tsps. (.6 oz.)	5	0.			
STRAWBERRY ICE CREAM (Sealtest)	¼ pt. (2.3 oz.)	38	5.3			
STRAWBERRY PIE:						
Home recipe, made with lard (USDA)[1]	$^1/_6$ of 9″ pie (5.6 oz.)	307	12.5	5.	8.	
Home recipe, made with vegetable shortening (USDA)[2]	$^1/_6$ of 9″ pie (5.6 oz.)	307	12.5	3.	9.	
Creme (Tastykake)	4-oz. pie		15.3			
Frozen:						
(Morton)	$^1/_6$ of 20-oz. pie	232	10.9			
Cream (Banquet)	2½-oz. serving		7.8			
Cream (Morton)	¼ of 14.4-oz. pie	186	13.9			
Cream (Mrs. Smith's)	$^1/_6$ of 8″ pie (2.3 oz.)	95	11.8			
STRAWBERRY PIE FILLING:						
(Comstock)	½ cup (5.4 oz.)	77	.2			
(Lucky Leaf)	8 oz.	150	.6			
STRAWBERRY PRESERVE or JAM:						
Sweetened (Bama)	1 T. (.7 oz.)	2	<.1			(0)
Dietetic or low calorie:						
(Diet Delight)	1 T. (.6 oz.)	<1	Tr.			(0)
(Kraft)	1 oz.	36	<.1			(0)
(S and W) *Nutradiet*	1 T. (.5 oz.)		<.1			(0)
(Slenderella)	1 T. (.7 oz.)	20	Tr.			(0)

(USDA): United States Department of Agriculture
*Prepared as Package Directs
[1]Principal sources of fat: lard & butter.
[2]Principal sources of fat: vegetable shortening & butter.

Food and Description	Measure or Quantity	Sodium (mg.)	Fats in grams Total	Satu- rated	Unsatu- rated	Choles- terol (mg.)
(Smucker's)	1 T. (.7 oz.)	Tr.	<.1			(0)
(Tillie Lewis)	1 T. (.5 oz.)	3	Tr.			0
STRAWBERRY RENNET MIX:						
Powder:						
Dry (Junket)	1 oz.	11	<.1			
*(Junket)	4 oz.	56	3.9			
Tablet:						
Dry (Junket)	1 tablet (<1 gram)	197	Tr.			
*& sugar (Junket)	4 oz.	98	3.9			
STRAWBERRY-RHUBARB PIE:						
(Tastykake)	4-oz. pie		14.6			
Frozen:						
(Morton)	⅛ of 46-oz. pie	382	16.5			
(Mrs. Smith's)	⅙ of 8″ pie (4.2 oz.)	395	14.2			
(Mrs. Smith's) old fashion	⅙ of 9″ pie (5.8 oz.)	470	22.7			
(Mrs. Smith's)	⅛ of 10″ pie (5.6 oz.)	509	18.4			
STRAWBERRY-RHUBARB PIE FILLING, canned (Lucky Leaf)	8 oz.	192	.4			
STRAWBERRY SOFT DRINK:						
Sweetened:						
(Canada Dry)	6 fl. oz.	14+	0.			0
(Clicquot Club)	6 fl. oz.	11	0.			0
(Cott)	6 fl. oz.	11	0.			0
(Fanta)	6 fl. oz.	7	0.			0
(Hoffman)	6 fl. oz.	14	0.			0
(Mission)	6 fl. oz.	11	0.			0
(Shasta)	6 fl. oz.	36	0.			0
(Yukon Club)	6 fl. oz.	14	0.			0
Low calorie:						
(Canada Dry) bottle or can	6 fl. oz.	14+	0.			0
(Clicquot Club)	6 fl. oz.	45	0.			0
(Cott)	6 fl. oz.	45	0.			0
(Hoffman)	6 fl. oz.	56	0.			0
(Mission)	6 fl. oz.	45	0.			0
(Shasta)	6 fl. oz.	47	0.			0
STRAWBERRY SYRUP, low calorie						
(No-Cal)	1 tsp.	<1	0.			(0)

(USDA): United States Department of Agriculture
*Prepared as Package Directs

Food and Description	Measure or Quantity	Sodium (mg.)	Total	— Fats in grams — Satu- rated	Unsatu- rated	Choles- terol (mg.)
STRAWBERRY TURNOVER,						
frozen (Pepperidge Farm)	1 turnover (3.3 oz.)	251	19.9			
STRUDEL, frozen (Pepperidge Farm):						
Apple	¹/₆ of strudel (2.5 oz.)	146	10.2			
Blueberry	¹/₆ of strudel (2.5 oz.)	156	9.8			
Cherry	¹/₆ of strudel (2.5 oz.)	160	9.8			
Pineapple-cheese	¹/₆ of strudel (2.3 oz.)	208	12.1			
STURGEON (USDA):						
Raw, section	1 lb. (weighed with skin & bones)		7.3			
Raw, meat only	4 oz.		2.2			
Smoked	4 oz.		2.0			
Steamed	4 oz.	122	6.5			
SUCCOTASH, frozen:						
Not thawed (USDA)	4 oz.	51	.5			0
Boiled, drained (USDA)	½ cup (3.4 oz.)	36	.4			0
(Birds Eye)	½ cup (3.3 oz.)	39	.5			0
SUCKER, CARP, raw (USDA):						
Whole	1 lb. (weighed whole)		5.7			
Meat only	4 oz.		3.6			
SUCKER, including **WHITE and MULLET,** raw (USDA):						
Whole	1 lb. (weighed whole)	109	3.5			
Meat only	4 oz.	64	2.0			
SUET, raw (USDA)	1 oz.		26.6			
SUGAR, beet or cane (There is no difference in values among brands):						
Brown:						
(USDA)	1 lb.	136	0.			0
Brownulated (USDA)	1 cup (5.4 oz.)	46	0.			0
Firm-packed (USDA)	1 cup (7.5 oz.)	64	0.			0
Firm-packed (USDA)	1 T. (.5 oz.)	4	0.			0

(USDA): United States Department of Agriculture
*Prepared as Package Directs

Food and Description	Measure or Quantity	Sodium (mg.)	Fats in grams Total	Satu- rated	Unsatu- rated	Choles- terol (mg.)
Confectioners':						
(USDA)	1 lb.	5	0.			0
Unsifted (USDA)	1 cup (4.3 oz.)	1	0.			0
Unsifted (USDA)	1 T. (8 grams)	<1	0.			0
Sifted (USDA)	1 cup (3.4 oz.)	<1	0.			0
Sifted (USDA)	1 T. (6 grams)	<1	0.			0
Stirred (USDA)	1 cup (4.2 oz.)	1	0.			0
Stirred (USDA)	1 T. (8 grams)	<1	0.			0
Granulated:						
(USDA)	1 lb.	5	0.			0
(USDA)	1 cup (6.9 oz.)	2	0.			0
(USDA)	1 T. (.4 oz.)	<1	0.			0
(USDA)	1 lump (1⅛″ x ¾″ x ⅜″, 6 grams)	<1	0.			0
Maple (USDA)	1 lb.	64	0.			0
Maple (USDA)	1¾″ x 1¼″ x ½″ piece (1.2 oz.)	4	0.			0
SUGAR APPLE, raw (USDA):						
Whole	1 lb. (weighed with skin & seeds)	22	.6			0
Flesh only	4 oz.	12	.3			0
SUGAR CHEX, cereal, dry	⅞ cup (1 oz.)	220	2.2			(0)
SUGAR FROSTED FLAKES, cereal (Kellogg's)	¾ cup (1 oz.)	170	0.			(0)
SUGAR JETS, cereal (General Mills)	1 cup (1 oz.)	218	1.1			(0)
SUGAR POPS, cereal (Kellogg's)	1 cup (1 oz.)	68	.1			(0)
SUGAR SMACKS, cereal (Kellogg's)	1 cup (1 oz.)	22	.5			(0)
SUGAR SPARKLED TWINKLES, cereal (General Mills)	1 cup (1 oz.)	230	.8			(0)
SUGAR SUBSTITUTE:						
(Adolph's)	1 tsp. (4 grams)	Tr.	0.			0
Superose (Whitlock)	1 packet (1 gram)	4	0.			(0)

(USDA): United States Department of Agriculture
*Prepared as Package Directs

Food and Description	Measure or Quantity	Sodium (mg.)	— Fats in grams —			Choles- terol (mg.)
			Total	Satu- rated	Unsatu- rated	
SUKI-YAKI MIX:						
(Durkee)	1.7-oz. pkg.	6271	.9			
*With meat & vegetables (Durkee)	6 cups (1.7-oz. pkg.)	7226	122.7			
SUNFLOWER SEED (USDA):						
In hulls	4 oz. (weighed in hull)	18	29.0	4.	25.	0
Hulled	1 oz.	9	13.4	2.	12.	0
SUNFLOWER SEED FLOUR (See **FLOUR**)						
SUPER ORANGE CRISP WHEAT PUFFS, cereal	1 cup (1 oz.)	146	.3			0
SUPER SUGAR CRISP WHEAT PUFFS, cereal (Post)	⅞ cup (1 oz.)	40	.2			0
SURINAM CHERRY (See **PITANGA**)						
SUZY Q (Hostess) 2 to pkg.	1 cake (2¼ oz.)	137	9.2			
SWAMP CABBAGE (USDA):						
Raw, whole	1 lb. (weighed untrimmed)		1.1			0
Boiled, trimmed, drained	4 oz.		.2			0
SWEETBREADS (USDA):						
Beef, raw	1 lb.	435	72.6			1134
Beef, braised	4 oz.	132	26.3			528
Calf, raw	1 lb.		9.1			
Calf, braised	4 oz.		3.6			
Hog (See **PANCREAS**)						
Lamb, raw	1 lb.		17.2			
Lamb, braised	4 oz.		6.9			
SWEET POTATO:						
Raw (USDA):						
All kinds, unpared	1 lb. (weighed whole)	37	1.5			0
All kinds, pared	4 oz.	11	.5			0
Firm-fleshed, Jersey types, pared	4 oz.	11	.8			0

(USDA): United States Department of Agriculture
*Prepared as Package Directs

318

Food and Description	Measure or Quantity	Sodium (mg.)	—Fats in grams—			Choles- terol (mg.)
			Total	Satu- rated	Unsatu- rated	
Soft-fleshed, Puerto Rico variety, pared	4 oz.	11	.3			0
Baked, peeled after baking (USDA)	3.9-oz. sweet potato (5″ x 2″)	13	.6			0
Boiled, peeled after boiling (USDA)	5-oz. sweet potato (5″ x 2″)	15	.6			0
Candied, home recipe (USDA)[1]	6.2-oz. sweet potato (3½″ x 2¼″)	74	5.8	4.	2.	0
Canned, regular pack:						
In syrup, solids & liq. (USDA)	4 oz.	54	.2			0
Vacuum or solid pack (USDA)	½ cup (3.8 oz.)	52	.2			0
Heavy syrup, solids & liq. (Del Monte)	½ cup (4.2 oz.)	52	Tr.			
Vacuum pack (Taylor's)	½ cup		.1			(0)
Canned, dietetic pack, without added sugar & salt (USDA)	4 oz.	14	.1			0
Dehydrated flakes, dry (USDA)	½ cup (2 oz.)	105	.3			0
*Dehydrated flakes, prepared with water (USDA)	½ cup (4.4 oz.)	57	.1			0
Frozen:						
Candied (Mrs. Paul's)	⅓ of 12-oz. pkg.		.2			
Sweets & Apples, candied (Mrs. Paul's)	⅓ of 12-oz. pkg.		.1			
Candied yams (Birds Eye)	⅓ of 12-oz. pkg.	10	.3			0
With brown sugar, pineapple glaze (Birds Eye)	½ cup (3.3 oz.)	36	2.5			0

SWEET POTATO PIE:

Food and Description	Measure or Quantity	Sodium (mg.)	Total	Satu- rated	Unsatu- rated	Choles- terol (mg.)
Home recipe, made with lard (USDA)[2]	⅙ of 9″ pie (5.4 oz.)	331	17.2	6.	11.	
Home recipe, made with vegetable shortening (USDA)[3]	⅙ of 9″ (5.4 oz.)	331	17.2	5.	13.	
(Tastykake)	4-oz. pie		15.0			

SWEETSOP (See SUGAR APPLE)

*SWISS BURGER, mix, dinner

Food and Description	Measure or Quantity	Sodium (mg.)	Total	Satu- rated	Unsatu- rated	Choles- terol (mg.)
(Jeno's) Add·n' Heat	30-oz. pkg.		78.2			

(USDA): United States Department of Agriculture
*Prepared as Package Directs
[1]Principal source of fat: butter.
[2]Principal sources of fat: lard & butter.
[3]Principal sources of fat: vegetable shortening & butter.

Food and Description	Measure or Quantity	Sodium (mg.)	— Fats in grams —			Choles- terol (mg.)
			Total	Satu- rated	Unsatu- rated	
SWISS STEAK, frozen:						
(Stouffer's)	10-oz. pkg.		33.0			
Dinner (Swanson)	10-oz. dinner	682	15.6	5.	10.	
SWORDFISH (USDA):						
Raw, meat only	1 lb.		18.1			
Broiled, with butter or margarine	3″ x 3″ x ½″ steak					
	(4.4 oz.)		7.5			
Canned, solids & liq.	4 oz.		3.4			
SYRUP (See also individual listings by kind, such as **PANCAKE & WAFFLE SYRUP** or by brand name, such as ***LOG CABIN***):						
All fruit flavors (Smucker's)	1 T. (.7 oz.)	Tr.	0.			(0)

T

TABASCO (McIlhenny)	¼ tsp. (1 gram)	6	Tr.			(0)
TACO SEASONING MIX:						
(French's)	1¾-oz. pkg.	3800	1.2			
(Lawry's)	1¼-oz. pkg.		2.1			
TAHITIAN TREAT, soft drink						
(Canada Dry) bottle or can	6 fl. oz.	15+	0.			(0)
TAMALE:						
Canned:						
(Armour Star)	15½-oz. can		31.6			
(Hormel) beef	1 tamale (2.1 oz.)	357	6.0	2.	3.	5
(Wilson)	15½-oz. can	2949	29.9	16.	14.	75
Frozen:						
(Banquet) cooking bag	2 tamales with sauce					
	(3 oz. each)		15.7			
(Banquet) buffet	2-lb. pkg.		83.6			
TAMARIND, fresh (USDA):						
Whole	1 lb. (weighed					
	with pods & seeds)	111	1.3			0
Flesh only	4 oz.	58	.7			0

(USDA): United States Department of Agriculture
*Prepared as Package Directs

Food and Description	Measure or Quantity	Sodium (mg.)	—Fats in grams—			Choles- terol (mg.)
			Total	Satu- rated	Unsatu- rated	
TANDY TAKE (Tastykake):						
Chocolate	.7-oz. cake		6.2			
Choc-o-mint	.6-oz. cake		5.2			
Dandy Kake	.6-oz. cake		5.2			
Karamel	.7-oz. cake		4.9			
Orange	.6-oz. cake		4.8			
Peanut butter	.7-oz. cake		6.5			
***TANG**, instant breakfast drink:						
Grape	½ cup (4.7 oz.)	8	Tr.			0
Grapefruit	½ cup (4.7 oz.)	40	Tr.			0
Orange	½ cup (4.7 oz.)	54	.1			0
TANGELO, fresh (USDA):						
Juice from whole fruit	1 lb. (weighed with peel, membrane & seeds)		.3			0
Juice	½ cup (4.4 oz.)		.1			0
TANGERINE or MANDARIN ORANGE, fresh:						
Whole (USDA)	1 lb. (weighed with peel, membrane & seeds)	7	.7			0
Whole (USDA)	4.1-oz. tangerine (2⅜" dia.)	2	.2			0
Peeled (Sunkist)	1 large tangerine (4.1 oz.)	2	Tr.			0
Sections, without membranes (USDA)	1 cup (6.8 oz.)	4	.4			0
TANGERINE JUICE:						
Fresh (USDA)	½ cup (4.4 oz.)	1	.2			0
Canned, unsweetened (USDA)	½ cup (4.4 oz.)	1	.2			0
Canned, sweetened (USDA)	½ cup (4.4 oz.)	1	.2			0
Frozen concentrate, unsweetened:						
Undiluted (USDA)	6-oz. can	4	1.5			0
*Prepared with 3 parts water by volume (USDA)	½ cup (4.4 oz.)	1	.2			0
Frozen concentrate, sweetened:						
*(Minute Maid)	½ cup (4.2 oz.)	1	<.1			0
*(Snow Crop)	½ cup (4.2 oz.)	1	<.1			0

(USDA): United States Department of Agriculture
*Prepared as Package Directs

Food and Description	Measure or Quantity	Sodium (mg.)	— Fats in grams —			Choles- terol (mg.)
			Total	Satu- rated	Unsatu- rated	
TAPIOCA, dry, quick cooking, granulated:						
(USDA)	1 cup (5.4 oz.)	5	.3			0
(USDA)	1 T. (10 grams)	<1	<.1			0
(Minute Tapioca)	1 T.	2	Tr.			
TAPIOCA PUDDING:						
Apple, home recipe (USDA)	½ cup (4.4 oz.)	64	.1			
Cream, home recipe (USDA)	½ cup (2.9 oz.)	128	4.2	2.	3.	80
Chilled (Sealtest)	4 oz.	229	3.5			
Canned (Hunt's)	5-oz. can	210	5.8	1.	5.	
Mix:						
*All flavors (Jell-O)	½ cup (5.1 oz.)	166	4.6			13
*Chocolate (Royal)	½ cup (5.1 oz.)	120	5.5			14
*Fluffy (Minute Tapioca)	½ cup (4.4 oz.)	76	5.9			82
*Vanilla (Royal)	½ cup (5.1 oz.)	120	4.6			14
TARO, raw (USDA):						
Tubers, whole	1 lb. (weighed with skin)	27	.8			0
Tubers, skin removed	4 oz.	8	.2			0
Leaves & stems	1 lb.		3.6			0
TARRAGON (Spice Islands)	1 tsp.	<1				(0)
TASTE AMERICA, frozen (Green Giant):						
Florida style	⅓ of 10-oz. pkg.	335	2.8			
New Orleans style	⅓ of 10-oz. pkg.	572	3.6			
Northwest style	⅓ of 10-oz. pkg.	454	3.3			
Pennsylvania Dutch style	⅓ of 10-oz. pkg.	340	2.8			
San Francisco Style	⅓ of 10-oz. pkg.	501	.9			
TAUTOG or BLACKFISH, raw:						
Whole (USDA)	1 lb. (weighed whole)		1.8			
Meat only (USDA)	4 oz.		1.2			
TEA:						
Bag (Lipton)	1 bag	0	0.			(0)
Bag (Tender Leaf)	1 bag	Tr.	Tr.			0
Canned (Lipton)	12 fl. oz.	12				(0)

(USDA): United States Department of Agriculture
*Prepared as Package Directs

Food and Description	Measure or Quantity	Sodium (mg.)	—Fats in grams—			Choles- terol (mg.)
			Total	Satu- rated	Unsatu- rated	
Instant:						
Dry powder, slightly sweetened (USDA)	1 tsp. (<1 gram)		Tr.			0
*Beverage, slightly sweetened (USDA)	1 cup (8.4 oz.)		Tr.			0
*(Lipton)	1 cup	0	0.			(0)
Nestea	1 tsp. (<1 gram)	7	0.			(0)
(Tender Leaf)	1 rounded tsp.	Tr.	Tr.			0
TEAM, cereal	1⅓ cups (1 oz.)	214	.4			(0)
TEA MIX, iced:						
*All flavors (Salada)	1 cup (.5 oz. dry)	6	<.1			(0)
*(Tender Leaf)	1 cup	80				0
Lemon-flavored:						
*Low calorie (Lipton)	1 cup	0	<.1			(0)
*Low calorie (Tender Leaf)	1 cup	50	Tr.			0
TEMPTYS (Tastykake):						
Butter creme	⅔-oz. cake		4.4			
Chocolate	⅔-oz. cake		2.5			
Lemon	⅔-oz. cake		2.6			

TENDERGREEN (See **MUSTARD SPINACH**)

TEQUILA (See **DISTILLED LIQUOR**)

Food and Description	Measure or Quantity	Sodium (mg.)	Total	Satu- rated	Unsatu- rated	Choles- terol (mg.)
TERRAPIN, DIAMOND BACK, raw (USDA):						
In shell	1 lb. (weighed in shell)		3.3			
Meat only	4 oz.		4.0			
TEXTURED VEGETABLE PROTEIN:						
Breakfast links, *Morningstar Farms*	1 link (.8 oz.)	227	3.6	<1.	2.	0
Breakfast patties, *Morningstar Farms*	1 pattie (1.3 oz.)	378	6.0	2.	5.	0
Breakfast slices, *Morningstar Farms*	1 slice (1 oz.)	468	3.1	1.	2.	0
Burger Builders (Betty Crocker)	¼ cup		1.0			
Pathmark Plus:						
Dry	⅓ oz.		0.			
*Prepared	4 oz.		7.9			

(USDA): United States Department of Agriculture
*Prepared as Package Directs

Food and Description	Measure or Quantity	Sodium (mg.)	—Fats in grams—			Cholesterol (mg.)
			Total	Satu-rated	Unsatu-rated	
THICK & FROSTY (General Foods)	1 cup (8.3 oz.)	144	14.0			13
THURINGER, sausage:						
(USDA)	1 oz.		6.9			
(Hormel) Old Smokehouse	1 oz.	292	8.9			
Summer sausage, all meat (Oscar Mayer)	.8-oz. slice	287	6.4			
Summer sausage, pure beef (Oscar Mayer)	.8-oz. slice	278	5.8			(0)
THYME (Spice Islands)	1 tsp.	<1				(0)
TIKI, soft drink (Shasta):						
Sweetened	6 fl. oz.	22	0.			(0)
Low calorie	6 fl. oz.	37	0.			(0)
TILEFISH (USDA):						
Raw, whole	1 lb. (weighed whole)		1.2			
Baked, meat only	4 oz.		4.2			
TOASTER CAKE:						
Corn Treats (Arnold)	1.1-oz. piece		3.7			
Toastee (Howard Johnson's):						
Blueberry	1 piece (1¼ oz.)		5.0			
Cinnamon raisin	1 piece (1.1 oz.)		4.0			
Corn	1 piece (1¼ oz.)		3.4			
Orange	1 piece (1 oz.)		4.8			
Pound	1 piece (1 oz.)		4.5			
Toastette (Nabisco):						
Apple	1 piece (1⅔ oz.)	159	5.4			
Blueberry	1 piece (1⅔ oz.)	148	4.9			
Brown sugar, cinnamon	1 piece (1⅔ oz.)	161	5.9			
Cherry	1 piece (1⅔ oz.)	162	4.9			
Orange marmalade	1 piece (1⅔ oz.)	175	5.0			
Peach	1 piece (1⅔ oz.)	175	4.8			
Strawberry	1 piece (1⅔ oz.)	145	5.1			
Toast-r-Cake (Thomas):						
Bran	1 piece (1.2 oz.)	335	3.1			
Corn	1 piece (1.2 oz.)	345	3.9			
Orange	1 piece (1.2 oz.)	335	3.8			

(USDA): United States Department of Agriculture
*Prepared as Package Directs

Food and Description	Measure or Quantity	Sodium (mg.)	—Fats in grams—			Cholesterol (mg.)
			Total	Saturated	Unsaturated	

TOASTERINO, frozen (Buitoni):

Cheese, grilled	4 oz.		9.3			
Pizzaburger	4 oz.		12.4			
Sloppy Joe	4 oz.		11.7			

TODDLER FOOD (See **BABY FOOD**)

TOFFEE KRUNCH BAR

(Sealtest)	3 fl. oz. (1.7 oz.)	29	10.4			

TOFU (See **SOYBEAN CURD**)

TOMATO:

Fresh, green, whole, untrimmed (USDA)	1 lb. (weighed with core & stem end)	12	.8			0
Fresh, green, trimmed, unpeeled (USDA)	4 oz.	3	.2			0
Fresh, ripe (USDA):						
Whole, eaten with skin	1 lb.	14	.9			0
Whole, peeled	1 lb. (weighed with skin, stem ends & hard core)	12	.8			0
Whole, peeled	1 med. (2″ x 2½″, 5.3 oz.)	4	.3			0
Whole, peeled	1 small (1¾″ x 2¼″, 3.9 oz.)	3	.2			0
Sliced, peeled	½ cup (3.2 oz.)	3	.2			0
Cooked without salt	½ cup (4.3 oz.)	5	.2			0
Canned, regular pack:						
Whole, solids & liq. (USDA)	½ cup (4.2 oz.)	155	.2			0
(Contadina)	½ cup (4 oz.)	225	Tr.			
Solids & liq. (Del Monte)	½ cup (4.2 oz.)	159	1.0			0
Diced, in puree (Contadina)	½ cup (4 oz.)	232	Tr.			(0)
Sliced (Contadina)	½ cup (4 oz.)	352	Tr.			(0)
Stewed (Contadina)	½ cup (4 oz.)	216	Tr.			(0)
Stewed (Del Monte)	½ cup (4.2 oz.)	342	.2			0
Wedges, solids & liq. (Del Monte)	½ cup (4.1 oz.)	336	.2			0
Whole, peeled (Hunt's)	½ cup (4.2 oz.)	368	.2			(0)
Whole, solids & liq. (Stokely-Van Camp)	½ cup (4.1 oz.)		.2			(0)

(USDA): United States Department of Agriculture
*Prepared as Package Directs

Food and Description	Measure or Quantity	Sodium (mg.)	Fats in grams — Total	Satu- rated	Unsatu- rated	Choles- terol (mg.)
Canned, dietetic pack, low sodium:						
Solids & liq. (USDA)	4 oz.	3	.2			0
Solids & liq. (Blue Boy)	4 oz.	5	.1			(0)
Solids & liq. (Tillie Lewis)	½ cup (4.3 oz.)	15	.2			0
Whole, peeled (Diet Delight)	½ cup (4.3 oz.)	9	.1			(0)
Whole, unseasoned (S and W)						
Nutradiet	4 oz.	10	.1			(0)
TOMATO JUICE:						
Canned, regular pack:						
(USDA)	6 fl. oz. (6.4 oz.)	364	.2			0
(USDA)	½ cup (4.3 oz.)	244	.1			0
(Campbell)	6 fl. oz.	618	.2			(0)
(Del Monte)	½ cup (4.3 oz.)	197	.1			0
(Heinz)	5½-fl.-oz. can	537	.3			(0)
(Hunt's)	5½-fl.-oz. can	566	.2			(0)
(Stokely-Van Camp)	½ cup (4.3 oz.)		.1			(0)
Canned, dietetic pack, low sodium:						
(USDA)	4 oz. (by wt.)	3	.1			0
(Blue Boy)	4 oz. (by wt.)	1	<.1			(0)
(Diet Delight)	½ cup (3.9 oz.)	8	<.i			(0)
Unseasoned (S and W) *Nutradiet*	4 oz. (by wt.)	7	.1			(0)
Concentrate, canned (USDA)	4 oz. (by wt.)	896	.5			0
*Concentrate, canned, diluted with 3 parts water by volume						
(USDA)	4 oz. (by wt.)	237	.1			0
Dehydrated (USDA)	1 oz.	1115	.6			0
*Dehydrated (USDA)	½ cup (4.3 oz.)	312	.1			0
TOMATO JUICE COCKTAIL:						
(USDA)	4 oz. (by wt.)	227	.1			0
Snap-E-Tom	6 fl. oz.	1066	.2			
TOMATO PASTE, canned:						
(USDA) regular, no salt added	6-oz. can	65	.7			0
(USDA) regular, no salt added	½ cup (4.6 oz.)	49	.5			0
(USDA) regular, no salt added	1 T. (.6 oz.)	61	.6			0
(USDA) salt added	6-oz. can	1343	.7			0
(USDA) salt added	½ cup (4.6 oz.)	1019	.5			0
(USDA) salt added	1 T. (.6 oz.)	126	.6			0
(Contadina)	1 T. (.5 oz.)	5	Tr.			(0)
(Del Monte)	1 T. (.6 oz.)	9	<.1			0
(Hunt's)	½ cup (4.6 oz.)	497	.5			(0)

(USDA): United States Department of Agriculture
*Prepared as Package Directs

Food and Description	Measure or Quantity	Sodium (mg.)	Fats in grams Total	Satu- rated	Unsatu- rated	Choles- terol (mg.)
(Hunt's)	1 T. (.6 oz.)	62	<.1			(0)
(Stokely-Van Camp)	½ cup (4.6 oz.)		.5			(0)

TOMATO PUREE:
Canned, regular pack:

(USDA)	1 cup (8.8 oz.)	998	.5			0
(Contadina)	1 cup	24	.7			(0)
(Hunt's)	1 cup (8.8 oz.)	497	.5			(0)
Canned, dietetic pack (USDA)	8 oz.	14	.5			0

TOMATO SALAD, jellied

(Contadina)	½ cup	592	Tr.			(0)

TOMATO SAUCE, canned:

(Contadina)	1 cup	1296	Tr.			(0)
(Del Monte) plain	1 cup (8.8 oz.)	1252	.3			(0)
(Del Monte) with mushrooms	1 cup (8.8 oz.)	1262	.3			
(Del Monte) with onions	1 cup (8.8 oz.)	1218	1.3			
(Del Monte) with tomato tidbits	1 cup (8.8 oz.)	1212	.2			
(Hunt's) plain	1 cup (8.7 oz.)	1662	.4			(0)
(Hunt's) herb[1]	1 cup (8.8 oz.)	942	8.5	2.	7.	
(Hunt's) special	1 cup (8.7 oz.)	825	.5			
(Hunt's) with bits	1 cup (8.7 oz.)	1647	.4			
(Hunt's) with cheese[2]	1 cup (8.8 oz.)	1672	1.7			
(Hunt's) with mushrooms	1 cup (8.8 oz.)	1669	4.2			
(Hunt's) with onions	1 cup (8.8 oz.)	1704	.5			

TOMATO SOUP:
Canned, regular pack:

Condensed (USDA)	8 oz. (by wt.)	1796	4.8			
*Prepared with equal volume water (USDA)	1 cup (8.6 oz.)	970	2.4			
*Prepared with equal volume milk (USDA)[3]	1 cup (8.8 oz.)	1055	7.0	2.	4.	
*(Campbell)	1 cup	801	1.8	Tr.	2.	3
*(Heinz) California	1 cup (8.5 oz.)	868	.7			
(Heinz) *Great American*	1 cup (8¾ oz.)	1585	5.7			
*(Manischewitz)	1 cup		1.7			
*Beef (Campbell) *Noodle-O's*	1 cup	727	3.3			
*Bisque (Campbell)	1 cup	988	2.4	1.	1.	

(USDA): United States Department of Agriculture
*Prepared as Package Directs
[1]Principal source of fat: cottonseed oil.
[2]Principal source of fat: cheese.
[3]Principal source of fat: milk.

Food and Description	Measure or Quantity	Sodium (mg.)	— Fats in grams — Total	Satu- rated	Unsatu- rated	Choles- terol (mg.)
*Rice, old fashioned (Campbell)	1 cup	752	2.8	<1.	2.	
*Rice (Manischewitz)	1 cup		2.3			
With vegetables (Heinz) *Great American*	1 cup (8¾ oz.)	1007	5.2			
Canned, dietetic pack:						
Low sodium (Campbell)	7¼-oz. can	30	2.3			
*With rice (Claybourne)	8 oz.	34	1.0			
*With rice (Slim-ette)	8 oz. (by wt.)	28	.1			
(Tillie Lewis)	1 cup (8 oz.)	35	1.4			
TOMATO SOUP MIX:						
(Lipton) *Cup-a-Soup*	1 pkg. (.8 oz.)	670	.5	Tr.	Tr.	1
Vegetable:						
With noodles (USDA)[1]	1 oz.	1740	2.3	<1.	2.	
*With noodles (USDA)[1]	1 cup (8 oz.)	974	1.4			
*With noodles (Lipton)	1 cup (8 oz.)	1304	1.6	Tr.	1.	10
TOMCOD, ATLANTIC, raw (USDA):						
Whole	1 lb. (weighed whole)		.7			
Meat only	4 oz.		.5			
TOM COLLINS MIX (Party Tyme)	½-oz. pkg.	66	0.			(0)
TOM COLLINS or COLLINS MIXER SOFT DRINK:						
(Canada Dry) bottle or can	6 fl. oz.	13†	0.			0
(Dr. Brown's)	6 fl. oz.	9	0.			0
(Hoffman)	6 fl. oz.	9	0.			0
(Kirsch)	6 fl. oz.	<1	0.			0
(Shasta)	6 fl. oz.	22	0.			0
(Yukon Club)	6 fl. oz.	9	0.			0
TONGUE (USDA):						
Beef, medium fat, raw, untrimmed	1 lb.		52.0			
Beef, medium fat, braised	4 oz.	69	18.9			
Beef, smoked	4 oz.		32.7			
Calf, raw, untrimmed	1 lb.		18.5			
Calf, braised	4 oz.		6.8			
Hog, raw, untrimmed	1 lb.		53.8			
Hog, braised	4 oz.		19.7			

(USDA): United States Department of Agriculture
*Prepared as Package Directs
[1]Principal sources of fat: vegetable shortening & egg.

Food and Description	Measure or Quantity	Sodium (mg.)	Total	—Fats in grams— Satu- rated	Unsatu- rated	Choles- terol (mg.)
Lamb, raw, untrimmed	1 lb.	50.7				
Lamb, braised	4 oz.	20.6				
Sheep, raw, untrimmed	1 lb.	72.2				
Sheep, braised	4 oz.	28.7				
TONGUE, CANNED:						
Pickled (USDA)	1 oz.	5.8				
Potted or deviled (USDA)	1 oz.	6.5				
(Hormel)	1 oz. (12-oz. can)	5.0				
TOPPING (See also **CHOCOLATE SYRUP**):						
Sweetened:						
Butterscotch (Kraft)	1 oz.	80	1.2			(0)
Butterscotch (Smucker's)	1 T. (.7 oz.)	35	.2	0.	Tr.	
Caramel:						
(Smucker's)	1 T. (.7 oz.)	55	<.1	Tr.	Tr.	
Chocolate (Kraft)	1 oz.	55	.6			
Vanilla (Kraft)	1 oz.	65	<.1			
Cherry (Smucker's)	1 T. (.6 oz.)	Tr.	0.			
Chocolate or chocolate flavored:						
(Kraft)	1 oz.	25	.4			
Fudge (Hershey's)	1 oz.	111	3.8			
Fudge (Kraft)	1 oz.	96	3.6			
Fudge (Smucker's)	1 T. (.7 oz.)	18	.6	Tr.	Tr.	
Fudge, mint (Smucker's)	1 T. (.6 oz.)	18	.6	Tr.	Tr.	
Milk (Smucker's)	1 T. (.7 oz.)	23	1.4	<1.	<1.	
Marshmallow creme (Kraft)	1 oz.	17	0.			
Pecan (Kraft)	1 oz.	4	7.8			
Pecan in syrup (Smucker's)	1 T. (.7 oz.)	Tr.	5.1	Tr.	4.	
Pineapple (Kraft)	1 oz.	<1	<.1			
Pineapple (Smucker's)	1 T. (.6 oz.)	Tr.	0.			
Spoonmallow (Kraft)	1 oz.	16	0.			
Strawberry (Kraft)	1 oz.	<1	<.1			
Strawberry (Smucker's)	1 T. (.6 oz.)	Tr.	0.			
Walnut (Kraft)	1 oz.	<1	6.3			
Walnut in syrup (Smucker's)	1 T. (.6 oz.)	Tr.	4.3	Tr.	4.	
Dietetic, chocolate (Diet Delight)	1 T. (.6 oz.)	13	Tr.			
TOPPING, WHIPPED:						
(USDA) pressurized	1 cup (2.5 oz.)		17.0	15.	2.	
(USDA) pressurized	1 T. (4 grams)		1.	<1.	Tr.	
(Birds Eye) *Cool Whip*	1 T. (4 grams)	1	1.2			0

(USDA): United States Department of Agriculture
*Prepared as Package Directs

Food and Description	Measure or Quantity	Sodium (mg.)	Fats in grams Total	Saturated	Unsaturated	Cholesterol (mg.)
(Kraft)	1 oz.	19	7.0			
(Lucky Whip)	1 T. (4 grams)	3	1.2	Tr.	1.	0
(Sealtest) *Big Top*	1.5 fl. oz. (.5 oz.)	1	1.4			
(Sealtest) *Zip Whipt,* real cream	1.5 fl. oz. (.4 oz.)	3	2.0			
TOPPING, WHIPPED, MIX:						
*(D-Zerta)	1 T.	6	.6			Tr.
*(Dream Whip)	1 T. (.2 oz.)	4	.8			Tr.
*(Lucky Whip)	1 T. (4 grams)	3	.6	Tr.	<1.	0
TORTILLA (USDA)	.7-oz. tortilla (5″)		.6			
TOTAL, cereal (General Mills)	1¼ cups (1 oz.)	48	.5			(0)
TOWEL GOURD, raw (USDA):						
Unpared	1 lb. (weighed with skin)		.8			0
Pared	4 oz.		.2			0
TREET (Armour)	1 oz.		7.4			
TRIPE, beef (USDA):						
Commercial	4 oz.	82	2.3			
Pickled	4 oz.	52	1.5			
TROPICAL PUNCH SOFT DRINK (Yukon Club)	6 fl. oz.	14	0.			0
TROUT:						
Brook, fresh, whole (USDA)	1 lb. (weighed whole)		4.7			
Brook, fresh, meat only (USDA)	4 oz.		2.4			
Lake (See **LAKE TROUT**)						
Rainbow (USDA):						
Fresh, meat with skin	4 oz.		12.9	3.	10.	62
Canned	4 oz.		15.2	5.	11.	
Frozen (1000 Springs):						
Boned	5-oz. trout	29	4.2			
Dressed	5-oz. trout	35	5.1			
Boned & breaded	5-oz. trout					
TUNA:						
Raw, bluefin, meat only (USDA)	4 oz.		4.6	1.	4.	

(USDA): United States Department of Agriculture
*Prepared as Package Directs

Food and Description	Measure or Quantity	Sodium (mg.)	—Fats in grams—			Choles- terol (mg.)
			Total	Satu- rated	Unsatu- rated	
Raw, yellowfin, meat only (USDA)	4 oz.	42	3.4	1.	2.	
Raw, yellowfin, meat only, brined (USDA)	4 oz.	498	3.4	1.	2.	
Canned in oil:						
Solids & liq.:						
(USDA)[1]	6½-oz. can	1472	37.7	9.	29.	100
(Breast O' Chicken)	6½-oz. can		36.8			
Chunk, light (Chicken of the Sea)	6-oz. can	1196	25.6	3.	22.	74
Chunk, light (Del Monte)	6½-oz. can	1150	32.8			
Chunk, light (Del Monte)	1 cup (4.7 oz.)	831	23.7			
White albacore (Del Monte)	6½-oz. can	1019	37.7			
White albacore (Del Monte)	1 cup (4.7 oz.)	528	27.3			
Drained solids:						
(USDA)[2]	6½-oz. can		12.9	5.	8.	102
Albacore (Del Monte)	1 cup (5.6 oz.)	858	13.1			
Chunk, light (Chicken of the Sea)	6½-oz. can	1072	12.5	2.	11.	36
Canned in water:						
Solids & liq., no salt added:						
(USDA)	6½-oz. can	75	1.5			116
Solids & liq., salt added:						
(USDA)	6½-oz. can	1610	1.5			116
(Breast O' Chicken)	6½-oz. can		1.8			
Drained, solid, light (Chicken of the Sea)	6½-oz. can	1105	3.4			
Drained, solid, white (Chicken of the Sea)	6½-oz. can	1105	1.7			
Canned, dietetic, drained, chunk, white (Chicken of the Sea)	6½-oz. can	74	2.6			
TUNA CAKE, frozen, thins (Mrs. Paul's)	10-oz. pkg.		43.3			
TUNA PIE, frozen:						
(Banquet)	8-oz. pie		27.0			
(Morton)	8-oz. pie	715	19.8			
TUNA SALAD, home recipe[3,4]						
(USDA)	4 oz.		11.9	3.	9.	

(USDA): United States Department of Agriculture
*Prepared as Package Directs
[1]Principal sources of fat: cottonseed oil & tuna.
[2]Principal sources of fat: tuna & cottonseed oil.
[3]Prepared with tuna, celery, mayonnaise, pickle, onion & egg.
[4]Principal sources of fat: cottonseed oil, soybean oil, corn oil, tuna & egg.

Food and Description	Measure or Quantity	Sodium (mg.)	—Fats in grams— Total	Satu- rated	Unsatu- rated	Choles- terol (mg.)
TUNA SOUP, Creole, canned (Crosse & Blackwell)	6½ oz. (½ can)		1.3			
TURBOT, GREENLAND:						
Raw, whole (USDA)	1 lb. (weighed whole)		19.8			
Raw, meat only (USDA)	4 oz.	64	9.5	2.	8.	
Frozen (Weight Watchers)	18-oz. dinner		25.1			
Frozen, with apple (Weight Watchers)	9½-oz. luncheon		16.4			
TURKEY:						
Raw, ready-to-cook (USDA)	1 lb. (weighed with bones)		48.7	14.0	35.	272
Raw, meat & skin only (USDA)	4 oz.					84
Raw, dark meat (USDA)	4 oz.	92	4.9			85
Raw, light meat (USDA)	4 oz.	58	1.4			68
Raw, skin only (USDA)	4 oz.		44.5			125
Roasted (USDA):						
Flesh, skin & giblets	From 13½-lb. raw, ready-to-cook turkey		603.5			3864
Flesh & skin	From 13½-lb. raw, ready-to-cook turkey		338.9	106.	233.	3283
Flesh & skin	4 oz.		10.9	3.	7.	105
Meat only:						
Chopped	1 cup (5 oz.)	183	8.6	3.	6.	
Dark	4 oz.	112	9.4	2.	7.	115
Dark	1 slice (2½" x 1⅝" x ¼", .7 oz.)	21	1.8	Tr.	1.	21
Diced	1 cup (4.8 oz.)	176	8.2	3.	6.	
Light	4 oz.	93	4.4	1.	3.	87
Light	1 slice (4" x 2" x ¼", 3 oz.)	35	1.7	Tr.	1.	32
Skin only	1 oz.		11.9	3.	9.	36
Giblets, simmered (USDA)	2 oz.		8.7			
Smoked, cooked, pressed (Oscar Mayer)	.8-oz. slice	266	.4			
Canned, boned:						
(USDA)	4 oz.		14.2	5.	10.	
Solids & liq. (Lynden Farms)	5-oz. jar	527	10.5	4.	6.	
(Swanson) with broth	5-oz. can	650	9.0			
Canned, roast (Wilson) *Tender Made*	4 oz.	480	2.4	<1.	2.	76

(USDA): United States Department of Agriculture
*Prepared as Package Directs

Food and Description	Measure or Quantity	Sodium (mg.)	— Fats in grams —			Choles- terol (mg.)
			Total	Satu- rated	Unsatu- rated	
TURKEY DINNER:						
Canned, noodle (Lynden Farms)	15-oz. can	254	25.2	9.	16.	
Frozen:						
Sliced turkey, mashed potato, peas (USDA)[1]	12 oz.	1360	10.2	3.	7	
(Banquet):						
Meat compartment	6.8 oz.		5.2			
Peas compartment	1.9 oz.		1.0			
Potato compartment	1.8 oz.		.6			
Complete dinner	11.5-oz. dinner		6.8			
(Morton)	12-oz. dinner	1397	20.2			
(Morton) 3-course	1-lb. 1-oz. dinner	1575	24.1			
(Swanson)	11½-oz. dinner	1058	12.5	4.	9.	
(Swanson) 3-course	16-oz. dinner	1747	17.0			
With gravy, dressing & potato (Swanson)	8¾-oz. pkg.	955	10.5			
(Weight Watchers)	18-oz. dinner		9.9			
TURKEY FRICASSEE, canned						
(Lynden Farms)	14.5-oz. can	1669	13.0	5.	8.	
TURKEY GIZZARD (USDA):						
Raw	4 oz.	66	8.3			164
Simmered	4 oz.	58	9.8			260
TURKEY PIE:						
Home recipe, baked (USDA)[2]	⅓ of 9″ pie (8.2 oz.)	633	31.3	9.	22.	72
Frozen:						
Commercial, unheated (USDA)[2]	8 oz.	837	23.6	7.	17.	20
(Banquet)	8-oz. pie		20.1			
(Banquet)	2-lb. 4-oz. pie		57.3			
(Morton)	8-oz. pie	1080	21.9			
(Swanson)	8-oz. pie	960	22.9			
(Swanson) deep-dish	1-lb. pie	1963	40.9			
TURKEY, POTTED (USDA)	1 oz.		5.4			
TURKEY SOUP, canned:						
(Campbell) *Chunky*	1 cup	848	4.0			
Broth (Lynden Farms)	1 cup (8 oz.)	953	0.			

(USDA): United States Department of Agriculture
*Prepared as Package Directs
[1]Principal sources of fat: turkey & butter.
[2]Principal sources of fat: vegetable shortening, cream, turkey & butter.

Food and Description	Measure or Quantity	Sodium (mg.)	—Fats in grams—			Cholesterol (mg.)
			Total	Saturated	Unsaturated	
Noodle:						
Condensed (USDA)	8 oz. (by wt.)	1887	5.4			
*Prepared with equal volume water (USDA)	1 cup (8.8 oz.)	1040	3.0			
*(Campbell)	1 cup	814	3.0	<1.	2.	
*(Heinz)	1 cup (8½ oz.)	1072	3.1			
(Heinz) *Great American*	1 cup (8¾ oz.)	1205	2.7			
Low sodium (Campbell)	7½-oz. can	40	3.2			
Rice, with mushrooms (Heinz) *Great American*	1 cup (8½ oz.)	1293	3.0			
*Vegetable (Campbell)	1 cup	854	3.2	<1.	2.	
Vegetable (Heinz) *Great American*	1 cup (8½ oz.)	1032	2.8			
*TURKEY SOUP MIX, noodle (Lipton)	1 cup (8 oz.)	933	2.1	<1.	2.	12
TURKEY TETRAZZINI, frozen (Stouffer's)	12-oz. pkg.	1345	32.4			
TURMERIC (Spice Islands)	1 tsp.	1				(0)
TURNIP (USDA):						
Fresh, without tops	1 lb. (weighed with skins)	191	.8			0
Fresh, pared, diced	½ cup (2.4 oz.)	33	.1			0
Fresh, pared, slices	½ cup (2.3 oz.)	31	.1			0
Boiled without salt, drained, diced	½ cup (2.8 oz.)	27	.2			0
Boiled without salt, drained, mashed	½ cup (4 oz.)	39	.2			0
TURNIP GREENS, leaves & stems:						
Fresh (USDA)	1 lb. (weighed untrimmed)		1.1			0
Boiled, in small amount water, short time, drained (USDA)	½ cup (2.5 oz.)		.1			0
Boiled, in large amount water, long time, drained (USDA)	½ cup (2.5 oz.)		.1			0
Canned, solids & liq.:						
(USDA)	½ cup (4.1 oz.)	274	.4			0
(Stokely-Van Camp)	½ cup (3.9 oz.)		.4			(0)
Frozen:						
Not thawed (USDA)	4 oz.	26	.3			0

(USDA): United States Department of Agriculture
*Prepared as Package Directs

Food and Description	Measure or Quantity	Sodium (mg.)	Total	Satu- rated	Unsatu- rated	Choles- terol (mg.)
			— Fats in grams —			
Boiled, drained (USDA)	½ cup (2.9 oz.)	14	.2			0
Chopped (Birds Eye)	½ cup (3.3 oz.)	22	.3			0

TURNOVER (See individual kinds)

TURTLE, GREEN (USDA):
Raw, in shell	1 lb. (weighed in shell)		.5			
Raw, meat only	4 oz.		.6			
Canned	4 oz.		.8			

TV DINNER (See individual listing such as **BEEF DINNER, CHICKEN DINNER, CHINESE DINNER, ENCHILADA DINNER,** etc.)

TWINKIE (Hostess):
2 to pkg.	1 cake (1.5 oz.)	241	4.3			
12 to pkg.	1 cake (1.3 oz.)	214	3.8			

U

UPPER-10, soft drink	6 fl. oz. (6.5 oz.)	10+	0.			0

V

VANILLA, bean (Spice Islands)	2″ piece	<1				(0)

VANILLA CAKE, frozen
(Pepperidge Farm)	⅙ of cake (3.1 oz.)	258	14.6			

VANILLA ICE CREAM (See also individual brand names):
(Borden) 10.5% fat	¼ pt. (2.3 oz.)	29	7.0			
Lady Borden 14% fat	¼ pt. (2.5 oz.)	26	10.0			
(Sealtest) *Party Slice*	¼ pt. (2.3 oz.)	48	6.7			
(Sealtest) 10.2% fat	¼ pt. (2.3 oz.)	48	6.7			
(Sealtest) 12.1% fat	¼ pt. (2.3 oz.)	51	7.8			
French (Prestige)	¼ pt. (2.6 oz.)	42	12.0			
Fudge royale (Sealtest)	¼ pt. (2.3 oz.)	51	5.6			

(USDA): United States Department of Agriculture
*Prepared as Package Directs

Food and Description	Measure or Quantity	Sodium (mg.)	—Fats in grams— Total	Saturated	Unsaturated	Cholesterol (mg.)
VANILLA ICE MILK						
(Borden) *Lite-line*	¼ pt.		1.9			
VANILLA PIE FILLING MIX						
(See **VANILLA PUDDING MIX**)						
VANILLA PUDDING:						
Blancmange, home recipe, with starch base (USDA)[1]	½ cup (4.5 oz.)	83	5.0	3.	2.	18
Canned (Betty Crocker)	½ cup	190	4.9			
Canned (Del Monte)	5-oz. can	321	5.1			
Canned (Hunt's)[2]	5-oz. can	186	12.4	2.	10.	
Canned (Thank You)	½ cup (4.5 oz.)		4.7			
Chilled (Breakstone)	5-oz. container	170	13.3			0
Chilled (Sealtest)	4 oz.	147	3.3			
VANILLA PUDDING or PIE FILLING MIX:						
Sweetened:						
*Regular, plain or French (Jell-O)	½ cup (5.2 oz.)	224	4.6			13
*Instant, plain or French (Jell-O)	½ cup (5.3 oz.)	406	4.7			13
*Regular (Royal)	½ cup (5.1 oz.)	230	4.8			14
*Instant (Royal)	¼ cup (5.1 oz.)	310	4.6			14
*Low calorie (D-Zerta)	½ cup (4.6 oz.)	142	4.5			
VANILLA RENNET MIX:						
Powder:						
Dry (Junket)	1 oz.	11	.1			
*(Junket)	4 oz.	56	3.9			
Tablet:						
Dry (Junket)	1 tablet (<1 gram)	197	Tr.			
*& sugar (Junket)	4 oz.	98	3.9			
VEAL, medium fat (USDA):						
Chuck, raw	1 lb. (weighed with bone)	327	36.0	17.	19.	258
Chuck, braised, lean & fat	4 oz.	91	14.5	7.	8.	115
Flank, raw	1 lb. (weighed with bone)	404	121.0	60.	61.	319

(USDA): United States Department of Agriculture
*Prepared as Package Directs
[1]Principal source of fat: milk.
[2]Principal source of fat: soybean oil.

Food and Description	Measure or Quantity	Sodium (mg.)	—Fats in grams—			Cholesterol (mg.)
			Total	Saturated	Unsaturated	
Flank, stewed, lean & fat	4 oz.	91	36.6	18.	18.	115
Foreshank, raw	1 lb. (weighed with bone)	212	19.0	10.	9.	168
Foreshank, stewed, lean & fat	4 oz.	91	11.8	6.	6.	115
Loin, raw	1 lb. (weighed with bone)	338	41.0	20.	21.	267
Loin, broiled, medium done, chop, lean & fat	4 oz.	91	15.2	8.	7.	115
Plate, raw	1 lb. (weighed with bone)	322	61.0	30.	31.	254
Plate, stewed, lean & fat	4 oz.	91	24.0	12.	12.	115
Rib, raw, lean & fat	1 lb. (weighed with bone)	314	49.0	23.	26.	248
Rib, roasted, medium done, lean & fat	4 oz.	91	19.2	9.	10.	115
Round & rump, raw	1 lb. (weighed with bone)	314	31.0	16.	15.	248
Round & rump, broiled, steak or cutlet, lean & fat	4 oz. (weighed without bone)	91	12.6	7.	6.	115
VEAL DINNER, frozen:						
Parmigiana (Kraft)	11-oz. dinner	1962	30.3			
Parmigiana (Swanson)	12¼-oz. dinner	1335	23.0			
Parmigiana (Weight Watchers)	9½-oz. luncheon		16.1			
Breaded veal with spaghetti in tomato sauce (Swanson)	8¼-oz. pkg.	974	11.5			
VEGETABLE BOUILLON CUBE:						
(Herb-Ox)	1 cube (4 grams)	900	.1			
(Steero)	1 cube (4 grams)		.1			
(Wyler's)	1 cube (4 grams)		.3			
VEGETABLE FAT (See **FAT**)						
VEGETABLE JUICE COCKTAIL, canned:						
(USDA)	4 oz. (by wt.)	227	.1			0
Unseasoned (S and W) *Nutradiet*	4 oz. (by wt.)	16	.1			(0)
V-8 (Campbell)	¾ cup	618	.2			(0)

(USDA): United States Department of Agriculture
*Prepared as Package Directs

337

Food and Description	Measure or Quantity	Sodium (mg.)	— Fats in grams —			Choles- terol (mg.)
			Total	Satu- rated	Unsatu- rated	

VEGETABLES, MIXED:
Canned, regular pack:

(Veg-All)	½ cup (4 oz.)		.1			0
Solids & liq. (Del Monte)	½ cup (4 oz.)	518	.5			(0)
Drained solids (Del Monte)	½ cup (2.8 oz.)	354	.4			(0)
Drained liq. (Del Monte)	4 oz.	536	.3			(0)
Solids & liq. (Stokely-Van Camp)	½ cup (3.8 oz.)		.4			(0)
Canned, Chinese,						
Chop Suey (Hung's)	4 oz.		.3			
Frozen:						
Not thawed (USDA)	4 oz.	67	.3			0
Boiled, drained (USDA)	½ cup (3.2 oz.)	48	.3			0
(Birds Eye)	½ cup (3.3 oz.)	46	.3			0
In butter sauce (Green Giant)	⅓ of 10-oz. pkg.	354	1.9			
Chinese (Birds Eye)	⅓ of 10-oz. pkg.	443	4.1			0
Jubilee (Birds Eye)	⅓ of 10-oz. pkg.	321	6.5			Tr.
With onion sauce (Birds Eye)	½ cup (2.7 oz.)	393	6.5			Tr.

VEGETABLE OYSTER (See SALSIFY)

VEGETABLE SOUP:
Canned, regular pack:

*(Campbell)	1 cup	776	1.6	Tr.	1.	3
(Campbell) *Chunky*	1 cup	904	3.0			
*(Campbell) old fashioned	1 cup	811	2.4	<1.	2.	
Beef, condensed (USDA)	8 oz. (by wt.)	1937	4.1			
*Beef, prepared with equal volume water (USDA)	1 cup (8.6 oz.)	1046	2.2			
*Beef (Campbell)	1 cup	858	2.6	<1.	2.	6
*Beef (Heinz)	1 cup (8½ oz.)	1005	1.5			
Beef (Heinz) *Great American*	1 cup (8¾ oz.)	1157	3.4			
With beef broth, condensed (USDA)	8 oz. (by wt.)	1565	3.2			
*With beef broth, prepared with equal volume water (USDA)	1 cup (8.8 oz.)	862	1.8			
With beef broth (Heinz) *Great American*	1 cup (8¾ oz.)	1116	3.7			
*With beef stock (Heinz)	1 cup (8½ oz.)	1049	2.1			
With ground beef (Heinz) *Great American*	1 cup (8¾ oz.)	1161	5.3			

(USDA): United States Department of Agriculture
*Prepared as Package Directs

| Food and Description | Measure or Quantity | Sodium (mg.) | —Fats in grams— | | Choles-terol (mg.) |
			Total	Satu-rated	Unsatu-rated	
*& *Noodle-O's* (Campbell)	1 cup	975	2.5			
Vegetarian:						
Condensed (USDA)	8 oz. (by wt.)	1551	3.9			
*Prepared with equal volume						
water (USDA)	1 cup (8.6 oz.)	838	2.0			
*(Campbell)	1 cup	585	1.7	Tr.	2.	7
*(Heinz)	1 cup (8¾ oz.)	1250	2.0			
(Heinz) *Great American*	1 cup (8½ oz.)	1326	3.5			
*(Manischewitz)	1 cup		1.8			
Canned, dietetic pack:						
Low sodium (Campbell)	7½-oz. can	40	1.8			
Beef, low sodium (Campbell)	7½-oz. can	40	2.7			
*(Claybourne)	8 oz.	45	.7			
*(Slim-ette)	8 oz. (by wt.)	16	.2			
(Tillie Lewis)	1 cup (8 oz.)	60	1.1			
Frozen:						
With beef, condensed (USDA)	8 oz. (by wt.)	1796	5.2			
*With beef, prepared with equal						
volume water (USDA)	8 oz. (by wt.)	898	2.7			

VEGETABLE SOUP MIX:

*(Wyler's)	1 cup		1.1			
*Beef (Lipton)	1 cup	1217	1.5	Tr.	1.	
*Chicken (Wyler's)	1 cup		.8			
*& noodle (Lipton) Country	1 cup	1169	1.1	Tr.	<1.	13
Spring (Lipton) *Cup-a-Soup*	½-oz. pkg.	1065	.8	Tr.	<1.	4

VEGETABLE STEW, canned

(Hormel) *Dinty Moore*	8 oz.	1021	8.4	4.	4.	11

"VEGETARIAN FOODS":

Canned or dry:						
Beans, rich brown (Loma Linda)	½ cup (3.7 oz.)	700	1.4			0
Bean, soy:						
Boston style (Loma Linda)	½ cup (3.7 oz.)	779	4.3			0
Green, drained (Loma Linda)	½ cup (3.7 oz.)	333	4.7			0
Tomato sauce (Loma Linda)	½ cup (3.7 oz.)	454	3.4			0
Big franks, drained (Loma Linda)	1 frank (1.6 oz.)	229	5.6			0
Breading meal (Loma Linda)	1 cup (3.7 oz.)	2625	5.4			0
Breading meal (Worthington)	¼ cup (1.1 oz.)	936	.5			
Burger aid (Worthington)	1 oz.		.2			
Cheze-O-Soy (Worthington)	½" slice (2.5 oz.)		7.1			
Chili (Worthington)	¼ can (5 oz.)		6.0			

(USDA): United States Department of Agriculture
*Prepared as Package Directs

Food and Description	Measure or Quantity	Sodium (mg.)	Fats in grams — Total	Satu- rated	Unsatu- rated	Choles- terol (mg.)
Chili with beans (Loma Linda)	½ cup (4.7 oz.)	498	3.3			0
Choplet (Worthington)	1 choplet (2.2 oz.)	350	1.2			
Choplet burger (Worthington)	⅓ cup (3.2 oz.)		1.9			
Cutlet (Worthington)	1 cutlet (2.2 oz.)		1.2			
Dinner bits, drained (Loma Linda)	1 bit (.5 oz.)	96	1.2			0
Dinner cuts, drained (Loma Linda)	1 cut (1.6 oz.)	252	.8			0
Dinner cuts, no salt added, drained (Loma Linda)	1 cup (1.6 oz.)	5	.8			0
Fry stick (Worthington)	1 piece (2.3 oz.)		3.4			
Garbanzo (Loma Linda)	½ cup (4.1 oz.)	534	1.3			0
GranBurger (Worthington)	1 oz.	798	.3			
Granola (Loma Linda)	½ cup (2 oz.)	107	8.5			
Gravy Quik, brown (Loma Linda)	⅙ of pkg. (6 grams)	194	.4			0
J-7901 or J-7901A (Worthington)	1 oz.		.3			
Jell Quik, dry (Loma Linda)	1 oz.	291	.8			0
Kaffir Tea (Worthington)	1 bag (1 gram)		Tr.			
Lentils, drained (Loma Linda)	¾ cup (3.5 oz.)	472	.7			0
Linketts, drained (Loma Linda)	1 link (1.3 oz.)	231	5.0			0
Little links, drained (Loma Linda)	1 link (.8 oz.)	120	2.8			0
Madison burger (Worthington)	⅓ cup (2 oz.)		2.3			
Meat-like loaf (Loma Linda):						
Beef, drained	¼" slice (2 oz.)	400	7.1			0
Chicken	¼" slice (2 oz.)	403	6.9			0
Luncheon	¼" slice (2 oz.)	591	7.3			0
Turkey	¼" slice (2 oz.)	554	7.5			0
Meat loaf mix (Worthington)	2 oz.	921	8.0			
Multigen powder (Loma Linda)	½ cup (2.1 oz.)		2.1			0
Non-meatballs (Worthington)	1 piece (.6 oz.)		1.9			
Non-meat with tomato (Worthington)	2⅓-oz.		11.6			
Numete (Worthington)	½" slice (2.3 oz.)		11.0			
Nuteena (Loma Linda)	½" slice (2.5 oz.)	331	11.8			0
Oven-cooked wheat (Loma Linda)	½ cup (2.6 oz.)	7	2.5			0
Peanuts & soya (USDA)	4 oz.		19.2			
Prime vegetable burger (Worthington)	½" slice (2.1 oz.)	444	0.			
Proteena (Loma Linda)	½" slice (2.6 oz.)	385	7.5			0
Protose (Worthington)	½" slice (2.7 oz.)		8.5			
Rediburger (Loma Linda)	½" slice (2.5 oz.)	259	9.3			0

(USDA): United States Department of Agriculture
*Prepared as Package Directs

Food and Description	Measure or Quantity	Sodium (mg.)	Total	Satu- rated	Unsatu- rated	Choles- terol (mg.)
				Fats in grams		
Redi-loaf mix, beef (Loma Linda)	⅓ cup (1.2 oz.)	770	8.0			0
Redi-loaf mix, chicken (Loma Linda)	⅓ cup (1.2 oz.)	490	7.5			0
Ruskets, biscuits (Loma Linda)	1 biscuit (.6 oz.)	47	.5			0
Ruskets, flakes (Loma Linda)	1 cup (1 oz.)	73	.8			0
Sandwich spread (Loma Linda)	1 T. (.6 oz.)	97	1.4			0
Sandwich spread (Worthington)	3 T. (1.2 oz.)		5.1			
Saucette (Worthington)	1 link (.6 oz.)		2.6			
Savita (Worthington)	1 tsp. (.4 oz.)	465	0.			
Savorex (Loma Linda)	1 oz.		<.1			0
Seasoning, chicken (Loma Linda)	1 T. (3 grams)	84	.2			0
Soyagen:						
A.P. & malt powder (Loma Linda)	¼ cup (1.4 oz.)	204	9.0			0
Carob, powder (Loma Linda)	¼ cup (1.4 oz.)	212	8.6			0
Liquid (Loma Linda)	1 cup (8.6 oz.)	303	8.5			0
Soyalac, concentrate, liquid (Loma Linda)	1 fl. oz. (1.1 oz.)	16	2.5			0
Soyalac, infant powder (Loma Linda)	1 oz.	78	7.4			0
Soyalac, ready-to-use (Loma Linda)	1 oz.	9	1.1			0
Soyameat (Worthington):						
Sliced beef	1 slice (1 oz.)	140	3.3			
Diced beef	1 oz.	174	1.4			
Diced chicken	1 oz.		1.7			<1
Fried chicken	1 piece (1.2 oz.)	230	4.9			0
Sliced chicken	1 slice (1.1 oz.)		2.2			<1
Salisbury steak	1 slice (2.5 oz.)	5530	9.0			
Soyamel, any kind (Worthington)	1 oz.		5.9			0
Soy flour (Loma Linda)	¼ cup (.9 oz.)	3	5.8			0
Stew pac, drained (Loma Linda)	½ cup (3 oz.)	481	7.7			0
Stripple Zips (Worthington)	1 oz.	943	6.8			
Tamales (Worthington)	1 oz.		<.1			
Tastee cuts (Loma Linda)	1 cut (1.5 oz.)	222	.7			0
Tenderbit (Loma Linda)	1 piece (.9 oz.)	126	1.2			0
Vegeburger (Loma Linda)	½ cup (3.9 oz.)	385	2.7			0
Vegeburger, no salt added (Loma Linda)	½ cup (3.9 oz.)	17	2.7			0
Vegechee (Loma Linda)	½" slice (2.5 oz.)	397	6.8			0
Vegelona (Loma Linda)	½" slice (3.3 oz.)	439	6.5			0

(USDA): United States Department of Agriculture
*Prepared as Package Directs

Food and Description	Measure or Quantity	Sodium (mg.)	Total	Satu-rated	Unsatu-rated	Choles-terol (mg.)
			— Fats in grams —			
Vegetarian burger (Worthington)	⅓ cup (2.5 oz.)		4.4			
Vegetable skallop (Worthington)	1 piece (.7 oz.)		.3			0
Vegetable steaks (Worthington)	1 piece (.7 oz.)		.3			
Veja-links (Worthington)	1 link (1.2 oz.)	225	5.6			<1
Wham, sliced or loaf (Worthington)	1 slice (1 oz.)		2.5			1
Wheat germ, natural or toasted (Loma Linda)	1 T. (.4 oz.)	21	1.2			0
Wheat protein (USDA)	4 oz.		.9			
Wheat protein, nuts or peanuts (USDA)	4 oz.		8.1			
Wheat protein, vegetable oil (USDA)	4 oz.		12.1			
Wheat & soy protein (USDA)	4 oz.		1.4			
Wheat & soy protein, soy or other vegetable oil (USDA)	4 oz.		6.4			
Worthington 209	1 slice (.5 oz.)	96	1.1			0
Yum (Worthington)	1 serving (1.5 oz.)		4.0			
Frozen (Worthington):						
Beef pie	1 pie (7.9 oz.)		28.4			
Beef style, loaf or sliced	1 slice (1 oz.)	156	2.7			
Chicken pie	1 pie (7.9 oz.)		26.9			
Chicken style, diced, roll or sliced	1 oz. or 1 slice	294	5.0			
Chic-Ketts	1 oz.		2.5			
Chili	¼ can (5 oz.)		5.9			
Corned beef, loaf or sliced	1 slice (.5 oz.)	179	2.3			0
Croquettes	1 croquette (1 oz.)		3.2			
FriPats	1 pat (2.6 oz.)	607	10.6			0
Holiday roast	1 slice (2 oz.)		9.1			0
Non-meatballs	1 piece (.6 oz.)		3.3			
Prosage	⅜" slice (1.2 oz.)	354	4.8			<1
Salisbury steak	1 slice (2 oz.)		5.3			
Smoked beef, roll or sliced	1 slice (7 grams)		.7			
Stripples	1 slice (7 grams)	95	1.0			
Turkey, smoked, loaf or sliced	1 slice (.7 oz.)	214	3.0			0
Wham, diced, sliced or loaf	1 slice (1 oz.)		2.7			1
VENISON, raw, lean meat only (USDA)	4 oz.		4.5	3.	1.	

(USDA): United States Department of Agriculture
*Prepared as Package Directs

Food and Description	Measure or Quantity	Sodium (mg.)	Fats in grams Total	Satu- rated	Unsatu- rated	Choles- terol (mg.)
VERMOUTH, dry or sweet (Great Western) 16% alcohol	3 fl. oz.	20				
VERNORS, soft drink:						
Regular	6 fl. oz. (6.2 oz.)	1	0.			0
Low calorie	6 fl. oz. (6.2 oz.)	3	0.			0
VICHYSSOISE SOUP (Crosse & Blackwell)	½ can (6½ oz.)		5.2			
VIENNA SAUSAGE, canned:						
(USDA)	1 oz.		5.6			
(USDA)	1 sausage (from 5-oz. can, .6 oz.)		3.2			
(Armour Star)	5-oz. can		36.1			
(Armour Star)	1 sausage (.6 oz.)		4.1			
(Hormel)	1 sausage (.6 oz.)	127	3.7			
(Van Camp)	1 oz.		5.6			
(Wilson)	1 oz.	228	7.9			13
VINEGAR:						
Cider:						
(USDA)	½ cup (4.2 oz.)	1	0.			0
(USDA)	1 T. (.5 oz.)	<1	0.			0
Distilled:						
(USDA)	½ cup (4.2 oz.)	1				0
(USDA)	1 T. (.5 oz.)	<1				0
Red or white wine (Regina)	1 T. (.5 oz.)	<1	Tr.			(0)
Red wine:						
Plain (Spice Islands)	2 T.	6				(0)
Eschalot (Spice Islands)	2 T.	10				(0)
Garlic (Spice Islands)	2 T.	9				(0)
Tarragon (Spice Islands)	2 T.	11				(0)
Rose (Spice Islands)	2 T.	10				(0)
White wine:						
Plain (Spice Islands)	2 T.	11				(0)
Basil (Spice Islands)	2 T.	10				(0)
Tarragon (Spice Islands)	2 T.	10				(0)
VINESPINACH or BASELLA, raw (USDA)	4 oz.		.3			0
VIN ROSE (See **ROSE WINE**)						
VIRGIN SOUR MIX (Party Tyme)	½-oz. pkg.	61	0.			(0)

(USDA): United States Department of Agriculture
*Prepared as Package Directs

343

Food and Description	Measure or Quantity	Sodium (mg.)	—Fats in grams—			Choles- terol (mg.)
			Total	Satu- rated	Unsatu- rated	

VODKA, unflavored (See
DISTILLED LIQUOR)

VODKA SOFT DRINK,
  sweetened (Shasta)

VODKA SOFT DRINK, sweetened (Shasta)	6 fl. oz.	22	0.			0

W

WAFER (See **COOKIE** or
CRACKER)

WAFFLE:

Food and Description	Measure or Quantity	Sodium	Total	Satu- rated	Unsatu- rated	Choles- terol
Home recipe (USDA)[1]	2.6-oz. waffle (7″ dia.)	356	7.4	2.	5.	
Frozen: (USDA)[2]	1.6-oz. waffle (8 in 13-oz. pkg.)	296	2.9	<.1	2.	
(USDA)[2]	.8-oz. waffle (6 in 5-oz. pkg.)	155	1.5	Tr.	1.	
Buttermilk (Aunt Jemima)	1 section (¾ oz.)	170	2.3			
Original (Aunt Jemima)	1 section (¾ oz.)	160	2.4			

WAFFLE MIX (USDA) (See also
PANCAKE & WAFFLE MIX):

Food and Description	Measure or Quantity	Sodium	Total	Satu- rated	Unsatu- rated	Choles- terol
Dry, complete mix[2]	1 oz.	291	5.4	1.	4.	
*Prepared with water[2]	2.6-oz. waffle (½″ x 4½″ x 5½″, 7″ dia.)	420	10.5	2.	8.	
Dry, incomplete mix	1 oz.	406	.5			
*Prepared with egg & milk[3]	2.6-oz. waffle (7″ dia.)	514	8.0	3.	5.	45
*Prepared with egg & milk[3]	7.1-oz. waffle (9″ x 9″ x ⅝″, 1⅛ cup batter)	1372	21.2	8.	13	120

WAFFLE SYRUP (See **SYRUP)**

WALNUT:
  Black:

Food and Description	Measure or Quantity	Sodium	Total	Satu- rated	Unsatu- rated	Choles- terol
In shell, whole (USDA)	1 lb. (weighed in shell)	3	59.2	4.	55.	0

(USDA): United States Department of Agriculture
*Prepared as Package Directs
[1]Principal sources of fat: vegetable shortening, egg & milk.
[2]Principal sources of fat: vegetable shortening & egg.
[3]Principal sources of fat: vegetable shortening, egg & milk.

Food and Description	Measure or Quantity	Sodium (mg.)	— Fats in grams —			Choles-terol (mg.)
			Total	Satu-rated	Unsatu-rated	
Shelled, whole (USDA)	4 oz.	3	67.2	5.	63.	0
Chopped (USDA)	½ cup (2.1 oz.)	2	35.6	2.	33.	0
Kernels (Hammons)	4 oz.		63.8			0
English or Persian:						
In shell, whole (USDA)	1 lb. (weighed in shell)	4	130.6	9.	122.	0
Shelled, whole (USDA)	4 oz.	2	72.6	4.5	68.	0
Chopped (USDA)	½ cup (2.1 oz.)	1	38.4	2.	36.	0
Chopped (USDA)	1 T. (8 grams)	<1	4.8	Tr.	4.	0
Halves (USDA)	½ cup (1.8 oz.)	1	32.0	2.	30.	0
(Diamond)	3-oz. bag (¾ cup)	2	54.7	5.	50.	(0)
(Diamond)	15 halves (.5 oz.)	<1	9.7	<1.	9.	(0)

*WALNUT CAKE MIX, BLACK,

(Betty Crocker)	$^{1}/_{12}$ of cake	271	5.6			

WATER (See page 353)

WATER CHESTNUT, CHINESE,
raw (USDA):

Whole	1 lb. (weighed unpeeled)	70	.7			0
Peeled	4 oz.	23	.2			0

WATERCRESS, raw (USDA):

Untrimmed	½ lb. (weighed untrimmed)	108	.6			0
Trimmed	½ cup (.6 oz.)	8	<.1			0

WATERMELON, fresh (USDA):

Whole	1 lb. (weighed with rind)	2	.4			0
Wedge	2-lb. wedge (4″ x 8″, measured with rind)	4	.9			0
Slice	½ slice (12.2 oz., ¾″ x 10″)	2	.3			0
Diced	1 cup (5.6 oz.)	2	3			0

WATERMELON RIND (Crosse & Blackwell)

Blackwell)	1 T. (.6 oz.)	210	0.			(0)

(USDA): United States Department of Agriculture
*Prepared as Package Directs

Food and Description	Measure or Quantity	Sodium (mg.)	—Fats in grams—			Choles-terol (mg.)
			Total	Satu-rated	Unsatu-rated	
WATERMELON SOFT DRINK, sweetened:						
(Hoffman)	6 fl. oz.	14	0.			0
(Nedick's)	6 fl. oz.	14	0.			0
WAX GOURD, raw (USDA):						
Whole	1 lb. (weighed with skin & cavity contents)	19	.6			0
Flesh only	4 oz.	7	.2			0
WEAKFISH (USDA):						
Raw, whole	1 lb. (weighed whole)	163	12.2			
Broiled, meat only, salt added	4 oz.	635	12.9			
WELSH RAREBIT:						
Home recipe (USDA)[1]	1 cup (8.2 oz.)	770	31.6	16.	15.	72
Canned (Snow)	4 oz.		11.4			
WEST INDIAN CHERRY (See **ACEROLA**)						
WHALE MEAT, raw (USDA)	4 oz.	88	8.5	1.	7.	
WHEAT CHEX, cereal (Ralston)	⅔ cup (1 oz.)	217	.3			(0)
WHEATENA, cereal	½ cup (.9 oz. dry)		.5			(0)
WHEAT FLAKES, cereal, crushed (USDA)	1 cup (2.5 oz.)	722	1.1			0
WHEAT GERM, crude, commercial, milled (USDA)	1 oz.	<1	3.1	<1.	3.	0
WHEAT GERM, CEREAL:						
(USDA)	¼ cup (1 oz.)	<1	3.2	<1.	3.	0
(Kretschmer)	¼ cup (1 oz.)	1	3.1			(0)
With sugar & honey (Kretschmer)	¼ cup (1 oz.)		2.3			(0)
WHEATIES, cereal (General Mills)	1¼ cups (1 oz.)	393	.5			(0)
WHEAT OATA, cereal, dry	¼ cup (1 oz.)	1	1.4			(0)

(USDA): United States Department of Agriculture
*Prepared as Package Directs
[1]Principal sources of fat: cheese, butter & milk.

Food and Description	Measure or Quantity	Sodium (mg.)	— Fats in grams —			Choles- terol (mg.)
			Total	Satu- rated	Unsatu- rated	
WHEAT, PUFFED, cereal:						
Added nutrients, without salt						
(USDA)	1 cup (.4 oz.)	<1	.2			0
Frosted with sugar and honey						
(USDA)	1 cup (.4 oz.)	19	.3			0
(Checker)	½ oz.	3	.3			(0)
(Quaker)	1⅓ cups (½ oz.)	<1	.2			(0)
(Sunland)	½ oz.	3	.3			(0)
(Whiffs)	½ oz.	3	.3			(0)
WHEAT, ROLLED (USDA):						
Uncooked	1 cup (3.1 oz.)	2	1.7			0
Cooked, salt added	1 cup (7.7 oz.)	640	.9			0
WHEAT, SHREDDED, cereal (See **SHREDDED WHEAT**)						
WHEY, fluid (USDA)	1 cup (8.6 oz.)		.7			
*__WHIP 'N CHILL__ (Jell-O):						
All flavors except chocolate	½ cup (3 oz.)	60	5.3			3
Chocolate	½ cup (3 oz.)	48	5.5			3
WHISKEY or WHISKY (See **DISTILLED LIQUOR**)						
WHISKEY SOUR MIX (Party Tyme)	½-oz. pkg.	87	0.			(0)
WHISKEY SOUR SOFT DRINK, sweetened (Shasta)	6 fl. oz.	22	0.			(0)
WHITEFISH, LAKE (USDA):						
Raw, whole	1 lb. (weighed whole)	111	17.5			
Raw, meat only	4 oz.	59	9.3			
Baked, stuffed, home recipe[1]	4 oz.	221	15.9			
Smoked	4 oz.		8.3			
WHITEFISH & PIKE (See **GEFILTE FISH**)						
WIENER (See **FRANKFURTER**)						

(USDA): United States Department of Agriculture
*Prepared as Package Directs
[1]Prepared with bacon, butter, onion, celery & bread crumbs.

Food and Description	Measure or Quantity	Sodium (mg.)	—Fats in grams— Total	Satu- rated	Unsatu- rated	Choles- terol (mg.)
WILD BERRY, fruit drink (Hi-C)	6 fl. oz. (6.3 oz.)	Tr.	Tr.			0
WILD RICE, raw (USDA)	½ cup (2.9 oz.)	6	.6			0
WINE (Most wines are listed by kind, brand, vineyard, region or grape name): Cooking, Sauterne or Burgundy						
(Regina)	½ cup (3.9 oz.)	657	0.			(0)
Cooking, Sherry (Regina)	½ cup (3.9 oz.)	657	0.			(0)
Dessert (USDA) 18.8% alcohol	3 fl. oz. (3.1 oz.)	4	0.			0
Table (USDA) 12.2% alcohol	3 fl. oz. (3.1 oz.)	4	0.			0
WINK, soft drink (Canada Dry) bottle or can	6 fl. oz.	15+	0.			(0)
WORCESTERSHIRE SAUCE (See **SAUCE,** Worcestershire)						
WRECKFISH, raw, meat only (USDA)	4 oz.		4.4			

Y

YAM (USDA): Raw, whole	1 lb. (weighed with skin)		.8			0
Raw, flesh only Canned & frozen (See **SWEET POTATO**)	4 oz.		.2			0
YAM BEAN, raw (USDA): Unpared tuber	1 lb. (weighed unpared)		.8			0
Pared tuber	4 oz.		.2			0
YEAST: Baker's:						
Compressed (USDA)	1 oz.	5	.1			0
Compressed (Fleischmann's)	³/₅-oz. cake	3	.1			
Dry (USDA)	1 oz.	15	.5			0
Dry (USDA)	1 pkg. (7 grams)	4	.1			0
Dry (Fleischmann's)	¼ oz. (pkg. or jar)	5	.1			0

(USDA): United States Department of Agriculture
*Prepared as Package Directs

Food and Description	Measure or Quantity	Sodium (mg.)	— Fats in grams —			Cholesterol (mg.)
			Total	Saturated	Unsaturated	
Brewer's dry, debittered (USDA)	1 oz.	34	.3			0
Brewer's dry, debittered (USDA)	1 T. (8 grams)	10	<.1			0
YELLOWTAIL, raw, meat only (USDA)	4 oz.		6.1			
YOGURT:						
Made from whole milk (USDA)	½ cup (4.3 oz.)	57	4.1	2.	2.	
Made from partially skimmed milk, plain or vanilla:						
(USDA)	½ cup (4.3 oz.)	62	2.1	1.	<1.	10
(USDA)	8-oz. container	116	3.9	2.	2.	18
Made from partially skimmed milk, fruit-flavored (USDA)	8-oz. container		2.7			15
Plain:						
(Borden) Swiss style	5-oz. container	101	1.4			
(Borden) Swiss style	8-oz. container	161	2.3			
(Breakstone)	8-oz. container	168	3.9			10
(Breakstone)	1 T. (.5 oz.)	10	.2			<1
(Dannon)	8-oz. container	170	3.6	2.	1.	11
Apple, Dutch (Dannon)	8-oz. container	125	2.5	2.	1.	8
Apricot:						
(Breakstone)	8-oz. container	128	2.9			8
(Breakstone) *Swiss Parfait*	8-oz. container	128	3.6			10
(Dannon)	8-oz. container	125	2.5	2.	1.	8
Black cherry (Breakstone) *Swiss Parfait*	8-oz. container	136	3.6			10
Blueberry:						
(Breakstone)	8-oz. container	136	2.9			8
(Breakstone) *Swiss Parfait*	8-oz. container	136	3.6			10
(Dannon)	8-oz. container	125	2.5	2.	1.	8
(Sealtest) *Light n' Lively*	8-oz. container	114	1.8			
(SugarLo)	8-oz. container	141	2.3	1.	1.	7
Boysenberry (Dannon)	8-oz. container	125	2.5	2.	1.	8
Cherry:						
(Dannon)	8-oz. container	125	2.5	2.	1.	8
Dark (SugarLo)	8-oz. container	141	2.3	1.	1.	7
Cinnamon apple (Breakstone)	8-oz. container	128	2.9			8
Coffee (Dannon)	8-oz. container	152	3.2	2.	1.	11
Danny (Dannon)	2½-oz. pop	51	5.0	3.	2.	
Honey (Breakstone) *Swiss Parfait*	8-oz. container	128	3.6			10
Lemon:						
(Breakstone) *Swiss Parfait*	8-oz. container	128	3.6			10
(Sealtest) *Light n' Lively*	8-oz. container	127	1.9			

(USDA): United States Department of Agriculture
*Prepared as Package Directs

| Food and Description | Measure or Quantity | Sodium (mg.) | —Fats in grams— | | | Choles-terol (mg.) |
			Total	Satu-rated	Unsatu-rated	
Lime (Breakstone) *Swiss Parfait*	8-oz. container	128	3.9			10
Mandarin orange:						
(Borden) Swiss style	5-oz. container	76	1.0			
(Borden) Swiss style	8-oz. container	121	1.6			
(Breakstone) *Swiss Parfait*	8-oz. container	128	3.6			10
Peach:						
(Borden) Swiss style	5-oz. container	74	1.0			
(Borden) Swiss style	8-oz. container	119	1.6			
(Breakstone) *Swiss Parfait*	8-oz. container	128	3.4			8
Melba (Breakstone) *Swiss Parfait*	8-oz. container	128	3.4			8
(Sealtest) *Light n' Lively*	8-oz. container	116	1.8			
(SugarLo)	8-oz. container	141	2.3	1.	1.	7
Pineapple:						
(Breakstone)	8-oz. container	128	2.9			8
(Sealtest) *Light n' Lively*	8-oz. container	120	1.8			
(SugarLo)	8-oz. container	141	2.3	1.	1.	7
Pineapple-orange (Dannon)	8-oz. container	125	2.5	2.	1.	8
Prune whip (Breakstone)	8-oz. container	128	2.9			8
Prune whip (Dannon)	8-oz. container	125	2.5	2.	1.	8
Raspberry:						
(Borden) Swiss style	5-oz. container	82	1.1			
(Borden) Swiss style	8-oz. container	132	1.8			
(Breakstone)	8-oz. container	136	2.9			8
Red (Breakstone) *Swiss Parfait*	8-oz. container	144	3.9			10
(Dannon)	8-oz. container	125	2.5	2.	1.	8
Red (Sealtest) *Light n' Lively*	8-oz. container	125	1.9			
(SugarLo)	8-oz. container	143	2.3	1.	1.	7
Strawberry:						
(Borden) Swiss style	5-oz. container	81	1.1			
(Borden) Swiss style	8-oz. container	129	1.8			
(Breakstone)	8-oz. container	128	2.9			8
(Breakstone) *Swiss Parfait*	8-oz. container	128	3.6			9
(Dannon)	8-oz. container	125	2.5	2.	1.	8
(Sealtest) *Light n' Lively*	8-oz. container	127	1.9			
(SugarLo)	8-oz. container	141	2.3	1.	1.	7
Vanilla:						
(Borden) Swiss style	5-oz. container	87	1.3			
(Borden) Swiss style	8-oz. container	139	2.0			
(Breakstone)	8-oz. container	208	3.4			10
(Dannon)	8-oz. container	152	3.2	2.	1.	11

YOUNGBERRY, fresh (See **BLACKBERRY,** fresh)

(USDA): United States Department of Agriculture
*Prepared as Package Directs

Food and Description	Measure or Quantity	Sodium (mg.)	— Fats in grams —			Choles- terol (mg.)
			Total	Satu- rated	Unsatu- rated	

Z

ZELLERSCHWARZE KATZ, wine
(Julius Kayser) 9% alcohol — 3 fl. oz. — 2 — 0. — — — (0)

ZING, cereal beverage, 0.4%alcohol — 12 fl. oz. (12 oz.) — 41 — — — — 0

ZITI, baked, with sauce, frozen
(Buitoni) — 4 oz. — — 1.6

ZUCCHINI (See **SQUASH, SUMMER**)

ZWIEBACK:
(USDA)[1] — 1 oz. — 71 — 2.5 — <1. — — 2.
(Nabisco) — 1 piece (7 grams) — 6 — .7

(USDA): United States Department of Agriculture
*Prepared as Package Directs
[1]Principal sources of fat: vegetable shortening & egg.

BIBLIOGRAPHY

Dawson, Elsie H., Gilpin, Gladys L., and Fulton, Lois H., *Average weight of a measured cup of various foods*. U.S.D.A. ARS 61-6, February 1969, 19 pp.

Durfor, C. N., and Becker, E., *Geological Survey Water Supply Paper* 1812, Washington, U. S. Government Printing Office, 1964, pp. 1–364.

Feeley, R. M., Criner, P. E., and Watt, B. K., "Cholesterol Content of Food," *Journal of the American Dietetic Association,* 61, August 2, 1972, pp. 134–149.

Merrill, A. L. and Watt, B. K., *Energy value of foods — basis and derivation*. U.S.D.A. Handb. 74, 105 pp., 1955.

Pecot, Rebecca K., Jaeger, Carol M., and Watt, Bernice K., *Proximate composition of beef from carcass to cooked meat: Method of derivation and tables of values*. U.S.D.A. Home Economics Research Report 31, 32 pp. 1965.

Pecot, Rebecca K. and Watt, Bernice K., *Food yields: Summarized by different stages of preparation*. U.S.D.A. Handb. 102, 93 pp., 1956.

U.S.D.A. *Nutritive value of foods*. Home and Garden Bul. 72, 36 pp., 1964, and revised edition, 1970, 41 pp.

U.S.D.A. Unpubl. Data 1969.

Watt, Bernice K., Merrill, Annabel L., *et al., Composition of foods: Raw, processed, prepared*. U.S.D.A. Agriculture Handb. 8, 190 pp., 1963.

APPENDIX

WATER SUPPLIES

This chart covers the largest cities in the United States and their principal sources of water supplies. The two types of resources are ground water (wells and infiltration galleries) and surface water (streams, lakes and reservoirs). Most cities use but one source; a second group always uses more than one source; a third group uses different sources only part of the year. Therefore, the column headed "% of water supply" includes both primary and auxiliary sources wherever used.

SODIUM CONTENT OF LOCAL WATER SUPPLIES

State, city & plant, reservoir or lake	% of water supply	Finished water or raw water	Sodium (mg.) 1 cup
ALABAMA:			
Birmingham:			
Inland Lake	10%	R	.5
Putnam Station filter plant	10%	F	.4
Cahaba River	90%	R	2.4
Shades Mountain filter plant	90%	F	2.3
Mobile:			
Big Creek	100%	R	.5
Treatment plant	100%	F	.6
Montgomery:			
Court Street plant, 18 wells	39%	R	12.1
Court Street treatment plant	39%	F	13.0
Day Street plant, 31 wells	61%	R	12.1
Day Street treatment plant	61%	F	12.6
ARIZONA:			
Phoenix:			
Verde River	12%	R	31.0
Verde River filter plant	12%	F	31.0
Squaw Peak filter plant	30%	F	32.7
Verde well field	17%	F	9.7
Scottsdale well 36	15%	F	22.8
Tucson:			
Composite of southside wells	33%	F	15.9

State, city & plant, reservoir or lake	% of water supply	Finished water or raw water	Sodium (mg.) 1 cup
Composite of northside wells	33%	F	8.1
Composite of Upper Santa Cruz wells	33%	F	7.8
CALIFORNIA:			
Fresno:			
Composite or wells	100%	R	4.0
Long Beach:			
Long Beach well water		R	16.8
Long Beach treatment plant	60%	F	17.5
La Verne treatment plant	40%	F	46.9
Los Angeles:			
Colorado River	20%	R	21.8
San Fernando Reservoir	60%	F	7.6
Weymouth treatment plant	20%	F	35.8
Oakland:			
Orinda filter plant	99%	F	.6
San Pablo filter plant		F	.2
Upper San Leandro filter plant		F	1.6
Chabot filter plant		F	5.5
Sacramento:			
Sacramento River	85%	R	2.4
Filter plant	85%	F	2.8
San Diego:			
Lake Hodges Reservoir	10%	R	23.5
Alvarado treatment plant	47%	F	23.5
Torrey Pines treatment plant	10%	F	25.4
Lower Otay Reservoir	4%	R	21.3
San Francisco:			
Calaveres Reservoir	18%	F	2.8
Crystal Springs Reservoir	8%	F	.8
San Andreas Reservoir	2%	F	1.2
Hetch Hetchy Reservoir	72%	R	.3
Hetch Hetchy treatment plant	72%	F	.7
San Jose:			
Wells	80%	F	6.9
COLORADO:			
Denver:			
Frazer River & Williams Fork	47%	R	.5
Moffat filter plant	46%	F	.6
South Platte River & Bear Creek	38%	R	5.7
Marston Lake, northside filter plant	31%	F	5.7
South Platte River	15%	R	8.1
Kassler filter plant	15%	F	7.3
Marston Lake, southside filter plant	8%	F	6.4
CONNECTICUT:			
Bridgeport:			
Easton Reservoir	21%	R	.9
Easton treatment plant	21%	F	.9

State, city & plant, reservoir or lake	% of water supply	Finished water or raw water	Sodium (mg.) 1 cup
Hemlocks Reservoir	51%	R	.9
Hemlocks treatment plant	51%	F	.9
Trap Falls Reservoir	26%	R	.9
Trap Falls treatment plant	26%	F	1.0
Hartford:			
West Hartford filter plant	97%	F	.7
New Haven:			
Whitney filter plant	13%	F	1.4
Lake Gaillard	41%	F	.7
Lake Saltonstall	12%	F	1.5
Woodbridge system	12%	F	.9
Lake Wintergreen	9%	F	.6
Lake Bethany	4%	F	.8
Beaver Brook Lake	4%	F	2.6
Maltby Lakes	2%	F	1.3
WASHINGTON, D.C.:			
Potomac River	100%	R	1.8
Dalecarlia filter plant	45%	F	1.9
McMillan filter plant	55%	F	2.4
FLORIDA:			
Jacksonville:			
Well	100%	R	3.3
Chlorination plant	100%	F	3.3
Miami:			
Hialeah well fields	40%	R	5.2
Hialeah treatment plant	40%	F	5.5
Orr well field	60%	R	2.8
Orr treatment plant	60%	F	3.1
St. Petersburg:			
Cosme well field	100%	R	1.4
Treatment plant	100%	F	1.4
Tampa:			
Hillsborough River	97%	R	1.9
Tampa waterworks	97%	F	1.9
GEORGIA:			
Atlanta:			
Chattahoochee River	100%	R	.5
Hemphill filter plant	100%	F	.5
Savannah:			
Abercorn Creek	100%	R	.8
Cherokee Hill plant	100%	F	1.0
Well 2	100%	R	2.1
HAWAII:			
Honolulu:			
Kaimuki pumping station	13%	R	15.2
Beretania pumping station	23%	R	8.3
Kalihi underground station	21%	R	8.1

State, city & plant, reservoir or lake	% of water supply	Finished water or raw water	Sodium (mg.) 1 cup
ILLINOIS:			
Chicago:			
Lake Michigan	100%	R	.9
Chicago Avenue station	46%	F	.9
Lake View station	20%	F	.9
South District filtration plant	34%	F	1.0
Rockford:			
Unit well 15	92%	F	.8
Composite of 6 group wells	8%	F	1.8
INDIANA:			
Evansville:			
Filtration plant	100%	F	3.1
Fort Wayne:			
St. Joseph River (impounded)	100%	R	2.3
Three Rivers filtration plant	100%	F	3.6
Gary:			
Gary-Hobart filter plant	100%	F	1.1
Indianapolis:			
Fall Creek purification plant	45%	F	1.9
White River purification plant	55%	F	3.3
South Bend:			
Coquillard Station well 3	18%	F	1.8
Pinhook Station well 3	18%	F	1.3
North Station wells 5 & 7	18%	F	2.0
Oliver Station well 4	20%	F	2.8
IOWA:			
Des Moines:			
Infiltration gallery	75%	R	3.1
Des Moines waterworks	77%	F	7.8
KANSAS:			
Kansas City:			
Missouri River	100%	R	6.2
Quindaro treatment plant	100%	F	5.9
Topeka:			
Kansas River	100%	R	1.7
Treatment plant	100%	F	26.3
Well 2	1%	R	15.6
Well 3	1%	R	16.6
Wichita:			
Wells in Equus Beds	100%	R	14.2
Treatment plant	100%	F	14.5
Well 4 near Bentley		R	35.1
KENTUCKY:			
Louisville:			
Ohio River	100%	R	3.8
Filtration plant	100%	F	6.2

State, city & plant, reservoir or lake	% of water supply	Finished water or raw water	Sodium (mg.) 1 cup
LOUISIANA:			
Baton Rouge:			
Lula Street plant, 7 wells	26%	F	17.1
Lafayette Street plant, 3 wells	10%	F	23.0
Government Street plant, 5 wells	37%	F	18.5
Bankston Street plant, 5 wells	27%	F	15.9
New Orleans:			
Mississippi River	100%	R	4.0
Algiers purification plant	3%	F	4.3
Carrollton purification plant	97%	F	4.3
Shreveport:			
Cross Lake	100%	R	5.7
Cross Lake treatment plant	74%	F	5.7
McNeill Street treatment plant	26%	F	5.7
MARYLAND:			
Baltimore:			
Loch Raven Reservoir	55%	R	.9
Montebello filtration plant	55%	F	.7
Liberty Reservoir	45%	R	.9
Ashburton filtration plant	45%	F	.9
MASSACHUSETTS:			
Boston:			
Quabbin Reservoir	100%	R	.4
Norumbega Reservoir	100%	F	.6
Springfield:			
Little River (Cobble Mountain Reservoir)	91%	F	.7
Ludlow Reservoir	9%	F	.7
Worcester:			
Holden Reservoir	35%	F	.5
Lynde Brook & Holden Reservoir	65%	F	.8
MICHIGAN:			
Detroit:			
Detroit River	100%	R	1.0
Water Works Park Station	100%	F	.9
Flint:			
Flint River (impounded)	100%	R	3.6
Filtration plant	100%	F	6.6
Grand Rapids:			
Lake Michigan	100%	R	1.0
Filtration plant	100%	F	1.2
MINNESOTA:			
Minneapolis:			
Mississippi River	100%	R	1.7
Fridley filtration plant	51%	F	1.6
Columbia Heights filtration plant	49%	F	1.5
St. Paul:			
Mississippi River	90%	R	1.6

State, city & plant, reservoir or lake	% of water supply	Finished water or raw water	Sodium (mg.) 1 cup
McCarron purification plant	90%	F	1.4
MISSISSIPPI:			
Jackson:			
Treatment plant	100%	F	.8
MISSOURI:			
Kansas City:			
Missouri River	100%	R	8.3
Treatment plant	100%	F	9.0
St. Louis:			
Missouri River	34%	R	4.0
Howard Bend purification plant	34%	F	5.2
Mississippi River	66%	R	4.0
Chain of Rocks purification plant	66%	F	5.2
NEBRASKA:			
Lincoln:			
Composite of Ashland wells, 6, 9, 54-1, -4, -7, -9, -11	96%	R	5.9
Lincoln wells	4%	R	5.9
Ashland purification plant	100%	F	5.9
Omaha:			
Missouri River	100%	R	15.4
Minne Lusa treatment plant	100%	F	15.4
NEW JERSEY:			
Jersey City:			
Boonton Reservoir	100%	F	1.1
Newark:			
Wanaque River	45%	R	.9
Wanaque Reservoir		F	.8
Cedar Grove treatment plant	55%	F	1.0
Paterson:			
Passaic River	47%	R	.5
Little Falls treatment plant	47%	F	1.1
Wanaque River	53%	R	.9
Wanaque Reservoir		F	.8
NEW MEXICO:			
Albuquerque:			
West Mesa Station, 40 wells	4%	F	25.4
Santa Barbara Station	34%	F	12.3
Eubank Station, 44 wells	50%	F	7.8
Thomas Station, 8 wells	6%	F	11.1
NEW YORK:			
Albany:			
Alcove Reservoir	92%	R	.4
Feura Bush filter plant	92%	F	.4
Buffalo:			
Lake Erie	100%	R	2.3

State, city & plant, reservoir or lake	% of water supply	Finished water or raw water	Sodium (mg.) 1 cup
Filtration plant	100%	F	2.2
New York City:			
Catskill & Delaware supplies	79%	F	.4
Croton supply	18%	F	1.0
Jamaica wells (8, 8A, 17A & 31)	3%	F	4.0
Rochester:			
Lake Ontario	21%	R	
Lake Ontario filter plant	21%	F	2.4
Hemlock Lake	79%	R	1.2
Upland supply	79%	F	1.1
Candice Lake		R	.7
Syracuse:			
Skaneateles Lake	100%	R	3.8
Treatment plant	100%	F	4.0
Yonkers:			
Saw Mill Reservoir	18%	F	4.7
Grassy Sprain Reservoir	17%	F	4.5
Catskill Aqueduct	65%	F	.4
NORTH CAROLINA:			
Charlotte:			
Catawba River	100%	R	.9
Hoskins treatment plant	100%	F	1.0
Greensboro:			
Lake Brandt	100%	R	.7
Filter plant	100%	F	.6
OHIO:			
Akron:			
Treatment plant	100%	F	1.5
Cincinnati:			
Ohio River	100%	R	
Treatment plant	100%	F	4.3
Cleveland:			
Nottingham filtration plant	100%	F	2.6
Columbus:			
Big Walnut Creek		R	
Morse Road treatment plant	55%	F	13.7
Scioto River		R	
Dublin Road treatment plant	45%	F	8.1
Dayton:			
Well water	100%	R	2.6
Ottawa Street treatment plant	100%	F	4.0
Toledo:			
Lake Erie	100%	R	2.8
Collins Park treatment plant	100%	F	2.8
Youngstown:			
Meander Creek treatment plant	100%	F	6.2

State, city & plant, reservoir or lake	% of water supply	Finished water or raw water	Sodium (mg.) 1 cup
OKLAHOMA:			
Oklahoma City:			
Lake Hefner	100%	R	21.3
Lake Hefner treatment plant	100%	F	19.9
Tulsa:			
Sapvinaw Creek		R	
Treatment plant	100%	F	1.0
OREGON:			
Portland:			
Bull Run Headworks	100%	F	.3
PENNSYLVANIA:			
Erie:			
Lake Erie	100%	F	1.9
Chestnut Street filtration plant	100%	F	2.8
Philadelphia:			
Delaware River		R	
Torresdale filter plant	50%	F	1.1
Schuykill River		R	
Belmont filter plant	50%	F	1.7
Pittsburgh:			
Allegheny River		F	
Aspinwall filter plant	60%	F	1.6
Aldrich filter plant	40%	F	4.0
RHODE ISLAND:			
Providence:			
Scituate Reservoir	92%	R	.7
Filter plant	92%	F	.8
TENNESSEE:			
Chattanooga:			
Tennessee River	100%	R	1.9
Treatment plant	100%	F	2.0
Memphis:			
Allen well field	28%	R	1.8
Allen filtration plant	28%	F	1.8
Sheadan well field	28%	R	2.8
Sheadan filtration plant	28%	F	2.8
McCord well field	16%	R	1.5
McCord filtration plant	16%	F	1.5
Parkway well field	28%	R	4.0
Parkway filtration plant	28%	F	4.0
Nashville:			
Cumberland River	100%	R	.7
Treatment plant	100%	F	.8
TEXAS:			
Amarillo:			
Composite of wells southwest of city	55%	F	5.9
Palo Duro well field		R	4.5
McDonald well 2		R	4.0

State, city & plant, reservoir or lake	% of water supply	Finished water or raw water	Sodium (mg.) 1 cup
Bush well 4		R	6.9
Westex well 3		R	6.4
Well 6, section 49		R	8.3
Composite of 25 wells in Carson County	45%	F	5.0
Austin:			
Colorado River	100%	R	8.1
Filter plant 1	53%	F	7.8
Filter plant 2	47%	F	7.8
Tap at 807 Brazos Street	100%	F	7.8
Corpus Christi:			
Nueces River	100%	R	14.7
Cunningham treatment plant	100%	F	14.7
Dallas:			
Garza-Little Elm Reservoir	67%	R	9.2
Elm Fork treatment plant	67%	F	9.7
Grapevine & Garza-Little Elm Reservoir	26%	R	8.1
Bachman plant	26%	F	9.7
Lake Lavon	7%	F	3.6
El Paso:			
Rio Grande	14%	R	39.1
Rio Grande treatment plant	14%	F	43.6
Wells at Mesa Station	16%	F	22.0
Canutillo Station, 6 wells	25%	F	20.9
Airport Station, 3 wells	13%	F	28.2
Nevins Stationn 7 wells	14%	F	20.1
Well V-70 in lower valley	15%	F	40.8
Fort Worth:			
Lake Worth	99%	R	4.7
North Holly treatment plant	99%	F	4.7
Lake Benbrook		R	3.6
Houston:			
Heights well field	14%	F	26.3
East End well field	5%	F	41.9
South Park well field	3%	F	38.2
South End well field	6%	F	23.7
Meyerland well	1%	F	20.6
Northeast well field	9%	F	25.1
Southwest well field	18%	F	20.4
Central well field	3%	F	33.9
San Jacinto River	24%	R	2.6
San Jacinto purification plant	24%	F	2.6
Lubbock:			
City well 62 at Northwest well field	7%	R	30.6
Composite of wells in South part Lubbock	4%	F	26.3
Northeast well field	13%	F	25.1
Composite of wells in Sand Hills well field	65%	F	6.4

State, city & plant, reservoir or lake	% of water supply	Finished water or raw water	Sodium (mg.) 1 cup
Well 102 in Shallowater well field	10%	R	24.2
San Antonio:			
Market Street Station, 4 wells		F	1.9
Wells of Bexar Metropolitan			
Water District		R	2.1
Artesia Station, 5 wells		F	1.8
Mission Station, 5 wells		F	2.4
Basin Station, 4 wells		F	1.7
34th Street Station, 3 wells		F	1.6
UTAH:			
Salt Lake City:			
Deer Creek Reservoir	24%	R	1.0
Little Cottonwood treatment plant	11%	F	2.8
Big Cottonwood Creek	33%	R	1.0
Big Cottonwood treatment plant	30%	F	1.3
Mountain Dell treatment plant	14%	F	5.2
City Creek treatment plant	9%	F	1.0
Artesian wells, 3rd East Station	4%	F	6.6
VIRGINIA:			
Norfolk:			
Lake Wright	62%	R	3.3
Moores Bridges treatment plant	62%	F	2.6
Lake Prince	38%	R	1.1
37th Street treatment plant	38%	F	1.4
Lake Burnt Mills		R	1.0
Richmond:			
James River	100%	R	.7
Douglasdale Road filtration plant	100%	F	.9
WASHINGTON:			
Seattle:			
Cedar River	100%	R	.5
Lake Youngs purification plant	100%	F	.4
Spokane:			
Parkway well 5	32%	F	.7
Electric well 2	45%	F	.7
Tacoma:			
Treatment plant	94%	F	.6
Treatment plant		F	.5
WISCONSIN:			
Madison:			
Main Station wells	25%	F	.9
Unit well 6	25%	F	.7
Unit well 11	25%	F	.7
Unit well 12	25%	F	.5
Milwaukee:			
Lake Michigan	100%	R	1.0
Linnwood Avenue purification plant	100%	F	1.0

Visual Meat Portions and Their
Sodium, Fat, and Cholesterol Content
(Source: USDA)

This Thick

One piece of *round steak* (lean only) of this size (3 oz., cooked medium) has approximately 5 gr. total fat, of which 2 gr. are saturated and 3 gr. are unsaturated; 77 mg. of cholesterol; and 51 mg. of sodium (See also p. 37).

This Thick

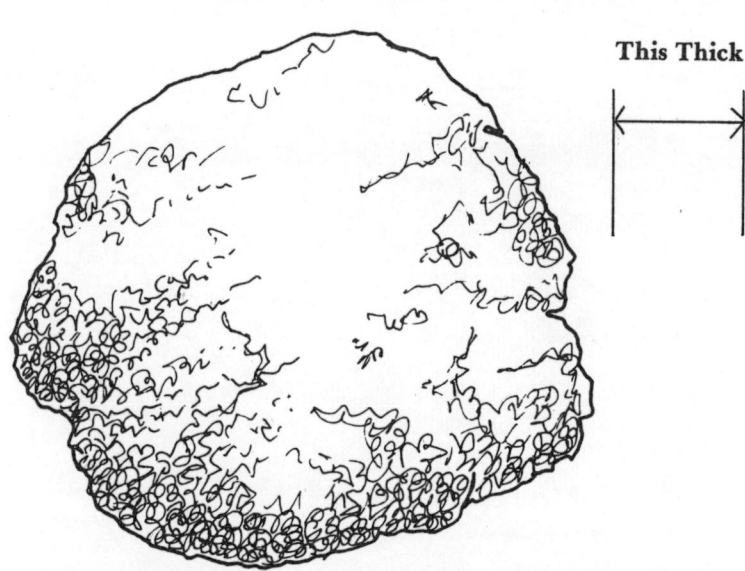

One *hamburger* (lean only) of this size (3 oz., cooked medium) has approximately 10 gr. total fat, of which 5 gr. are saturated and 5 gr. are unsaturated; 77 mg. of cholesterol; and 41 mg. of sodium (See also p. 36).

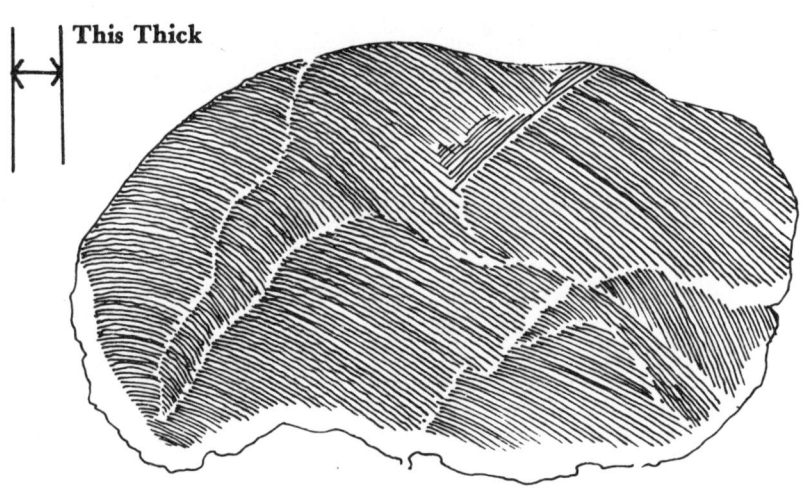

Two slices of *roast beef round* (lean only) of this size (3 oz., cooked medium) have approximately 5 gr. total fat, of which 2 gr. are saturated and 3 gr. are unsaturated; 77 mg. of cholesterol; and 51 mg. of sodium (See also p. 37).

This Thick

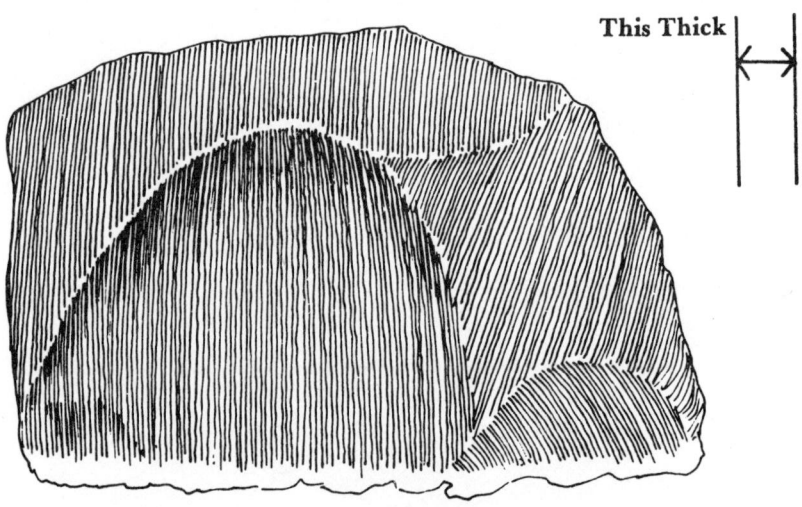

Two slices of *cured ham* (lean only) of this size (3 oz., cooked) have approximately 8 gr. total fat, of which 3 gr. are saturated and 5 gr. are unsaturated; 75 mg. of cholesterol; and 790 mg. of sodium (See also p. 259).

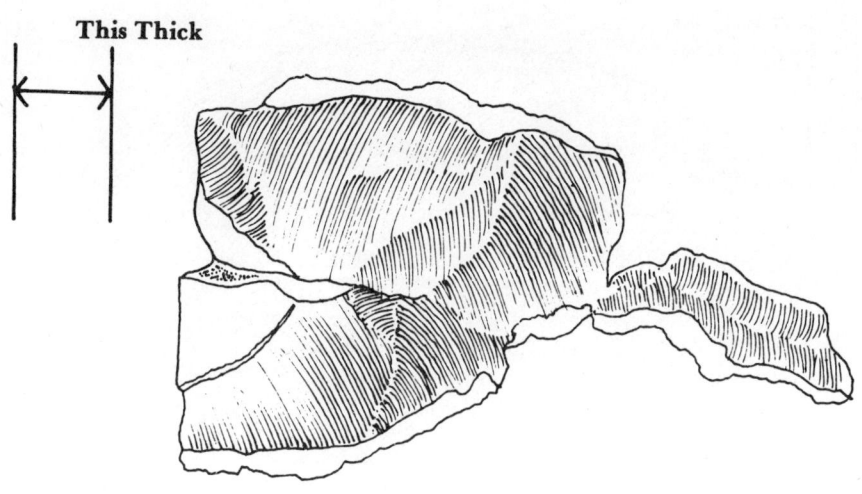

Two *lamb chops* (lean only) of this size (3 oz., cooked medium) have approximately 7 gr. total fat, of which 4 gr. are saturated and 3 gr. are unsaturated; 85 mg. of cholesterol; and 60 mg. of sodium (See also p. 190).

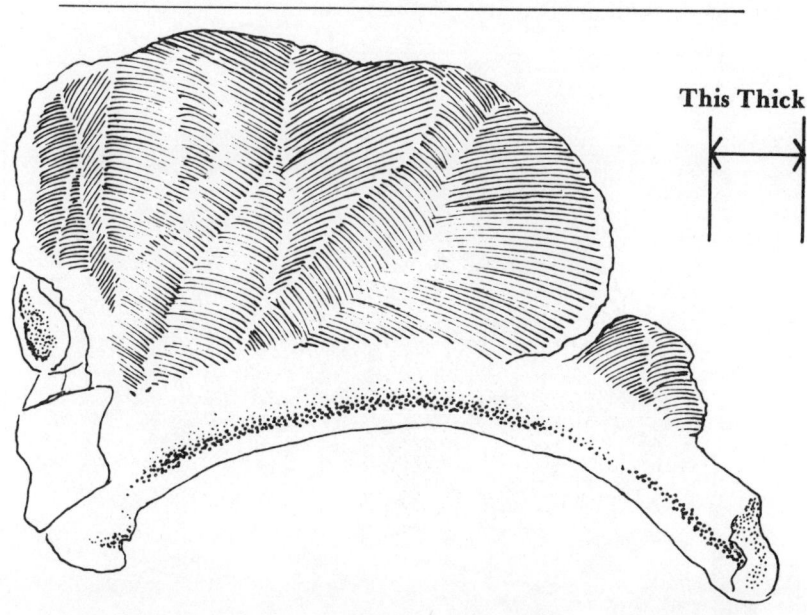

Two *pork chops* (lean only) of this size (3 oz., cooked) have approximately 13 gr. total fat, of which 5 gr. are saturated and 8 gr. are unsaturated; 75 mg. of cholesterol; and 55 mg. of sodium (See also p. 259).

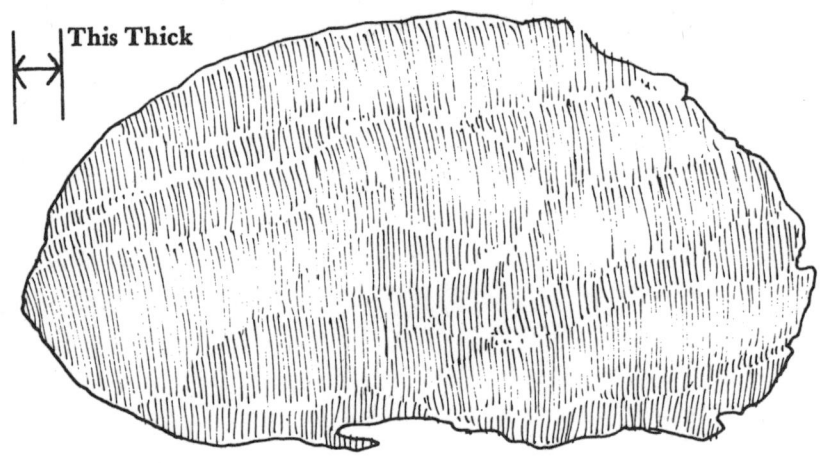

This Thick

Two slices of this size (3 oz., cooked) of the light meat of a *roast turkey* have approximately 3 gr. total fat, of which 1 gr. is saturated and 2 gr. are unsaturated; 55 mg. of cholesterol; and 70 mg. of sodium. Two slices of this size (3 oz., cooked) of the dark meat of a *roast turkey* have approximately 7 gr. total fat, of which 2 gr. are saturated and 5 gr. are unsaturated; 73 mg. of cholesterol; and 84 mg.of sodium (See also p. 332).

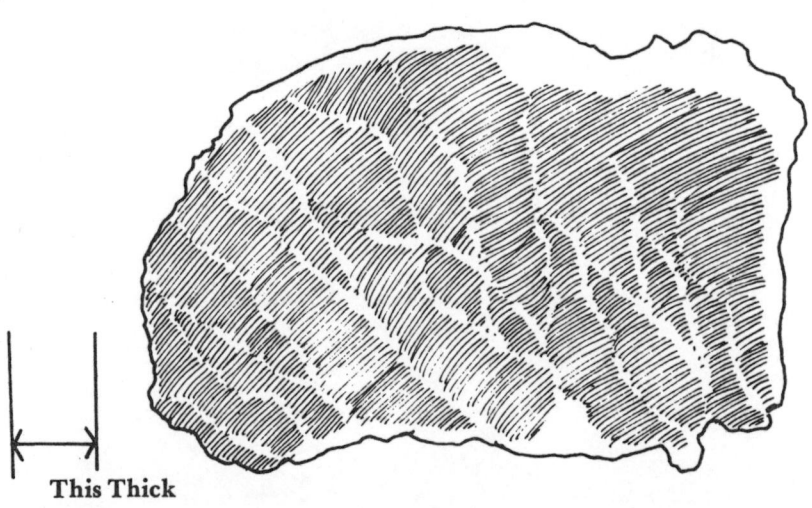

This Thick

One *veal cultlet* (trimmed) of this size (3 oz., cooked medium) has approximately 10 gr. total fat, of which 5 gr. are saturated and 5 gr. are unsaturated; 84 mg. of cholesterol; and 68 mg. of sodium (See also p. 337).